Peter A. M. Weiss
Donald R. Coustan (eds.)

Gestational Diabetes

Springer-Verlag Wien New York

Prof. Dr. Peter A. M. Weiss
Department of Obstetrics and Gynecology
University of Graz, Austria

Prof. Dr. Donald R. Coustan
Director of Maternal-Fetal Medicine
Women and Infants Hospital of Rhode Island
Providence, Rhode Island, U.S.A.

With 51 Figures

Cover drawing: H. Rosegger, M.D.

Library of Congress Cataloging-in-Publication Data. Gestational diabetes. Includes index. 1. Diabetes in pregnancy. I. Weiss, Peter A. M., 1934—   . II. Coustan, Donald R. [DNLM: 1. Pregnancy in Diabetes. WQ 248 G393]. RG580.D5G47. 1988. 618.3. 87-32098

ISBN-13: 978-3-7091-8927-6   e-ISBN-13: 978-3-7091-8925-2
DOI: 10.1007/978-3-7091-8925-2

Foreword

It is an honor and a pleasure to have been asked by Professor Weiss to co-edit this book on gestational diabetes. The tremendous increase in interest and investigation in this area in recent years has precipitated a pressing need for the interchange of ideas and approaches among investigators. We are particularly pleased with the international nature of the contributors, because European and North American researchers often work in relative isolation from each other, despite the advanced state of communication currently available. I would like to express my appreciation to Professor Weiss, and to his colleague Dr. Hannes Hofmann, for their excellent organizational abilities and hard work, which has made collaboration both pleasant and easy.

Donald R. Coustan

Preface

The incidence of gestational diabetes in developed countries lies between 1 and 8%. The general decrease in perinatal mortality and especially in perinatal morbidity has made the proportion of each due to gestational diabetes increasingly recognizable and more significant. Moreover, a growing number of investigations make us realize that a disturbed intrauterine environment in patients with unrecognized or insufficiently treated gestational diabetes may produce fetal jeopardy leading to a non-genetic fuel-mediated disposition towards obesity nad Typ II diabetes later in life.

The increased relevance has, thus, greatly stimulated research in this field. Since gestational diabetes has been at the center of research at the University of Graz-Department of Obstetrics and Gynecology since 1975, it appeared appropriate to deal with it comprehensively—the more so as the problem has been invariably underestimated by many a specialist. Thanks to personal contact with diabetologists and specialists in the field of gestational diabetes, we succeeded in enrolling the cooperation of many authorities on the subject. They contributed greatly to the outcome of this book by describing their latest investigations in gestational diabetes and related fields. This spectrum of opinion guarantees that all facets of the subject are dealt with.

I express sincere thanks to all the authors and coauthors for their kind assistance. My special thanks go to Professor Donald R. Coustan for his invaluable editorial cooperation. Without his advice and help this book could have never been issued in so short a time.

Graz, November 1987   **Peter A. M. Weiss**

Contents

Selected Topics on Gestational Diabetes

A. The Significance of Gestational Diabetes

B. On Physiologic and Pathophysiologic Features of Gestational Diabetes

C. On Screening and Diagnostics in Gestational Diabetes

15 Macrosomia and Birth Trauma in Infants of Diet Treated Gestational Diabetic Women 160

By J. L. Kitzmiller, L. A. Hoedt, E. P. Gunderson, T. S. Theiss, C. L. Ceresa, and A. M. Kitzmiller

16 Fetal Hemoglobin in the Infant of Gestational Diabetic Mothers (IGDM) 167

By P. Doménech, X. Pastor, O. Cruz, F. Botet, R. Jiménez, and J. L. Aguilar

17 Somatometric Study in the Infant of Gestational Diabetic Mother 171

By X. Pastor, P. Doménech, J. M. Jorba, A. Martínez-Gutierrez, J. Figueras, R. Jiménez, and R. Casamitjana

18 The Impact of Diet-induced Ketosis During Pregnancy on the Offspring 176

By R. Steldinger, B. Weber, J. Kneer, and H. Hättig

F. On Oral Contraception in Gestational Diabetes

19 Oral Contraceptives in Women with Previous Gestational Diabetes: Influence on Glucose Metabolism 185

By S. O. Skouby and O. Andersen

G. On the Further Fate of Women Who Had Gestational Diabetes

20 Follow-up Studies in Women with Gestational Diabetes Mellitus. The Experience at Los Angeles Country/University of Southern California Medical Center 191
By J. H. Mestman

21 The Emergence of Diabetes and Impaired Glucose Tolerance in Women Who Had Gestational Diabetes 199
By J. N. Oats, N. A. Beischer, and P. T. Grant

List of Contributors

Bellmann Otto, M.D.
Professor, Bonn University Clinic of Obstetrics and Gynecology, Federal Republic of Germany.
Postal address: Krankenanstalten „Florence Nightingale" Frauenklinik, Kreuzbergstrasse 79, D-4000 Düsseldorf 31 (Kaiserswerth), Federal Republic of Germany.

Co-authors:
Lang Norbert, M.D.
Professor, Bonn University Clinic of Obstetrics and Gynecology.
Postal address: Universitäts-Frauenklinik Bonn, Sigmund-Freud-Strasse 25, D-5300 Bonn 1, Federal Republic of Germany.
Schlebusch Harald, Ph.D.
Site of research and postal address as above.
Niesen Mathilde, M.D.
Site of research and postal address as above.
Schönhardt Rüdiger, M.D., Dr. rer. nat.
Site of research and postal address as above.

Burkart Wolfgang C., Ph.D., M.D.
Münster University Clinic of Obstetrics and Gynecology, Federal Republic of Germany.
Postal address: Albert-Schweitzer-Strasse 33, D-4400 Münster, Federal Republic of Germany.

Co-authors:
Circel Ulrich, M.D.
Site of research and postal address as above.
Hanker Jürgen P., M.D.
Site of research and postal address as above.
Schneider Herrmann P. G., M.D.
Professor, Chairman of the Department of Gynecology and Obstetrics.
Site of research and postal address as above.

Coustan Donald R., M.D.
Professor of Obstetrics and Gynecology, Brown University Program in Medicine, Director of Maternal-Fetal Medicine, Women and Infants Hospital of Rhode Island, U.S.A.
Postal address: Woman and Infants Hospital, 101 Dudley Street, Providence, RI 02905, U.S.A.

Czekelius Peter A. M., M.D.
Associate Professor (Doz.), Center of Obstetrics and Gynecology, Philipps-University, Marburg/Lahn, Federal Republic of Germany.
Postal address: Universitäts-Frauenklinik Marburg, Pilgrimstein 3, D-3550 Marburg/Lahn, Federal Republic of Germany.

Co-author:
Rollmann Johannes, C.M.
Site of research and postal address as above.

Desoye Gernot, Ph.D.
Staff Biochemist, Hormone Laboratory, Department of Obstetrics and Gynecology, University of Graz, Austria.
Postal address: Geburtshilflich-gynäkologische Universitätsklinik, Auenbruggerplatz 14, A-8036 Graz, Austria.

Doménech Pere, M.D.
Department of Pediatrics, Hospital Clinic, Faculty of Medicine, University of Barcelona, Spain.
Postal address: Villaroel, 170, E-08036 Barcelona, Spain.

Co-authors:
Pastor Xavier, M.D.
Associate Professor.
Site of research and postal address as above.
Cruz Ofelia, M.D.
Site of research and postal address as above.
Botet Francesc, M.D.
Associate Professor, Staff Neonatologist.
Site of research and postal address as above.
Jinénez Rafael, M.D.
Chairman of Pediatrics-Neonatology, Chief of the Department of Pediatrics.
Site of research and postal address as above.
Aguilar Josep Lluis, M.D.
Hematologist, Staff physician in the Laboratory of Hematology, Hospital Clinic, School of Hematology, Faculty of Medicine, University of Barcelona, Spain.
Postal address: as above.

Fuhrmann Kurt, M.D.
Associate Professor (M.R., Doz.), Diabetes Center "Gerhard Katsch", Karlsburg, and Berlin Kaulsdorf Medical Center, German Democratic Republic; Head of the Department of Obstetrics and Gynecology.
Postal address: Alt-Kaulsdorf, Frauenklinik, DDR-1144 Berlin, German Democratic Republic.

Goldman Jack, M.D.
Professor, Tel Aviv Medical School, Golda Meir Medical Center (Hasharon).
Postal address: Department of Obstetrics and Gynecology Golda Meir Medical Center, Petah-Tikva, Israel.
Co-authors:
Dicker Dov, M.D.
Instructor, Tel-Aviv Medical School.
Site of research and postal address as above.
Feldberg Dov, M.D.
Instructor, Tel-Aviv Medical School.
Site of research and postal address as above.

Yeshaya Arie, M.D.
Site of research and postal address as above.
Karp Moshe, M.D.
Professor, Tel-Aviv University Medical School, The Institute of Pediatric and Adolescent Endocrinology, Division of Juvenile Diabetes, Beilinson Medical Center.
Postal address: Division of Juvenile Diabetes, Beilinson Medical Center, Petah-Tikva, Israel.

Hoet Joseph J., M.D.
Professor, Correspondant of the Royal Medical Academy of Belgium, President of the International Diabetes Federation, Department of Medicine, Catholic University of Louvain, Brussels, Belgium.
Postal address: Avenue Hippocrate 10, B-1200 Bruxelles, Belgium.

Hofmann Hannes M. H., M.D.
University of Graz, Department of Obstetrics and Gynecology Graz, Austria.
Postal address: Geburtshilflich-gynäkologische Universitätsklinik, Auenbruggerplatz 14, A-8036 Graz, Austria.

Co-authors:
Weiss Peter A. M., M.D.
Associate Professor.
Site of research and postal address as above.
Kainer Franz, M.D.
Site of research as above.

Kitzmiller John L., M.D.
Professor in Residence, Chief of Obstetrics; Director, House Officer Training, Department of Obstetrics, Gynecology and Reproductive Sciences, University of California, San Francisco, U.S.A.
Postal address: UCSF, Dept. Ob/Gyn, Box 10132, 513 Parnassus Ave, San Francisco, CA 94143, U.S.A.

Co-authors:
Hoedt Lisa A.
Coordinator Perinatal Testing Service, Childrens Hospital, San Francisco.
Postal address: 37000 California St., San Francisco, CA 94118, USA.
Gunderson Erica P.
Nutrition Coordinator, California Diabetes and Pregnancy Program, Childrens Hospital, San Francisco.
Postal address: as above.
Theiss Trudy S.
Nutritionist.
Site of research and postal address: as above.
Ceresa Carol L.
Director Dietary Services, Childrens Hospital, San Francisco.
Postal address: 3700 California St., San Francisco, CA 94143, USA.
Kitzmiller Angela M.
Student, Children Hospital, San Francisco.
Postal address: 91 Glen Road, Brookline, MA 02146, USA.

Kühl Claus, M.D.
Vice President, Consultant Diabetes Center, Rigshospitalet, Copenhagen, Department of
Obstetrics and Gynecology, Rigshopsitalet, University of Copenhagen and Hvidöre
Hospital Klampenborg, Denmark.
Postal address: Novo Industri a/s, Novo Alle, DK-2880 Bagsvaerd, Denmark.

Co-author:
Andersen Ole, M.D.
Site of research as above.
Postal address: Pediatric Department G, Rigshospitalet, DK-2100 Copenhagen, Denmark.

Lupo Virginia R., M.D.
Assistant Professor of Obstetrics and Gynecology, University of Minnesota, Hennepin
County Medical Center, Minneapolis, Minnesota; Department of Obstetrics and Gyne-
cology University of Cincinnati Hospitals, Cincinnati, OH 45263, U.S.A.
Postal address: Department of Obstetrics-Gynecology, Hennepin County Medical Center,
101 Park Avenue, South Minneapolis, MN 55415, U.S.A.

Co-author:
Stys Stanley J., M.D.
Associate Professor of Obstetrics and Gynecology, Dartmouth Medical School.
Postal address: Hanover, NH 03156, U.S.A.

Mestman Jorge H., M.D.
Clinical Professor of Medicine and Obstetrics and Gynecology, Director of the Section of
Endocrinology and Metabolism, University of Southern California, School of Medicine.
Postal address: The Hospital of the Good Samaritan, 616 South Witmer Street, Los Angeles,
CA 90017-2395, U.S.A.

Oats Jeremy N., M.D.
First Assistant, MBBS, DM (Nottingham) MRCOG FRACOG, Department of Obstetrics
and Gynecology, University of Melbourne and Senior Obstetrician Diabetes Unit, Mercy
Maternity Hospital, Australia.
Postal address: 126-158 Clarendon Street, East Melbourne, Victoria 3002, Australia.

Co-authors:
Beischer Norman A., M.D.
Professor and Chairman, MD, MGO FRCS (Edin) FRACS, FRCOG, FRACOG, De-
partment of Obstetrics and Gynecology, University of Melbourne, Mercy Maternity Ho-
spital, Australia.
Postal address: as above.

Grant Peter T., M.D.
Registrar, MBBS, MRACOG, Department Obstetrics and Gynecology, University of Mel-
bourne, Mercy Maternity Hospital East Melbourne, Australia.
Postal address: as above.

Pastor Xavier, M.D.
Associate Professor, Department of Pediatrics, Hospital Clinic, Faculty of Medicine, Uni-
versity of Barcelona.
Postal address: Villaroel 170, E-08036 Barcelona, Spain.

Co-authors:
Doménech Pere, M.D.
Site of research and postal address as above.
Jorba Josep Maria, M.D.
Site of research and postal address as above.
Martinez-Gutierrez, M.D.
Site of research and postal address as above.
Figueras Josep, M.D.
Professor in Pediatrics, Staff Neonatologist.
Site of research and postal address as above.
Jiménez Rafael, M.D.
Chairman of Pediatrics-Neonatology.
Chief of the Department of Pediatrics.
Site of research and postal address as above.
Casamitjana Roser
Biologist, Staff Laboratory Specialist.
Laboratory of Hormonology, Hospital Clinic.
Postal address: as above.

Peterson Charles M., M.D.
Director of Research, Columbia University, College of Physicians and Surgeons, Sansum Medical Research Foundation.
Postal address: 2219 Bath street, Santa Barbara, CA 93105, U.S.A.

Co-author:
Jovanovic Lois, M.D.
Albert Einstein College of Medicine.
Senior Scientist at the Sansum Medical Research Foundation.
Postal address: as above.

Skouby Sven O., M.D.
Diabetes Center, Department of Obstetrics and Gynecology, Rigshospitalet, Copenhagen, Denmark.
Postal address: Y 4042, Rigshospitalet, Blegdamsvej 9, DK-2100 Copenhagen O, Denmark.

Co-author:
Andersen Ole, M.D.
Institution and postal address as above.

Steldinger Rainer, M.D.
Department of Obstetrics and Gynecology, Free University of Berlin.
Postal address: Universitäts-Frauenklinik, Pulsstrasse 4–14, D-1000 Berlin 19.

Co-authors:
Weber Bruno, M.D.
Professor, Department of Pediatrics, Free University of Berlin.-
Postal address: Universitäts-Kinderklinik, Heubnerweg 6, D-1000 Berlin 19.
Kneer Johannes, Ph.D.
Institution and site of research as above.
Postal address (at present): Abt. PKF-PK, Raum 121, Grenzacherstrasse 70, CH-4002 Basel, Switzerland.

Hättig Hein, Psychologist
Institution and site of research as above.
Postal address (at present): Klinikum Charlottenburg der Freien Universität Berlin, Neurologische Klinik, Spandauer Damm 130, D-1000 Berlin 19.

Van Assche André F., M.D.
Professor, Head of the Department of Obstetrics and Gynecology, Universitaire Ziekenhuizen, Sint-Rafaël-Gasthuisberg, Belgium.
Postal address: Herestraat 49, B-3000 Leuven, Belgium.

Co-authors:
Aerts Leona M., Ph.D.
Verhaeghe Johan, M.D.
Postal address as above.

Weiss Peter A. M., M.D.
Associate Professor, Vice-President of the Styrian Diabetes Association; Department of Obstetrics and Gynecology, University of Graz, Graz, Austria.
Postal address: Geburtshilflich-gynäkologische Universitätsklinik, Auenbruggerplatz 14, A-8036 Graz, Austria.

1

Gestational Diabetes: A Survey and the Graz Approach to Diagnosis and Therapy

P. A. M. Weiss

Department of Obstetrics and Gynecology, University of Graz, Graz, Austria

Gestational diabetes mellitus is a heterogenous clinical entity with substantial phenotypic and genotypic diversity in the mother (FREINKEL et al., 1985). It is most common in obese women and in those over 30 years old. Typically, the insulin response to glucose intake is delayed, while the basal plasma insulin level is normal or even elevated (KÜHL et al., 1985; PERSSON and LUNELL, 1975). The increase of maternal plasma insulin response to an increase in blood glucose, however, is diminished (FREINKEL et al., 1985), leading to a relative insulin deficiency in the gravida and consequently to fetopathy. HLA antigens DR 3 and DR 4 are often associated with gestational diabetes mellitus as are cytoplasmic islet-cell antibodies (FREINKEL et al., 1985).

The prevalence of gestational diabetes shows geographic, ethnic, and racial differences. It has been reported to be between 0.15% and 12.3% (Table 1). Comparisons are difficult since screening methods and diagnostic criteria vary greatly. Under-reporting and under-recording are likely as long as there is no general screening in pregnancy (SEPE et al., 1985).

The fetal risks in disorders of maternal carbohydrate metabolism have been evaluated differently — usually underestimated. This is both because definitions of gestational diabetes vary and because intensive monitoring and specific therapy can prevent high perinatal mortality once the disorder has been diagnosed. However, it is misleading to calculate the risk of a *recognized* complication of pregnancy from the perinatal mortality. We found no perinatal loss in a series of approximately 200 consecutive pregnancies of *overt* diabetics (White Class B–R) followed at the University Graz Department of Obstetrics between 1978 and 1985. However this does not mean that insulin dependent diabetes poses no risk or even protects the pregnancy. The same is true for gestational diabetes.

O'SULLIVAN (1974) reported a perinatal mortality rate associated with gestational diabetes of up to 16.1%. In another study SALZBERGER and LIBAN (1975) analyzed 1,000 perinatal deaths and found that a disorder of carbohydrate metabolism was implicated as the probable cause of death in 28%. However, insulin-dependent diabetes had been diagnosed in only 4.8%. JACKSON and WOOLF (1958) reported perinatal losses of 29% in prediabetic women during the 5-year period before they manifested diabetes. ROVERSI (1979) identified a perinatal mortality rate of 24.5% in the obstetric histories of White Class-A diabetic women.

In 1975 we began determining the insulin concentration in the amniotic fluid of every patient who underwent amniocentesis at our institution. Initially these were fundamental studies of fetal insulin homeostasis and so no diagnostic or therapeutic conclusions were drawn. Abnormally high amniotic fluid insulin levels were found retrospectively in 60 patients after 28 weeks gestation. In 13

Table 1. Prevalence of "gestational diabetes"

Investigator	No. screened	% Gestational diabetes
Lind, 1983; Newcastle	ND	0.15
*Csaba, 1986; Pecs	6604	0.96
Macafee and Beischer, 1974; Melbourne	1000	1.00
Chen et al., 1972; New York	1269	1.10
*Fuhrmann, 1986; Karlsburg	2510	1.10
Stangenberg, 1984; Stockholm	ND	1.30
Beard et al., 1980; London	3317	1.50
Lavin, 1985; Ohio	2077	1.50
Guttorm, 1975; Copenhagen	514	1.70
Hadden, 1980; Belfast	30300	2.20
Freinkel et al., 1985; Chicago	8300	2.40
O'Sullivan, 1975; Boston	725	2.50 (—7.5)
Merkatz et al., 1980; Cleveland	2225	3.10 (—11.5)
*Jovanovic, 1986; New York	300	3.20
Carpenter and Coustan 1982; New Haven	381	3.40
Sutherland, 1984; Aberdeen	ND	4.00
*Sketelj, 1986; Ljubljana	306	5.60 (—7.8)
*Weiss and Hofmann, 1986; Graz	4090	8.80
*Irsigler et al., 1986; Wien	1172	11.40
Mestman et al., 1971; Los Angeles	658	12.30

ND no data.

Source: 1. Proceedings of the second international workshop-conference on gestational diabetes 1984 (Diabetes 34 [Suppl] 2: 2. *Proceedings of the first international Graz Symposium on disturbance of carbohydrate metabolism 1986 (Probleme der perinatalen Medizin), Maudrich, Wien München Bern.

of these patients the elevated amniotic-fluid insulin level was probably due to tocolytic agents (ritodrine) (Weiss et al., 1984), but in 47 patients (11 primiparae and 36 multiparae) the elevation was most likely due to unrecognized gestational diabetes. Eleven of these multiparae (30.6%) had an obstetric history of one or more perinatal losses. The perinatal mortality of all previous pregnancies was 16.7% (Table 2); that of the index pregnancy still as high as 6.4% (Table 3). These results ilustrate the main risks of gestational diabetes:

– It is diagnosed too late or not at all.

– Patients with diagnosed diabetes are often treated insufficiently. Perinatal mortality rates can be markedly elevated in patients diagnosed late or not at all. Undertreatment, however, increases primarily the incidence of obstetric complications (preterm delivery, preeclampsia, urinary tract in-

fection, polyhydramnios) and of obstetric interventions (induction of labor, cesarean section). While obstetric monitoring and intensive neonatal care can almost eliminate excess perinatal deaths, the neonatal morbidity, however, still remains exaggerated.

The specific problem of diabetic fetopathy is hyperinsulinism, leading to hypoglycemia, overly high birth weight, and enlarged and immature organs in the offspring. Researchers increasingly implicate in utero damage and exhaustion of fetal beta-islet cells in the occurence of diabetes mellitus later in life, so-called "fuel-mediated teratogenesis" (Freinkel et al., 1985; Martin et al., 1985). Van Beek (1939) and Potter (1941) already reported inflammatory and degenerative changes in pancreatic islet cells of newborns with diabetic mothers, and they suspected the possibility of resulting diabetes. Similarly Hultquist and Olding (1975) de-

Table 2. Obstetric history of 90 pregnancies in 36 multiparae with high amniotic fluid insulin content (WEISS et al. 1984)

	Control	Raised amniotic fluid insulin contant	Multiplication factor	Difference
Age (years)	28.2	29.1		
Amniotic fluid insulin content	6.9 µU/ml	23.2 µU/ml	3.4-fold	p < 0.0001 (paired t test)
Infants over 4000 g (%)	7.3	12.2	1.7-fold	NS (x^2 test)
Malformation (%)	0.5	3.3	6.7-fold	p < 0.025 (Fischer exact test)
Perinatal mortality (%) in previous pregnancies	1.3	16.7	12.8-fold	p < 0.0001 (x^2 test)

NS not significant.
Reprinted with permission from The American College of Obstetricians and Gynecologists (WEISS et al. [1984] Obstetrics and Gynecology 63: 76).

Table 3. Fetal outcome and complications of the pregnancy in 47 women (36 multiparae, 11 primiparae) with high amniotic fluid insulin content (WEISS et al. 1984)

	Control	Raised amniotic fluid insulin content	Multiplication factor	Difference (x^2 test)
Premature births (%)	7.5	17.0	2.3-fold	p < 0.05
Infants more than 4000 g (%)	7.3	6.4	0.9-fold	NS[d]
Rate of cesarean sections (%)	7.0	12.8	1.8-fold	NS
Morbidity (%)	8.7	27.7[a]	3.2-fold	p < 0.001
Neonatal intensive care (%)	4.2	19.1[b]	4.6-fold	p < 0.001
Perinatal mortality (%)	1.3	6.4[c]	4.9-fold	p < 0.05[d]

[a] Intervention of the neonatologist required.
[b] Five cases of hyperbilirubinemia with blood exchange transfusion, two cases of infantile respiratory distress syndrome: male 1900 g, 45 cm; female 2050 g, 45 cm: both artificially ventilated; female 2800 g, 47 cm: diabetogenic fetal disease; male 2860 g, 49 cm: cheilognato-palatoschisis.
[c] Male 5340 g, 56 cm: infratentorial hemorrhage; female 1120 g, 37 cm: stillbirth (unexplained); female 1100 g, 35 cm: stillbirth (unexplained).
[d] Fischer exact test.
Reprinted with permission from The American College of Obstetricians and Gynecologists (WEISS et al. [1984] Obstetrics and Gynecology 63: 76).

scribed fibrosis and hyalinization of the pancreatic islets of newborns of diabetic mothers. SCHEIBENREITER and THALHAMMER (1966) voiced the same suspicion after observing that a high percentage of adolescents with diabetes had high birth weights. FARQUHAR (1969) found a higher incidence of diabetes mellitus in children of Type-II diabetic mothers than in those of diabetic fathers. PETTITT et al. (1982) observed a higher incidence of diabetes in offspring of women who developed diabetes during pregnancy than in those who developed the disease after pregnancy. Finally VAN ASSCHE et al. (1983) reported beta cell degranulation in children of diabetic mothers, and they demonstrated in animals that fetal hyperinsulinism can have a prospective diabetogenic effect (Van Assche and Aerts, 1985).

Considering these investigations it is not surprising that offspring with diabetogenic fetopathy tend towards obesity at school age, in puberty, and in adolesence (Voiir et al., 1980; Pettitt et al., 1983, 1985); 25% show an exaggerated insulin response to glucose administration, and 18% have abnormal results in a glucose tolerance test (Amendt et al., 1976; Kohlhoff and Roth, 1986). A delayed insulin response, similar to that of patients with Type-II diabetes, has also been described (Rosenkranz et al., 1979).

Thus it must be emphasized that a low or absent perinatal mortality rate in association with gestational diabetes is not a measure of the quality of metabolic management, but only of obstetric care. Considering the long-term sequelae, optimal care must also prevent fetal hyperinsulinism with or without macrosomia. Van Assche et al. describe their research in this very important area in a separate section.

Nomenclature and classification

Diabetes mellitus is the most common serious metabolic disorder of humans, affecting about 2% to 5% of the general population worldwide (Schernthaner, 1980). The National Diabetes Data Group (1979) and the WHO Expert Comittee on Diabetes Mellitus (1980) classified the "heterogenous syndrome" diabetes mellitus according to its ethiopathology (Table 4), but other classifications persist.

Type I diabetes is also called *overt diabetes, insulin-dependent diabetes (IDD)*, or *juvenile-onset diabetes (JOD)*. The peak age of manifestation is 9 years (Schernthaner, 1980). *Type-II diabetes* is also called *non-insulin-dependent (NIDD)* or *maturity-onset diabetes (MOD)*. Type-II disease has a subgroup: *maturity-onset diabetes of young people (MODY)*, also called *Type III diabetes*. Type II disease is more common with increasing age, but it has been found as early as in a 12-year-old gravida (Hollingsworth, 1983). Thus the discriminations according to age are not absolute, and overlapping does occur.

Impaired glucose tolerance, diagnosed by an abnormal glucose-tolerance test and without other evidence of diabetes, is also called *subclinical, chemical* or *latent diabetes*. *Gestational diabetes* is defined as carbohydrate intolerance that begins during and ends with pregnancy (Essex et al., 1973). Since it is usually diagnosed by the glucose-tolerance test, it is often called by the terms applied to impaired glucose tolerance outside pregnancy.

Impaired glucose tolerance in pregnancy can be a result of a number of pathophysiologic mechanism. Normal-weight women with this disorder usually have a relative insulin deficiency whereas hyperinsulinemia and insulin resistance are more common in obese

Table 4. Classification of diabetes mellitus

- *Type I:* insulin-dependent diabetes mellitus (IDD)
- *Type II:* noninsulin-dependent diabetes mellitus (NIDD)
  Obese
  Lean
- *Secondary diabetes mellitus:* pancreatoprival diabetes, diabetes in endocrinopathy
- *Impaired glucose tolerance*
  Obese
  Lean
- *Gestational diabetes*

Table 5. Potential diabetes

History	Clinical signs
Neonates over 4000 g	hydramnios
Unexplained perinatal losses	large fetal placental unit (ultrasonography)
Repeated unexplained premature deliveries	repeated glucosuria
Repeated unexplained miscarriages	obesity
Diabetes in the family	age over 30 years
Malformation	EPH-gestosis (preeclampsia)
	recurrent urinary infections

women. The fasting and the postprandial blood glucose level may be high in both lean and obese patients, as are delayed insulin response to nutritional stimuli and reduced insulin binding to erythrocyte receptors (CHENEY et al., 1985; FREINKEL et al., 1985; KÜHL and ANDERSON, 1986).

The term *potential diabetes* is applied to a spectrum of anamnestic or clinical traits that indicate an increased risk of diabetes (Table 5). To recognize impaired glucose tolerance or manifest diabetes as early as possible, potential diabetics should be evaluated yearly in the non pregnant state and especially when they become pregnant.

The risk of fetal malformation, perinatal mortality, and neonatal morbidity are determined by the quality of metabolic management, by the mother's age at the manifestation of her diabetes, by diabetic vasculopathy, and by other obstetric complications, especially preeclampsia. Therefore a classification of diabetes in pregnancy must take these factors into account, in order to evaluate the child's prognosis, and to compare obstetric results of different centers.

PRISCILLA WHITE proposed the most widely used classification in 1959, and later enlarged it (1974, 1978). Classifications by PEDERSEN and MOLSTED PEDERSEN (1965), TYSON and HOCK (1976), and PYKE (ESSEX et al., 1973) are also used.

PEDERSEN and MOLSTED PEDERSEN (1965) base their classification on prognostically bad signs in pregnancy (PBSP). Among these are:

− Clinical pyelonephritis (positive urine culture and fever).

− Precoma or severe acidosis (venous bicarbonate concentration below 10 mEq/l, or 10–17 mEq/l, respectively).

− Hypertension of pregnancy (preeclampsia, gestosis, toxemia), and

− Neglectors

Neglectors are pregnant women whose prenatal care was inadequate through their own fault (refusal of treatment, presentation in the last trimester or in labor, psychopathy, low IQ),

Pyke's classification differentiates three groups:

I) Gestational diabetes mellitus. The metabolic disturbance begins during and ends with pregnancy.

II) Pregestational diabetes. Preexistant diabetes that continues beyond the pregnancy.

A) Without complications.

B) With complications (retinopathy, nephropathy, macroangiopathy).

Each of these groups can be divided into subgroups:

− With good metabolic control.

− Without optimal metabolic control (JOVANOVIC and PETERSON, 1982).

FREINKEL's group has instituted a further stratification of gestational diabetes according to the patient's fasting blood glucose (FREINKEL et al., 1985). They designated gravidae with abnormal oGTT results but

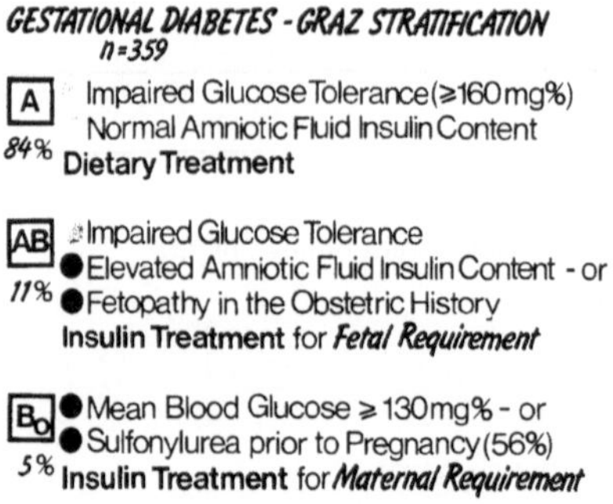

GESTATIONAL DIABETES - GRAZ STRATIFICATION
n=359

[A] Impaired Glucose Tolerance(≥160mg%)
     Normal Amniotic Fluid Insulin Content
84% Dietary Treatment

[AB] ≥ Impaired Glucose Tolerance
     ● Elevated Amniotic Fluid Insulin Content - or
11% ● Fetopathy in the Obstetric History
     Insulin Treatment for Fetal Requirement

[Bo] ● Mean Blood Glucose ≥ 130mg% - or
     ● Sulfonylurea prior to Pregnancy (56%)
5% Insulin Treatment for Maternal Requirement

Fig. 1. The Graz stratification of gestational diabetes. The distribution of the Classes A, AB, and Bo was calculated from 359 patients between 1980 and 1985. Additionally when the fasting blood glucose value exceeds 110 mg/dl on three consecutive days we also consider the patient as Class Bo

with fasting blood glucose values below 105 mg/dl (5.8 mmol/1) as Class A_1, gravidae with fasting values between 105 mg/dl and 130 mg/dl (7.2 mmol/1) as Class A_2, and gravidae with values over 130 mg/dl as Class B_1.

We use a modification of White's classification. The White Classes A to R are supplemented by our classes AB and Bo (Fig. 1). Group AB comprises pregnant women who, according to their own metabolic condition, could be managed by diet alone. However, these patients have an elevated amniotic-fluid insulin level indicating fetal hyperinsulinemia. Insulin treatment is therefore initiated only for the benefit of the fetus — the women themselves are not real insulin-re-

Table 6. Comparison of the stratification according to Freinkel (A_1, A_2, B_1) with the Graz classification (A, AB, Bo) in 359 gestational diabetic women (1980 to 1985)

	A	AB	Bo	Total (Freinkel)
A_1	302	26	2	330
A_2	0	11	5	16
B_1	0	2	11	13
Total (Graz)	302	39	18	

quiring White Class-B diabetics; they belong to a group between A and B.

Group Bo is a stratification of White Class B. It comprises women who develop insulin-requiring diabetes during pregnancy — with or without postpartal remission. Previously known diabetics who were treated by diet alone or with oral antidiabetic agents before pregnancy also fall into this group (first insulinization in pregnancy).

Today the most severe cases of diabetogenic fetal morbidity stem from Groups AB and Bo. Especially in these patients, the metabolic disorder is diagnosed late or not at all, or its implications are underestimated. In two of 18 Group Bo patients in 1980 to 1985, the diagnosis of diabetes mellitus was made only after the pregnant patient was admitted to the hospital with severe ketoacidosis.

Table 6 compares our subclassification with FREINKEL's. It can be seen that insulin demand is not invariably linked to elevated fasting blood glucose levels.

Physiology and pathophysiology

Treatment of diabetes mellitus in pregnancy aims to correct metabolic abnormalities and to imitate physiologic regulation through measures of restriction (diet) and of substitution (insulin administration). Knowledge of basic pathophysiologic concepts is thus a prerequisite to diagnosis and therapy. We summarize here important aspects of carbohydrate metabolism; some are discussed

in later sections or in connection with clinical problems.

Metabolic changes in pregnancy (Tables 7 to 9)
(References are given in the tables)

The *fasting blood glucose level* (Table 7) sinks to 60 to 70 mg/dl (3.3 to 3.9 mmol/l) during the course of normal pregnancy. This is a decrease of 10 to 20% (about 15 mg/dl, or 0.83 mmol/1). It is caused by improved peripheral utilization of glucose and by the demand of the fetus which syphons off 30 to 50 g glucose per day by the last trimester. The fasting blood glucose level in *gestational diabetes* is 70 to 110 mg/dl (3.9 to 6.1 mmol/l). Not every patient shows a significant increase. Elevated blood glucose levels may be caused by a relative insulin deficiency (usually in normal-weight women) or by peripheral insulin resistance (usually in obese women).

The *postprandial blood glucose level* in pregnancy is elevated to 130 to 140 mg/dl (7.2 to 7.8 mmol/l). This increase is most likely a result of placental anti-insulin hormones. The mean amplitude of glycemic excursions (MAGE) (SERVICE and NELSON, 1980) is about 45 mg/dl (2.5 mmol/l). Patients with *gestational diabetes* have postprandial glycemia of 150 to 160 mg/dl (8.3 to 8.9 mmol/l) as a result of placental anti-insulin hormones, insulin resistance, and delayed insulin secretion after nutritional stimuli (Fig. 20). The MAGE is about 60 mg/dl (3.3 mmol/l), reflecting greater fluctuations of the blood glucose value. *The mean blood glucose* (MBG) in normal pregnancy is 90 to 100 mg/dl (5.0 to 5.6 mmol/l), the same as in the non-pregnant state. The lower fasting values balance the higher postprandial levels. In *gestational diabetes* the MBG is about 100 mg/dl (5.6 mmol/l), or more. The difference from the MBG of normal pregnancy is often minimal. *Betamimetic agents* for tocolysis (terbutaline, hexoprenaline, ritodrine) enhance glycogenolysis and lipolysis. Within 30 minutes they increase MBG by about 30 mg/dl (1.7 mmol/l) in metabolically healthy subjects and by up to 70 mg/dl or more (3.9 mmol/l) in diabetic patients. *Glucose tolerance* improves early in normal pregnancy. This may be an effect of human chorionic gonadotropin (HCG). After the 20th week placental anti-insulin hormones progressively decrease glucose tolerance (Table 10, Figs. 7 and 9). The maximum blood glucose after an oral challenge peaks about 20 minutes later in late pregnancy than in early pregnancy (BELLMANN, 1984). *Insulin secretion* (Table 8) is increased during normal pregnancy. It peaks in the morning hours. Both the fasting and the stimulated plasma insulin level increase with advancing gestation. When betamimetic agents are given, insulin secretion rises three-fold within 60 minutes. Insulin secretion in *gestational diabetes* may be normal, increased, or decreased. High fasting values are often found, especially in obese patients, as is a sluggish (and sometimes decreased) insulin response to glucose stimulation (Fig. 20). In particular cases no correlation can be discerned between insulin secretion and glucose tolerance. The insulin resistance that is the main feature of gestational diabetes initiates the vicious circle: lack of receptors – hyperglycemia – hyperinsulinism – down-regulation of the receptors.

The effect of insulin is enhanced by insulinotropic hormones before 20 weeks gestation, but diminished by anti-insulin hormones thereafter (Table 10).

The placenta has been suspected of participating in the *breakdown of insulin*. Two such enzymatic pathways have been demonstrated *in vitro*. However, it is unlikely that this is the reason for the progressive insulin requirement during pregnancy since the serum insulin level is elevated in normal pregnancy and since both insulin-kinetics and insulin half-life are unchanged. Neither is fetal insulin catabolized by the placenta. There is no measurable arteriovenous difference in the cord blood insulin content. In normal pregnancy *insulin receptors* in lymphocytes are unchanged as compared to the non-pregnant state, while they are reduced in *gestational diabetes*.

Plasma *free fatty acids and glycerol* are decreased in the first half of pregnancy (Table 9), but increased by accelerated mobilization and utilization of fat in the second half. *Gestational diabetic* patients have significantly higher levels than healthy gravidae.

Table 7. Variations of the metabolism in normal pregnancy (NP) and gestational diabetes (GD)

Metabolic parameter	Behaviour in pregnancy	Probable cause of Variation in pregnancy	Comments	References
Fasting blood glucose				
in NP 60–70 mg/dl	→	peripheral glucose-utilization ↑. Syphoning by the fetoplacental unit.	Decrease 10–20% ($\sim$ 15 mg/dl, $\sim$ 0.83 mmol/l). At the last trimester the fetus syphons off 30–50 g glucose in 24 hours.	Bellmann, 1978a; Fuhrmann, 1982a; Service and Nelson, 1980; Persson and Lunell, 1975; Hollingsworth, 1983; Fredholm et al., 1978; Rathgen, 1980; Lipshitz and Vinik, 1978; van Lierde et al., 1982.
in GD 70–110 mg/dl	←	peripheral insulin resistance mainly; relative insulin deficiency.	Mostly in obese women. With repeatedly FBG values $\geqslant$ 110 mg/dl IDD must be considered.	
Postprandial blood glucose				
in NP 130–140 mg/dl	←	antiinsulin hormones (Table 10).	MAGE (mean amplitude of glycemic excursions) < 45 mg/dl.	
in GD 150–160 mg/dl	←	peripheral insulin resistance; delayed insulin secretion after nutritional stimuli.	MAGE $\sim$ 60 mg/dl.	

Mean blood glucose			
in NP 80–100 mg/dl	—		Fasting blood glucose ↓. Postprandial blood glucose ↑.
in GD 90–130 mg/dl	↑	as above.	Often minimal difference to normal pregnancy (see also Table 20)
with betamimetic agents (terbutalwie, hexoprenalin, ritodrine)	↑	glycogenolysis ↑. Lipolysis ↑.	NP↑ ~ 30 mg/dl. GD↑ 40–70 mg/dl. Onset of metabolic effect after approximately 30 minutes. The metabolic effect decreases in long-term tocolysis.
Glucose tolerance	[↑]↓	(insulinotropic and antiinsulin hormones see Table 10).	i. v. GTT: Glucose tolerance in early pregnancy ↑; in late pregnancy ↓. oGTT: The same as above (see Figs. 7 and 9). Later peaking with advancing pregnancy.

Table 8. Variations of the metabolism in normal pregnancy (NP) and gestational diabetes (GD)

Metabolic parameter	Behaviour in pregnancy	Probable cause of variation in pregnancy	Comments	References
Insulin-secretion				
in NP	↑	Hyperplasia of the islet beta cells (Table 10).	Maximum secretion in the early morning. Both the fasting and the postprandial levels increase with advancing pregnancy. 3-fold output 60 minutes after administration of betamimetic agents.	Bellmann, 1978; Burt and Davidson, 1974; Daweke and Hüter, 1970; Persson and Lunell, 1975; Hollingsworth, 1983; Fredholm et al. 1978; Rathgen, 1980; Lipshitz and Vinik, 1978; Turner et al., 1977; Steel et al., 1979; Bellmann and Hartmann, 1975; Pedersen et al., 1982;
in GD	↑[↑][↓]	Vicious circle: lack of receptors-hyperglycemia down-regulation of the receptors.	Elevated fasting levels, delayed response to glucose stimuli, shifting of the insulin to glucose ratio.	Kahn, 1982; Omori et al., 1982; Toyoda, 1982; Exon and Dixon, 1974; Mylvaganam et al., 1983; Schernthaner, 1980; Muck and Hommel, 1977; Weber et al., 1978.
with betamimetic agents		Direct effect on islet beta cells.	3-fold increase of insulin plasma level.	
Insulin-effect				
before the 20th gestational week	↑	HCG (?) (see Table 10).	Valid for NP, GD, IDD.	
after the 20th gestational week	→	Antiinsulin hormones (see Table 10). Postreceptor defect (?).	Decreased peripheral effect.	

Insulin-degradation			
in the placenta (?)	Two enzymatic pathways detected in vitro.	—	Its importance for the enhanced insulin requirement during pregnancy is contradicted by: 1. Elevated plasma-insulin levels during pregnancy. 2. Unchanged insulin-kinetics and insulin half-life. 3. No arteriovenous difference of insulin in the cord blood.
Insulin-receptors			
in NP			300,000 to 1,000,000 molecular weight of the membrane-protein. Circadian fluctuation. High binding at midnight, and in the early morning. Low binding in the afternoon.
in GD	Down regulation.	→	In obesity ↓. Dietary treatment ↑.

Table 9. Variations of the metabolism in normal pregnancy (NP) and gestational diabetes (GD)

Metabolic parameter	Behaviour in pregnancy	Probable cause of variation in pregnancy	Comments	References
Free fatty acids and glycerol *in NP*				
First half of pregnancy	→	Enhanced mobilization and utilization of fat.	↑ In gestational diabetes and with administration of betamimetic agents.	PERSSON and LUNELL, 1975; HOLLINGSWORTH, 1983; FREDHOLM et al., 1978; LIPSHITZ, 1978; FUHRMANN, 1982; PHELPS et al., 1983;
Second half of pregnancy	←			
Ketones	—		Distinctly elevated in GD.	WIDNESS et al., 1980; O'SHAUGNESSY et al., 1979; KJAERGAARD and DITZEL, 1979.
Glycohemoglobins				
prior to the 24th gestational week	→	Reciprocal course compared to insulin requirement and glucose tolerance during pregnancy.	In GD according to the metabolic control.	
after the 24th gestational week	←			

Ketones are also higher in diabetic than in non-diabetic subjects.

Glycosylated hemoglobin decreases until the 24th week of a normal pregnancy and rises thereafter. Thus it runs counter to insulin sensitivity (WEISS and HOFMANN, 1984 b) and to glucose tolerance (KAINER et al., 1986) during the course of pregnancy.

Detailed descriptions of the homeostasis of *insulin, proinsulin, and glucagon* as well as of *insulin-resistance* and *-degradation* are provided in the section "Pathophysiological background for gestational diabetes" by KÜHL and ANDERSON.

Endocrine effects of pregnancy on carbohydrate metabolism

The placenta is a highly potent endocrine organ that strongly influences maternal metabolism (Table 10 a, b, c). Its synthesis of steroid and protein hormones causes a diabetogenic effect, making pregnancy the most favorable time to uncover latent diabetes.

Insulin homeostasis in the fetoplacental unit

Through the integration of the cybernetic systems of mother and fetus, an imbalance of maternal metabolism can induce a fetal endocrine disturbance. A special feature is that glucose can freely pass through the placenta, but maternal or fetal insulin cannot. The fetal disorder can be recognized early by demonstrating increased fetal insulin production; then it can be treated specifically.

Insulin homeostasis in the fetoplacental unit of normal pregnancy

Insulin biosynthesis in humans is stimulated mainly by D-glucose (dextrose). The cyclic-AMP system serves as a second messenger (CERASI and GRILL, 1977; SCHATZ, 1977). Glucagon and growth hormone enhance the effect of glucose on the beta-islet cells. However, a high blood insulin level does not in-

hibit insulin production (ZILKER et al., 1977). Exogenous insulin inhibits endogenous insulin production indirectly by decreasing blood glucose.

The fetus begins to produce insulin by the 11th gestational week. Stimulation of increased insulin biosynthesis *in vitro* is possible only at abnormally high glucose concentrations in the culture medium (720 mg/dl, 40 mmol/l) (REIHER et al., 1982). Stimulation of increased insulin production in the fetus of a metabolically healthy pregnant woman is practically impossible at normal *in vivo* glucose concentrations, even at the end of gestation (PATERSON et al., 1968; FEIGE et al., 1977). Normally, the insulin dynamics of the fetus are immature (BELLMANN, 1978).

Insulin reaches the amniotic fluid via the fetal urine at 12 weeks' gestation, but only trace amounts (1.3 to 2.5 µU/ml) are detectable between the 12th and 16th week (WEISS et al., 1984 c). In the 16th week the amniotic-fluid insulin level surges to about 4.3 µU/ml. Between the 16th and the 42nd week, the mean insulin level rises from 4.3 µU/ml to 9.1 µU/ml. In the last five weeks of gestation, the ratio of cord-blood insulin to fetal-urine insulin to amniotic-fluid insulin is 100 to 92 to 77.

Follow-up measurements at intervals of weeks or months show that the amniotic fluid insulin level stays stable, *i.e.* the values in the individual pregnant woman stay in the same percentile range (WEISS 1979). Measurements at hourly intervals show no significant circadian fluctuations (WEISS et al., 1985) (Fig. 2). Figure 2 shows that neither food intake nor insulin administration exerts any immediate impact on the amniotic-fluid insulin concentration.

The amniotic-fluid insulin level is significantly decreased with major fetal malformations, intrauterine fetal death, preeclampsia, placental insufficiency, and intrauterine growth retardation. It is increased with fetal hemolytic disease, as well as with administration of glucocorticoids, beta-mimetic agents, or both (WEISS et al., 1984 c) (Fig. 3).

Table 10 a. Effect of various hormones on carbohydrate metabolism in pregnancy

Hormone	Change in pregnancy	Origin	Effect
Insulin	Increase peaking in the 2nd and 3rd trimester. Increased effect until the 17th week. thereafter accelerating decline until the 36th week, than renewed increase.	Pancreatic islet-organ beta cells.	Decreased blood glucose.
Glucagon	Increase in the 2nd–3rd trimester. 36th week: + 30%.	Pancreatic islet-organ alpha cells.	Diabetogenic.
Thyroxine	Continuous increase.	Thyroid.	Diabetogenic.
Glucocorticoids (cortisol)	Continuous increase.	Adrenal cortex placenta?	Antiinsulin.

Insulin homeostasis in the fetoplacental unit of the diabetic woman

Constant glucose oversupply to the fetus stimulates the fetal pancreatic islet cells to increase insulin production and gradually induces hypertrophy and hyperplasia (GABBE and QUILLIGAN, 1977). Especially the mass of the islet cells is increased. In fetuses of diabetic women with poor metabolic control, the relative islet cell volume in the 11th to 15th gestational week is still normal, *i.e.* 10%. But in the 21st to 26th week it increases to 27% − twice the norm (REIHER et al., 1983). As gestation proceeds, the insulin response to a standard glucose stimulus becomes quicker and more pronounced. Glucose stimuli that elicit no response in a normal fetus cause increased insulin secretion by such a hyperdynamic fetal islet organ (OBENSHAIN et al., 1970; FEIGE et al., 1977). With a maternal glucose load during delivery, FEIGE observed a nine-fold increase of the fetal insulin level (mature insulin secretion dynamics) in gestational diabetes. Our own studies showed that newborns with diabetogenic fetopathy produce five times as much insulin as healthy neonates in response to intravenous glucose administration (1.5 g/kg). Since the placental barrier prevents excess fetal insulin from leaving the fetoplacental unit, a disproportionate amount of the fetomaternal glucose pool is metabolized by the fetus.

Fetal insulin production may be considerable. We found an amniotic-fluid insulin level of 625.0 µU/ml (almost 100 times the norm) in a patient with gestational diabetes in the 37th gestational week; calculated from the birthweight of 4,050 g, fetal insulin production exceeded maternal production by far. Generally, the amounts of insulin produced by hyperdynamic fetal pancreatic islets must have an effect on maternal glucose tolerance, especially since the fetus takes up

Mechanism	Remarks	References
Peripheral glucose utilization ↑. Glycogen-synthesis ↓. Lipolysis ↓.	Proinsulin to insulin at a constant ratio. Earlier insulin peak. after glucose stimulation. Insulin-glucose index ↑.	BELLMANN, 1978 a; RATHGEN, 1980; PERSSON and LUNELL, 1975
Gluconeogenesis ↑. Glycogenolysis ↑. Hepatic glucose output ↑. Lipolysis ↑	Sharp drop after glucose. loading. NIDD: normal range. IDD: significantly ↓. Hexoprenaline ↑	BELLMANN, 1978 a; HOLLINGSWORTH, 1983; RATHGEN, 1980; LIPSHITZ and VINIK, 1978; TURNER et al., 1977
Intestinal glucose-absorption ↑. Glucose break-down in fat tissue ↑. Lipolysis ↑.	Thyroxine-binding globulin ↑. Free thyroxine remains constant.	BELLMANN, 1978; DAWEKE and HÜTER, 1970.
Peripheral glucose-utilization ↓. Gluconeogenesis ↑.	Serum level rises 3-fold in the 3rd trimester. Cortisol in the amniotic fluid: steep increase in the 36th week, moderate increase in IDD: unfavorable sign.	BELLMANN, 1978; DAWEKE and HÜTER, 1970; HOLLINGSWORTH, 1983; BRAZY et al., 1978; RATHGEN, 1980.

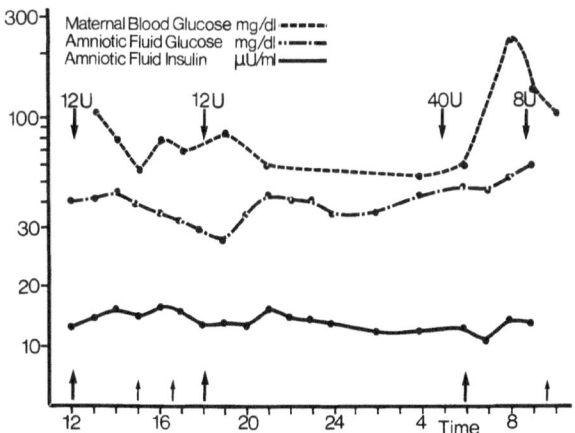

Fig. 2. The circadian course of the amniotic-fluid insulin level in an overt diabetic woman (hourly measurements). Top curve: maternal blood glucose (*mg/dl*); middle curve: amniotic fluid glucose (*mg/dl*); bottom curve; amniotic-fluid insulin (*μU/ml*). Arrows pointing upwards: meals; arrows pointing downwards: insulin injections. The scale applies to insulin as well as to glucose. Neither food intake nor insulin administration exerts any immediate impact on the amniotic-fluid insulin concentration

Table 10 b. The impact of hormones on carbohydrate metabolism in pregnancy

Hormone	Change in pregnancy	Origin	Effect
Adrenaline. noradrenaline	Steady increase during labor.	Adrenal gland.	Diabetogenic.
Growth hormone (STH, HGH)	Decrease (?)	Anterior pituitary.	Antiinsulin.
Prolactin	Continuous increase.	Anterior pituitary.	Impaired glucose tolerance.
Human chorionic gonadotropin (HCG)	1st peak in the 1st trimester. 2nd (lower) peak in the 3rd trimester.	Trophoblast.	Decreased blood glucose.
Human placental lactogen (HPL, HCS)	Increasing from the 6th to the 36th–38th week. Thereafter plateau or decrease	Placenta	Antiinsulin.

Table 10 c. The effect of estrogens and progesterone on carbohydrate metabolism in pregnancy

Hormone	Change in Pregnancy	Origin	Effect
Estrogens	Continuous increase.	Trophoblast.	Insulin enhancing.
Progesterone	Continuous increase.	Trophoblast.	Anti-cortisol.

Mechanism	Remarks	References
Glycogenolysis ↑. Lactic acid-production ↑. Gluconeogenesis from lactic acid Lipolysis ↑. Free fatty acids and glycerine ↑.	No systematic studies in pregnancy.	BELLMANN, 1978; RATHGEN, 1980.
Glucose-utilization ↓. Insulin-secretion ↑. Lipolysis ↓.	STH is inhibited by HPL, progesterone, cortisol. free fatty acids.	BELLMANN, 1978; DAWEK and HÜTER, 1970; RATHGEN, 1980.
Lipolysis ↑. Free fatty acids ↑. Glycerine ↑.	10-fold increase at the end of gestation. NIDD: ↓. IDD: ↓.	BELLMANN, 1978; HOLLINGSWORTH, 1983; RATHGEN, 1980; KIVINEN et al., 1979.
Insulinotrophic, beta-cell stimulating. Inhibition of insulin breakdown.	2nd peak in female fetus. Increased mass of islet organ and increased beta cell : alpha cell ratio (animal experiment).	BELLMANN, 1978; DAWEKE and HÜTER, 1970 HINCKERS, 1978; RATHGEN, 1980; FUHRMANN, 1982.
Glucose utilization ↓. Insulin effect ↓. Free fatty acids ↑. Lipolysis ↑	Glucose tolerance ↓. Despite insulin secretion ↑. Direct stimulatory effect on beta-cells	BELLMANN, 1978; DAWEKE and HÜTER, 1970; PERSSON and LUNELL, 1975; HOLLINGSWORTH, 1983; RATHGEN, 1980; FUHRMANN, 1982

Mechanism	Remarks	References
Sensitivity to insulin in fat muscle tissue ↑. Intestinal resorption of glucose ↑.	♂ Estrogens increase until birth. ♀ increase until the 35th–36th week.	BELLMANN, 1978; HOLLINGSWORTH, 1983; HINCKERS, 1978; RATHGEN, 1980; FUHRMANN, 1982.
Glycogen storage ↑. Peripheral glucose utilization ↑. Insulin synthesis ↑.		BELLMANN, 1978; HOLLINGSWORTH, 1983; HINCKERS, 1978; RATHGEN, 1980; FUHRMANN, 1982.

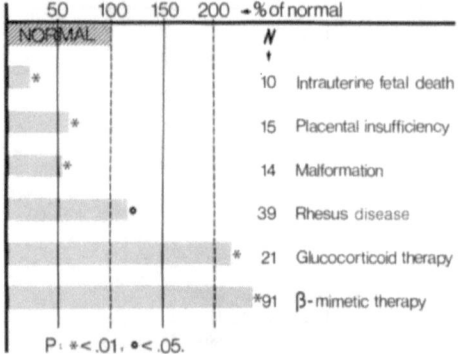

AMNIOTIC FLUID INSULIN CONTENT

50   100   150   200   → % of normal

	N
	10	Intrauterine fetal death
	15	Placental insufficiency
	14	Malformation
	39	Rhesus disease
	21	Glucocorticoid therapy
	*91	β-mimetic therapy

P: * < .01, • < .05.

Fig. 3. Amniotic-fluid insulin levels of 180 patients with complications of pregnancy or drug treatment. The mean period of gestation was approximately the 34 week (WEISS et al., 1984)

glucose twice as rapidly as the adult (BELL-MANN 1978). Investigations by SPELLACY et al. (1980) also indicate that the fetus may influence maternal carbohydrate metabolism. In women with twin pregnancies, he found significantly decreased blood glucose levels under loading despite reduced insulin biosynthesis in these pregnant women.

As a rule, maternal glucose tolerance worsens in the course of a normal pregnancy. However, in pregnant gestational diabetics we observed an apparent improvement of glucose tolerance by an average of 20 mg/dl (1.1 mmol/l) after the 35th gestational week in patients with exaggerated glucose syphoning by the fetus due to fetal hyperinsulinism (WEISS et al., 1984). The metabolic situation of the mother is thus seemingly improved, but that of the fetus is actually worsened. This "biochemical fetopathy" is unnoticed; it can be detected only by the demonstration of an elevated insulin concentration in the amniotic fluid. If allowed to continue biochemical fetopathy gradually induces "somatic fetopathy" with overweight newborns, and increased lipogenesis, visceromegaly, and the typical cushingoid features of the offspring.

SUSA et al. (1984) showed that continuous insulin administration to the fetuses of Rhe-

sus monkeys induces macrosomia. Increasing the fetal insulin level (normally 28 µU/ ml) to approximately 340 µU/ml for three weeks raised the fetal weight to 23% above normal. Increasing the fetal insulin level to 3625 µU/ml increased body weight by 27%. Despite the ten-fold higher insulin level, the difference of macrosomia was only marginal. This indicates that diabetic fetopathy occures above a critical fetal insulin level, and that a further increase causes only gradual differences in fetopathy. Our own studies indicate that "biochemical fetopathy" (hyperinsulinism, hypoglycemia, biochemical dysregulation in the newborn) occurs when the mean amniotic-fluid insulin concentration exceeds 17 µU/ml, and that "somatic fetopathy" (overly high birth weight in addition to the above) occurs above 20 µU/ml. This applies to both insulin dependent and gestational diabetes. Since the absolute values vary according to the methods used in different laboratories, relative values should be preferred. Thus amniotic-fluid insulin values of fetuses with biochemical or somatic fetopathy are on average 2.4 and 2.9 times higher, respectively, than those of healthy offspring.

The course of the amniotic-fluid insulin concentration during pregnancy is also instructive. Elevated levels in the 28th to 30th gestational week that decrease to normal by the 36th week most probably are not associated with any fetal jeopardy while elevated amniotic-fluid insulin values that do not decrease (or even increase) are associated with biochemical or somatic fetopathy (Fig. 4). The critical limit above which fetal insulinemia causes fetopathy is probably lower in humans than in animal experiments. In gestational diabetes hyperinsulinism may last up to 20 weeks and, furthermore, the substrate supply to the fetus is increased, while it is decreased in animal studies of exogenous insulin administration (SUSA et al., 1984).

Until the 24th gestational week, the glucose concentrations possible *in vivo* can only moderately stimulate the fetal beta cells. Only in diabetic pregnancies with poor met-

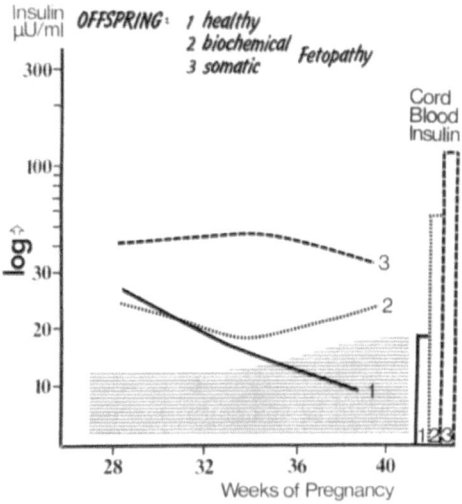

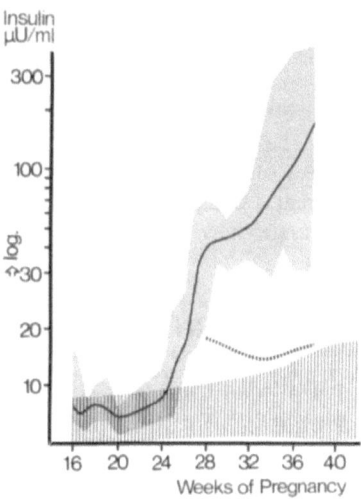

Fig. 4. The course of the mean amniotic-fluid insulin level during pregnancy (45 diabetic women, 183 determinations). *1* healthy offspring; *2* biochemical fetopathy and *3* somatic fetopathy in the newborn. The bars at the right represent the corresponding mean cord blood insulin concentration

Fig. 5. Amniotic-fluid insulin content in 84 diabetic women (White Class A–R) during *conventional* insulin therapy. Solid line: mean, dotted area: zone of dispersion; hatched zone: normal range of the amniotic-fluid insulin content between the 3rd and 97th percentile (WEISS et al., 1984). Broken line: mean amniotic-fluid insulin content in 91 metabolically healthy women during tocolytic therapy (ritodrine)

abolic control after the 26th gestational week can amniotic-fluid insulin levels be found clearly above the two-sigma range of normal (Fig. 5). This agrees with REIHER et al. (1983) who studied fetuses of diabetic mothers and found primarily a volume increase of the islet cells and then an increase of their insulin content only after the 21st to 26th gestational week. Full blown fetal hyperinsulinism and consequent onset of "somatic" fetopathy thus do not occur before approximately the 28th week of pregnancy (Fig. 5). The nutritive placental function considerably influences the fetal glucose supply and hence the stimulation of the fetal pancreatic islet beta cells. Placental insufficiency can result in a fetal undersupply of glucose despite maternal hyperglycemia (ABELL et al., 1976; SARLES and ADAMSONS, 1978). Another important factor in the genesis of diabetic fetopathy appears to be the individual responsiveness of the fetal islet cells to glucose stimulation. Even in twins, one child can have diabetic fetopathy while the other is healthy (BURKE et al., 1979). We have found insulin concentration differences of up to 27 µU/ml in the cord blood of homozygous twins. Additionally, HINCKERS (1978) reported a sex-specific disposition of female fetuses to diabetogenic fetopathy.

Conventional insulin treatment of diabetes in pregnancy (insulin once or twice daily, mean blood glucose > 120 mg/dl, > 6.6 mmol/l) almost certainly leads to fetal hyperinsulinism (Fig. 5). Intensified conventional insulin therapy (three of more insulin doses daily, mean blood glucose < 120 mg/dl, < 6.6 mmol/l) leads to fetal hyperinsulinism in about 30% of pregnancies. There is no threshold value of maternal glycemia that accurately proves or excludes diabetic fetopathy. TALLARIGO et al. (1986) recently reported that glucose tolerance is related to complications of pregnancy even in *nondiabetic* women, *i.e.* women who had normal oral glucose-tolerance test results according to accepted criteria (O'SULLIVAN and MAHAN, 1964). The highest two-hour plasma glucose levels were associated with a significant increase in the incidence of macrosomia (27.5%), congenital abnormalities (5.0%), and toxemia, cesarian section, or both (40%). These data indicate that even limited degrees of maternal hyperglycemia that are currently considered to be within the normal range, may affect the outcome

of pregnancy. Fetal hyperinsulinism can occur with apparently good maternal metabolic control (Persson et al., 1982; Weiss et al., 1978; Weiss, 1979) and with normal maternal glycohemoglobin levels (Burkart et al., 1984). But it can also be missing in fetuses of patients with poor metabolic control.

Screening for disorders of carbohydrate metabolism

Target groups

A general screening of all pregnant women by glucose loading would be ideal and should be aimed at (Freinkel and Josimovic, 1980; Jovanovich and Peterson, 1985). For practical reasons this is not always possible. However, at least all pregnant women with potential diabetes (Table 5), should categorically be screened, even though a recent report suggests that the incidence of gestational diabetes in this group is not as high as usually claimed (Lavin, 1985). In our own patients gestational diabetes was associated especially with obesity and age over 30 years (Table 11). The significance of risk factors is presented in the contribution by Fuhrmann later on.

The fasting blood glucose concentration can also be used to pick candidates for an oral glucose-tolerance test (Pederson, 1977; Fuhrmann, 1982a; Miller and Steinhoff, 1982). If the limit is set at 80 mg/dl (4.4 mmol/l), only every 6th or 7th gravida need undergo the test; this procedure uncovers 88% of gestational diabetics.
Prescreening by the 1-hour blood glucose concentration after a 50 g glucose load has also been suggested (see the contribution by Kitzmiller et al.); a complete test is recommended if prescreening values exceed 130 mg/dl (7.2 mmol/l) (O'Sullivan, 1973) or 140 mg/dl (7.7 mmol/l) (Gillmer, 1980). Lind and Anderson (1984) consider a random blood glucose check as sufficient for prescreening. A tolerance test is recom-

Table 11. Gestational diabetes. Demographic data. Mean (standard deviation)

	Control	Class A	Class AB	Class Bo
Number	28, 126	302	39	18
Age	25 (5)	31 (6)*	31 (6)*	32 (7)*
> 30 years	18%	59%*	69%*	56%*
Parity	1.0 (1.3)	2.2 (1.8)	2.1 (1.8)	1.7 (2.2)
Hight, cm	164 (6)	164 (6)	164 (6)	163 (6)
Prepregnancy weight, kg	61 (11)	67 (13)*	75 (15)*	76 (22)*
Body mass according to Jelliffe	102%	113%*	127%*	134%*
Broca	95%	105%*	117%*	124%*
Weight gain, kg	13.6	11.8	12.0	8.8

* < 0.01 (vs. controls).

mended if the postprandial blood glucose concentration exceeds 110 mg/dl (6.1 mmol/l) within 2 hours, or 101 mg/dl (5.6 mmol/l) thereafter.

Glycohemoglobins are not suited for screening for gestational diabetes since they yield false-positive results in 41% and false negatives in 26% (ARTAL et al., 1984; COUSINS et al., 1984). Detailed data of investigations of glycosylated proteins and hemoglobins by PETERSON and JOVANOVIC as well as CZEKELIUS and ROLLMAN are presented in their contributions.

Period of screening

Carbohydrate metabolism changes dynamically throughout pregnancy (BELLMANN, 1978 a). In the last trimester the fasting blood glucose concentration is 10% to 20% lower than in the non-pregnant state. In contrast, glucose tolerance is decreased (PEDERSEN, 1977), despite increasing insulin biosynthesis (KÜHL, 1975). According to JOVANOVIC and PETERSON (1985), the probability of uncovering gestational diabetes is highest between the 27th and 31st gestational week. They recommend repeating a negative test in obese (> 120% of ideal body weight) or elderly (> 32 years old) pregnant women.

We consider the period between the 20th and 28th gestational week as the best for routine screening, since, in the case of gestational diabetes, enough time for treatment remains. The evaluation before the 28th week of pregnancy must consider the dynamics of glucose tolerance during pregnancy (see p. 22, 24). After the 28th week, a hyperinsulinemic fetus may increasingly distort the results by siphoning off maternal glucose (see p. 18). We repeat glucose tolerance testing after 2 to 4 weeks in cases of borderline results (elevated fasting or 2-hour values only) and in obese and elderly women (> 30 years).

Methods of glucose tolerance testing

Various protocols for intravenous- and oral-glucose tolerance testing during pregnancy have been published, and there is no single universally agreed upon standard. In oral loading 50 g, 75 g, 100 g, 1 g/kg, 1.5 g/kg, etc. glucose are administered (HEISIG et al., 1975; LANG et al., 1978; MILLER and STEINHOFF, 1982). Moreover glucose is measured in capillary blood, venous blood or serum – and by different methods. Evaluation by different criteria impedes the comparison of results. As long as an established standard does not exist, the clinican should use the one with which he is most familiar. The glucose tolerance test by O'SULLIVAN and MAHAN (1966) is the most widely used. We, however, consider a load of 100 g glucose unphysiologically high. Also, researchers in the United States suggest reducing the glucose load to 75 g in the future (Summary and Recommendations of the Second International Workshop Conference on Gestational Diabetes Mellitus 1985). The WHO (1980) also recommends a 75 g glucose load. However, the WHO defines gestational diabetes by a fasting blood glucose value over 140 mg/dl (7.8 mmol/l) and a 2-hour value over 200 mg/dl (11.1 mmol/l). According to ANDERSON et al. (1986), a fasting blood glucose of 140 mg/dl corresponds to the mean plus the 10.3-fold standard deviation of the normal range in healthy pregnant women, while 200 mg/dl at 2 hours corresponds to the mean plus the 4.2-fold standard deviation.

We are throughly convinced that the WHO criteria for gestational diabetes are too high. In a joint study in Vienna and Graz (IRSIGLER et al., 1986), 2260 pregnant women were evaluated. No single case of gestational diabetes was found according to the WHO limits but 22 were identified by O'SULLIVAN's criteria.

Since the weight of pregnant women varies widely we in Graz load with 1 g glucose per kg body weight. We cannot imagine that gravidae weighing 37 kg or 138 kg (a ratio of 1 to 3.7) should be tested with equally large glucose loads. These plain reflections are quite familiar. The internist and the neonatologist frequently calculate dosages according to body surface-area or weight.

Table 12. Glucose tolerance test in 1005 healthy pregnant women. Loading with 50 g, 1 g/kg weight and 100 g glucose

Hours	Weeks of pregnancy	A: 50 g			B: 1 g/kg BW			C: 100 g			Difference (p)		
		x̄	SD	(N)	x̄	SD	(N)	x̄	SD	(N)	A:B	A:C	B:C
0	16–20	70.9	7.9	(62)	70.4	7.7	(70)	71.5	9.3	(72)			
	20–24	69.6	8.0	(119)	69.8	8.3	(91)	70.4	7.6	(93)			
	24–28	70.3	7.0	(75)	70.2	7.3	(71)	70.3	7.9	(84)			
	28–32	70.2	10.4	(45)	69.7	7.4	(60)	70.4	8.9	(47)			
	32–36	69.1	7.5	(34)	71.6	7.3	(43)	70.9	6.7	(39)			
1	16–20	108.6	16.5	(62)	114.0	25.2	(70)	116.0	25.3	(72)	NS	NS	NS
	20–24	108.7	19.6	(119)	107.0	21.5	(91)	115.5	26.3	(93)	NS	NS	NS
	24–28	112.8	24.6	(75)	117.5	27.4	(71)	117.2	20.7	(84)	NS	NS	NS
	28–32	116.1	22.2	(45)	120.7	21.8	(60)	116.0	23.3	(47)	NS	NS	NS
	32–36	121.3	31.9	(34)	121.3	27.1	(43)	127.0	23.4	(39)	NS	NS	NS
2	16–20	82.6	17.8	(62)	91.0	23.9	(70)	99.8	17.1	(72)	<0.05	<0.05	<0.05
	20–24	81.6	14.0	(119)	91.0	23.8	(91)	98.3	20.0	(93)	<0.005	<0.05	0.05
	24–28	82.8	18.9	(75)	93.7	17.1	(71)	97.2	17.9	(84)	<0.05	<0.05	NS
	28–32	88.5	19.3	(45)	94.5	24.7	(60)	103.0	18.3	(47)	NS	<0.05	NS
	32–36	89.6	20.6	(34)	98.9	24.7	(43)	106.0	23.0	(39)	NS	<0.05	NS

x̄ mean, SD standard deviation, BW body weight, NS nonsignificant, N number of cases.

In a randomized pilot study in 1,005 healthy pregnant women we found no significant differences of blood glucose at one hour after loading with 50 g, 1 g/kg, or 100 g glucose (KAINER et al., 1986) (Table 12). The mean 2-hour value showed small differences that diminished as pregnancy advanced (Table 12). Thus, at first glance it seems to make little difference whether the glucose load is 50 g or 100 g or in between. However, using a glucose load corresponding to the patients weight has the following apparent advantages:

– The fasting, 1-hour, and 2-hour blood glucose values are most clearly separated from each other. With a 50 g glucose load, the fasting value and the 2-hour value overlap, indicating underloading. On the other hand, using 100 g the 1-hour and 2-hour values overlap, indicating overloading (Fig. 6).

– The adequacy of the test using 1 g glucose/kg is further shown by the dynamics of glucose tolerance over the course of pregnancy (Figs. 7, 9). Glucose tolerance improves until the 22nd gestational week, and worsens thereafter. This behavior parallels

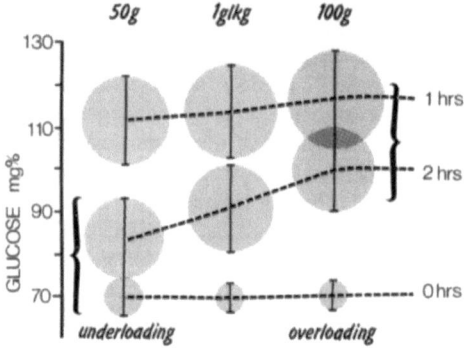

Fig. 6. oGTT results (mean ± SD) in 1005 metabolically healthy pregnant women at the 24 ± 5 gestational week. Groups of 335 probands were tested with 50 g, 1 g/kg, and 100 g glucose loads, respectively. The fasting, 1-hour, and 2-hour blood glucose values are most clearly separated from each other with a 1 g/kg glucose load

both the insulin needs of insulin-dependent diabetics during the course of pregnancy (WEISS and HOFMANN, 1984 b), and the course of glycosylated hemoglobin in healthy gravidae (WIDNESS et al., 1980; PHELPS et al., 1983). This apparently phys-

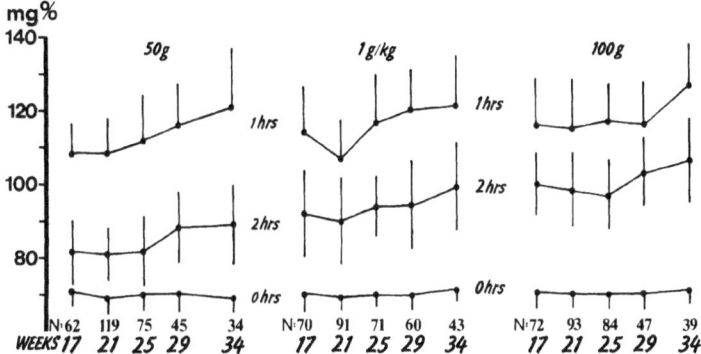

Fig. 7. The course of glucose tolerance during pregnancy under loading with 50 g, 1 g/kg and 100 g glucose. The mean values and standard deviations at 0, 1, and 2 hours were determined at about the 17th, 21st, 25th, 29th, and 34th gestational week

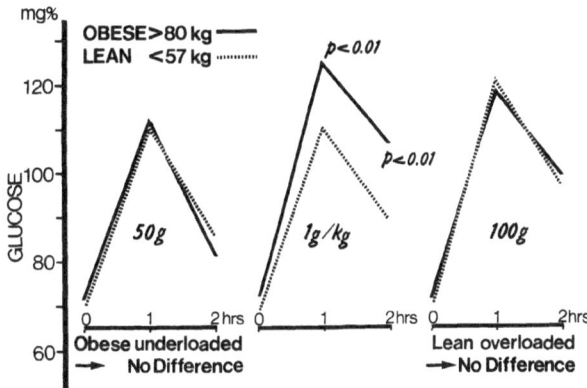

Fig. 8. Oral glucose-tolerance test in 159 lean (< 57 kg) and 146 obese (> 80 kg) pregnant women. Only a glucose load of 1 g/kg distinguishes these two groups significantly

iological pattern is weak with a 50 g glucose load and eliminated by a 100 g load.

– Obesity impairs glucose tolerance (BENTORP et al., 1983). This should become apparent when testing low-weight (< 57 kg) and overweight (> 80 kg) patients. When the test is performed with a glucose load of 1 g/kg, these groups do show significant differences in glucose tolerance (Fig. 8). This difference is lost at a 50 g load since overweight women are underloaded, and at a 100 g load since low-weight women are overloaded.

No specific diet precedes the test, which is carried out in the early morning (about 8 o'clock) after an overnight fast. Weighing glucose doses is tiresome. The glucose dosage is facilitated with 50 g, 10 g, and 1 g dispensers; producers are also willing to supply glucose in 50 g, 10 g, and 1 g tablets. The glucose is disolved in about 250 ml water and drunk within a few minutes; citric acid can be added for taste. Vomitting occured in 50 of over 8000 subjects (0.6%). During the test, the patient is instructed to reduce her activity to a minimum, and not to smoke.

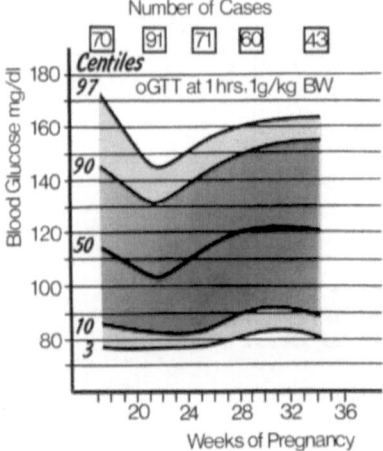

Fig. 9. Blood glucose percentiles 1 hour after a 1 g/kg glucose load. During the course of pregnancy at first an increase and then a decrease of glucose tolerance can be observed

The glucose concentration in the fasting state, and 1 and 2 hours after glucose intake is measured in capillary blood with a hexokinase method.

We use the highest blood glucose value for evaluation. This is usually the 1-hour, sometimes the 2-hour value (delayed gastric emptying?). Between the 26th and 30th gestational week, a value over 160 mg/dl (8.9 mmol/l) is considered borderline; a value over 200 mg/dl (11.1 mmol/l) is considered pathologic. 160 mg/dl corresponds to the 97th percentile of the blood glucose level in healthy pregnant women under loading in the 28th gestational week; 200 mg/dl represents the mean plus 3 standard deviations.

In testing prior to the 26th gestational week glycemia should be evaluated according to the percentiles in Fig. 9. Early testing leads to early therapy which prevents fetopathy. With elevated fasting blood glucose values (> 90 mg/dl, 5.0 mmol/l) which do not peak above normal limits we are rather skeptical concerning the fasting state of the test person and repeat testing.

Further diagnostic procedures

Monitoring blood glucose

A gravida with borderline or pathologic glucose tolerance constitutes a high-risk pregnancy. Consequently, the patient should be seen at least every 2 weeks. After an initial blood glucose profile (see p. 42) the fasting blood glucose level must be determined at these examinations. If it exceeds 110 mg/dl (6.1 mmol/l), a further glucose profile (see p. 43) should be drawn, and the fasting blood glucose concentration should be checked more often. Insulin-dependent diabetes with first manifestation during pregnancy (WHITE Bo, see p. 6) is assumed when the fasting blood glucose concentration repeatedly exceeds 110 mg/dl (6.1 mmol/l), or when the mean blood glucose level of one profile exceeds 130 mg/dl (7.2 mmol/l) (Fig. 1).

Amniotic fluid insulin determination

While maternal glucose threshold values are generally arbitrary, the fetal insulin production reveals whether a disturbance of maternal carbohydrate metabolism is relevant from a fetal point of view. Thus measurement of amniotic-fluid insulin levels identifies those pregnant women who really need insulin to prevent fetopathy. Since the gravida herself can almost always be managed by diet alone, only the fetus, which is most at risk, should present the indication for insulin treatment. In the 1970 s, three pilot studies in our center with 53, 88, and 33 gestational diabetic patients led to a number of fundamental observations. The first study showed that one fourth to one fifth of the newborns of gestational diabetic mothers

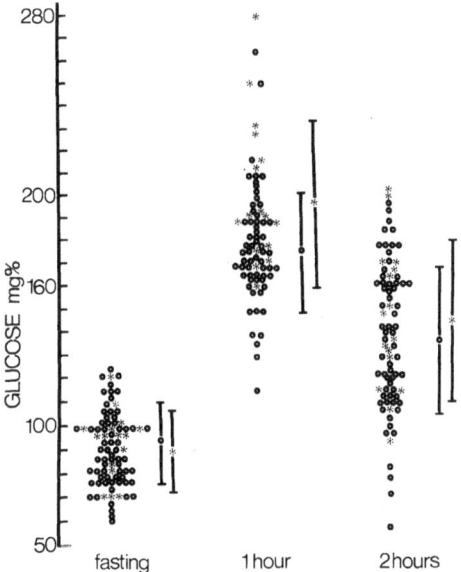

Fig. 10. oGTT results (1 g/kg) of 69 gestational diabetic women with subsequent healthy offspring (circles) and of 19 with fetopathy (overly high birth weight, hyperinsulinism of the neonate and additionally various disorders as shown in Fig. 12) in the newborn (stars). Bars: mean and standard deviation. A considerable overlap of the single values can be observed

showed fetopathy (overly high birth weight, fetal hyperinsulinism, and consequently often various disorders additionally described on page 52 ff.) despite maternal dietary treatment. The probability of fetopathy increases with the degree of glucose intolerance: 11% of the newborns had fetopathy when maternal blood glucose peaked between 160 to 200 mg/dl (8.9 to 11.1 mmol/l) after glucose loading and 36% when maternal blood glucose exceeded 200 mg/dl. In the individual patient, the prediction of fetopathy is not possible from the glucose-tolerance test results, since the ranges of blood glucose values in gestational diabetes with and without fetopathy overlap considerably (Fig. 10).

A further study demonstrated that other maternal parameters such as fasting blood glucose, mean blood glucose, or glycohemoglobin also fail to predict fetopathy accurately (Table 13). However, amniotic-fluid insulin was highly significantly elevated in every pregnancy with fetopathy (Table 13, Fig. 11).

The third pilot study showed in matched groups that in gestational diabetics with elevated amniotic-fluid insulin levels, insulin treatment prevented fetopathy, whereas dietary treatment did not (Fig. 12). Diet alone prevented diabetic fetopathy only in those patients with normal amniotic fluid insulin levels (WEISS et al., 1987).

These studies led us to the following conclusions:

● Maternal parameters cannot be used to accurately predict or exclude fetal diabetogenic harzards.
● Fetopathy is associated with elevated amniotic-fluid insulin.
● Insulin treatment is indicated by an elevated amniotic fluid insulin level.

The resulting diagnostic and therapeutic procedures are outlined in Fig. 13.

Table 13. Maternal and fetal data in gestational diabetes (mean values)

	Without fetopathy	With fetopathy	Difference
Number of cases	69	19	
Fasting blood glucose	93.0 mg/dl	91.8 mg/dl	NS
Mean blood glucose	87.0 mg/dl	93.0 mg/dl	NS
HbA$_1$	5.8%	6.2%	NS
Amniotic fluid insulin	7.1 µU/ml	20.7 µU/ml	< 0.001

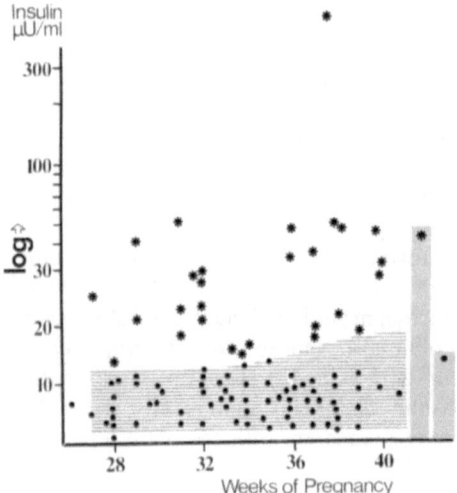

Fig. 11. 103 amniotic-fluid insulin values from 75 gestational diabetic women. Hatched zone: normal range (3rd to 97th percentile); dots: 75 values in 61 patients with subsequent delivery of a healthy infant; stars: 28 values of 14 women with the subsequent birth of a child with diabetogenic fetal disease; bars: the corresponding mean cord blood insulin concentration

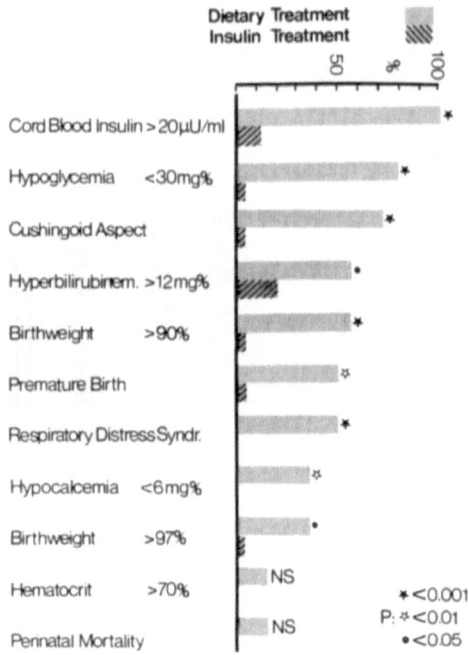

Fig. 12. Fetal outcome in gestational diabetics with *elevated* amniotic-fluid insulin concentration (Class AB). Dietary versus insulin treatment. Hatched bars: after diet alone (n = 14); dark bars: after combined dietary and insulin therapy (n = 19). The superior effect of insulin therapy is obvious (Weiss et al., 1987)

The time of the first amniotic-fluid insulin examination is determined by the onset of ability of the fetal islet organ to be hyperstimulated, approximately in the 28th gestational week (see p. 19). If gestational diabetes is recognized late, *i.e.* in the 32nd gestational week or later, amniotic insulin should be determined as soon as possible. According to our experience, fetal insulin homeostasis should be rechecked in the 35th to 36th week. Adequate insulin treatment can be demonstrated by normal amniotic-fluid insulin levels. Should amniotic-fluid insulin be elevated, maternal insulin dose must be increased. On the other hand, we have seen patients developing fetal hyperinsulinism – and thus require insulin treatment – late in pregnancy. During adequate insulin therapy, the amniotic-fluid insulin typically declines to normal levels, while staying high in inadequately treated patients. Figure 14 shows the typical course of amniotic fluid insulin levels in patients with adequate insulin treatment and healthy neonates who underwent 3 or more aminocenteses for additional reasons (suspected Rh hemolytic disease, determination of lung maturity, etc.). It also shows the course of amniotic-fluid insulin in patients who declined insulin treatment or who reduced therapeutic insulin doses of their own accord (neglectors). The offspring of these patients had marked fetopathy.

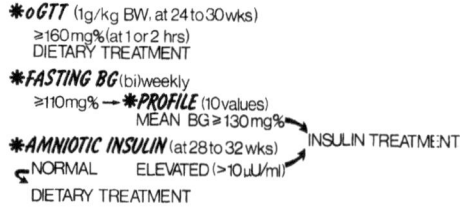

Fig. 13. Diagnostic and therapeutic plan to uncover and treat gestational diabetes

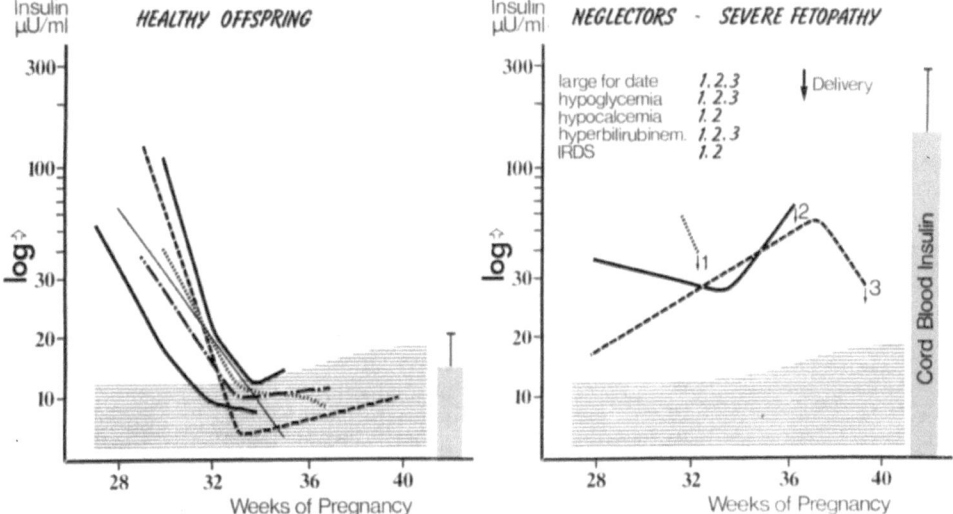

Fig. 14. The course of the amniotic fluid insulin concentration in 6 appropriately treated gestational diabetic women (left) and 3 neglectors refusing insulin therapy (right). While the newborns of the neglectors had severe fetopathy (overly high birth weight, hypoglycemia, respiratory dystress syndrome, hyperbilirubinemia, etc.), the others were completely healthy. Hatched zone: normal range of the amniotic fluid insulin concentration (3rd to 97th percentile); dotted bars: cord blood insulin level (mean, standard deviation)

Amniocentesis poses neglectible risks in experienced hands, especially after the 28th gestational week. In about 2,000 procedures on insulin-dependent and gestational diabetics, the only complication was a 3 cm hematoma of the abdominal wall. The procedure also seems justified in light of the 11% of tests that reveal fetal hyperinsulinism and thus the need for maternal insulin treatment. Genetic amniocentesis, which, since done earlier, has a higher rate of complications, shows an abnormal result in 1% to 5% only (WINTER and HOFMANN, 1985).

Methods of determing amniotic fluid insulin

Insulin was first demonstrated in the amniotic fluid in the early 1970s (CASPER and BENJAMIN, 1970; SPELLACY et al., 1973; DRAISY et al., 1977), but only later was it used as a prognostic fetal parameter in pregnancies complicated by maternal diabetes mellitus (WEISS et al., 1975; NEWMAN and TUTERA, 1976; WEISS et al., 1978, 1979, 1984; BURKART et al., 1984).

The reliability of the determination of insulin in the amniotic fluid and fetal urine (see p. 54, 55) depends on the methods used. A study at the University of Graz Department of Obstetrics compared five commercially available radioimmunoassays (RIA). The low protein content of the amniotic fluid (1/16th that of the serum; BENZIE et al., 1974) causes precipitation methods (PEG) and double-antibody methods to yield poorly reproducible results (WEISS et al., 1984 c). Especially the precipitation methods vary with the protein content of the samples since they are adjusted to the serum protein level. In contrast, well reproducible results were obtained by a solid-phase method on a sephadex basis (Phadebas®, Phadeseph®, Pharmacia Uppsala).

Because of the adsorption of insulin to glass, amniotic fluid should be stored in plastic vials. Freezing and thawing the samples once has no effect on the test results, and neither do small amounts of blood or meconium in the amniotic fluid. Amniotic fluid must not be filtered since most of its insulin is adsorbed by the filter.

Table 14a. Insulin levels in amniotic fluid of healty pregnant women between the 13th and 25th weeks of gestation (N = 530) (WEISS et al., 1964)

Week	Case no.	Insulin levels in amniotic fluid (μU/ml)	
		Mean values	SD
13	4	1.3	1.9
14	8	2.0	3.2
15	11	2.5	2.0
16	121	4.3	3.3
17	126	3.5	2.6
18	146	3.9	2.5
19	60	3.7	2.3
20	38	3.9	2.5
21	8	4.8	1.8
22	4	4.9	1.5
*			
25	4	5.1	1.4

SD standard deviation.
* During the 23rd and 24th weeks, only values from disturbed pregnancies were available.
Reprinted with permission from The American College of Obstetricians and Gynecologists (WEISS et al. [1984] Obstetrics and Gynecology 63: 371).

In gestational diabetes, no insulin antibodies are expected in the amniotic fluid. But they are also irrelevant in pregnancies of insulin-dependent diabetic mothers. FUHRMANN and KEILACKER (1983) found no insulin antibodies in 100 amniotic fluid samples of insulin-dependent diabetics (White Class B to R).

PERSSON et al. (1982) and TCHOBROUTSKY et al. (1980) found insulin antibodies in 7% and 2% of amniotic fluid samples, respectively. However, the insulin-binding capacity of these antibodies is only 0.08–0.22 μU/ml – causing an error of < 1% – and thus negligible for the analysis of the amniotic fluid insulin level. Tables 14a and b shows the normal amniotic insulin content of metabolically healthy gravidae, but it must be remembered that every such definition depends on the technique used. Consequently, every laboratory should determine its own normal values.

Fetal insulin production is also reflected by the C-peptide concentration in the amniotic fluid. In fetal hyperinsulinism, the insulin level correlates linearly with the C-peptide concentration (LIN et al., 1981; PERSSON et al., 1982; BURKART et al., 1984). However, we prefer the determination of amniotic fluid insulin level to that of C-peptide.

It is not yet possible to assess the accuracy of C-peptide RIAs since some of the different kits lack immunochemical identity between the standard and the specimen. Reports in the literature about C-peptide concentrations vary by a factor of up to four (BEISCHER, 1982). Cross-reactions with proinsulin and C-peptide fragments as well as C-peptide antibodies have been reported (BEISCHER, 1982; SOSENKO et al., 1979; PEA-

Table 14b. Insulin levels in amniotic fluid of healthy pregnant women between the 27th and 42nd weeks of gestation. (N = 458) (WEISS et al., 1984)

Week of gestation	Case no.	Insulin levels (μU/ml)			Percentiles rounded off				
		Mean	SD	Range	3	10	50	90	97
27/28	45	6.0	2.9	1.5–12.8	2.0	3.2	6.0	10.4	11.2
29/30	36	6.5	3.0	1.8–13.5	2.0	3.2	6.1	10.8	11.7
31/32	63	7.2	3.4	1.0–14.5	2.0	3.2	6.2	11.3	12.2
33/34	86	6.7	3.4	0.3–14.4	1.9	3.0	6.3	11.7	12.8
35/36	80	6.9	3.7	1.2–15.5	1.8	2.9	6.4	11.8	13.5
37/38	70	7.0	3.9	1.0–20.0	1.5	2.8	6.6	12.2	14.9
39/40	60	7.3	4.4	1.0–20.0	1.2	2.8	6.8	13.4	17.2
41/42	18	9.1	5.7	1.0–23.0	1.1	3.4	7.5	15.3	18.0

Reprinted with permission from The American College of Obstetricians and Gynecologists (WEISS et al.[1984] Obstetrics and Gynecology 63: 371).

COCK et al., 1983). As for the insulin measurements, the low amniotic fluid protein content can be assumed to negatively influence the results of the C-peptide analysis.

At low amniotic fluid insulin levels, *i.e.* beneath the tenth percentile, the fetus is being deprived of glucose – especially when pre-eclampsia or signs of prenatal dystrophy (ultrasound biometry, oligohydramnios) are also present. Its insulin synthesis is depressed by the decreased glucose stimulation (DE-PRINS et al., 1983; VORHERR, 1982; ABELL et al., 1976; PHILLIPS et al., 1969; VAN ASSCHE et al., 1977; GABBE and QUILLIGAN, 1977). This under-supply can be linked to or independent of maternal glycemia (SARLES and ADAMSONS, 1978). An undersupply of glucose to the fetus significantly increases perinatal mortality (ABELL et al., 1976; PHILLIPS et al., 1968), and may be detrimental to the development of the fetal brain (STEHBENS et al., 1977; ZUPPINGER et al., 1981). In cases of prenatal dystrophy, too tight metabolic regulation with insulin should be avoided since it promotes fetal dystrophy. We recommend adjusting the maternal insulin dose to raise the mean blood glucose by about 30 mg/dl (1.7 mmol/l) in cases of fetal growth retardation and amniotic fluid insulin levels below the 10th percentile.

The determination of glucose in the amniotic fluid

If an amniotic fluid sample of a diabetic gravida is available, the measurement of its glucose content provides additional clinical information. Reports about the glucose concentration in the amniotic fluid are based mainly on the last trimester (SEEDS et al., 1979; WOOD and SHERLINE, 1975; CASSADY et al., 1977; ARCHIMAUT et al., 1974; SPELLACY et al., 1973; PEDERSEN, 1954; NEWMAN and TUTERA, 1976). However, since the glucose content of the amniotic fluid is dynamic over the course of a pregnancy, a normal range can be established only by large series. Between 1980–1984, these considerations led us to determine the glucose level in about 2,500 amniotic fluid samples from gravidae in the 14th to 42nd week (WEISS et al., 1985). Table 15 shows the normal values and percentiles.

The origin of the glucose in the amniotic fluid is not quite clear. It is thought to reach the amniotic fluid via the fetal urine, at least in the second half of the gestation (WOOD and SHERLINE, 1976; SPELLACY et al., 1973). We have found that the glucose concentration in the urine of newborns and in the amniotic fluid at the end of pregnancy are virtually identical (WEISS et al., 1985).

Table 15. Statistical data of 1655 amniotic fluid glucose values of healthy pregnant women (WEISS et al., 1985)

Gestational week	Case no.	Mean values (mg/dL)	SD	Range	Percentiles				
					3	10	50	90	97
14/15	26	44.8	11.8	26–78	26	31	42	55	64
16/17	428	45.9	9.8	13–86	29	35	45	57	67
18/19	537	44.2	9.2	21–79	28	33	44	57	64
20/21	117	41.3	10.4	18–92	24	29	41	52	59
22–27	46	36.1	10.0	15–70	15	23	35	47	50
28/29	80	32.4	9.1	5–53	16	21	33	43	49
30/31	44	30.1	10.4	7–57	12	18	29	41	49
32/33	66	30.2	12.0	3–61	4	15	32	45	50
34/35	88	27.5	9.9	3–60	11	14	26.5	39	46
36/37	76	26.1	9.8	6–50	11	15	26	39	46
38/39	54	22.2	9.4	7–47	8	11	20	34	40
40–42	93	15.8	6.6	1–37	4	9	15	24	27

Table 16. Mean glucose values in amniotic fluid in pathological pregnancies (Weiss et al., 1985)

	Sample no.	Mean gestation period (week)	Mean glucose level (mg/dl) (normal mean value)	Percent of normal mean value
Malformations	50	28.7	18.4 (32.4)	56.8[a]
IDD AFI ↑	102	32.2	62.2 (30.2)	206.0[a]
IDD AFI ⊥	128	31.6	46.4 (30.2)	153.6[a]
oGTT ↑ AFI ↑	54	32.2	50.5 (30.2)	167.2[a]
oGTT ↑ AFI ⊥	192	33.9	35.6 (27.5)	129.5[a]
Hydramnios	115	31.7	28.8 (30.2)	95.4[b]

IDD insulin-dependent diabetics; *AFI* amniotic fluid insulin; ↑ elevated; ⊥ normal; *oGTT* oral glucose tolerance test.
[a] $p < 0.001$.
[b] Not significant.
Reprinted with permission from The American College of Obstetricians and Gynecologists. (Weiss et al. [1985] Obstetrics and Gynecology 65: 333).

The maternal blood glucose level correlates both with fetal blood glucose and amniotic fluid glucose. The maternal/fetal/amniotic fluid glucose gradient at the end of the gestation is approximately 12/8/2 (Feige et al., 1979). The amniotic fluid glucose level follows maternal glycemic excursions at an interval of about 250 minutes but with markedly lower amplitudes (Weiss et al., 1984 b; see also Fig. 2). Thus amniotic fluid glucose represents a "memory-like system" for the maternal metabolic situation over the past few hours (Archimaut et al., 1974).

The glucose concentration in amniotic fluid is increased in the diabetic state if metabolic control is not normoglycemic. This applies equally to IDDM and to disorders of glucose tolerance in pregnancy (Table 16). Low levels of amniotic fluid glucose have been associated with closure defects of the neural tube (Pettit et al., 1977; Guibaud et al., 1978). In 29 cases of neural tube defects and 16 of other fetal malformations, we found the amniotic fluid glucose to be beneath the tenth percentile in 68% and beneath the third in 44% (Weiss et al., 1984 b) (Tables 15, 16). Thus, the possibility of a major fetal malformation should be considered if the amniotic fluid glucose level is beneath the tenth percentile (Table 15). In hydramnios without fetal malformation, the amniotic fluid glucose level is not decreased.

Treatment

Dietary treatment

Dietary treatment of gestational diabetes is indicated when glucose tolerance tests are borderline or pathologic. The diet is the same as that prescribed for insulin-dependent diabetics but it has evolved considerably in the last decade.

Caloric restriction to 1,800 to 2,000 kcal (7,530 to 8,370 kJ) per day is the fundamental dietary principle (Matzkies et al., 1982). A rule of thumb assumes a caloric requirement of 30 to 35 kcal (125 to 146 kJ) per kg of ideal body weight. Ideal body weight is calculated most easily by the Broca index:
(Height in cm − 100) × 0.9 = ideal weight in kg.
Comparing the ideal weight with the reference standard of body mass according to Jelliffe (1966), which is based on comprehensive Canadian-Amercian insurance company data, shows that the Broca index determines an ideal weight too low for short women and too high for tall women, but for hights between 155 cm and 175 cm, the correlation between the reference standard and

the Broca index is close enough. Since the Broca index is easy to use and since most women are between 155 and 175 cm tall, it is commonly used.

In pregnancy, the patient's ideal weight must be corrected by the physiologic weight gain of pregnancy, which is largest between the 5th and 7th lunar month (Scientific Tables, GEIGY, 1982). The 50th percentile of the weekly weight gain in healthy gravidae is 0.42 kg between the 13th and 20th week; 0.48 kg between the 20th and 30th week; 0.42 kg between the 30th and 36th week, and 0.37 kg between the 36th and 40th week of gestation (HYTTEN, 1982).

A complete pregnancy requires 25,000 kcal (104,600 kJ) over its entire course (EMERSON, 1975). This amounts to an average additional requirement of only 90 kcal (376 kJ) per day. Thus a pregnant woman need not eat for two, as the saying goes. It is especially important for the diabetic gravida to avoid excess food intake and weight gain because, apart from leading to metabolic difficulties, the risk of preeclampsia may be increased.

Carbohydrate restriction in diabetes mellitus is obsolete. Cross-over studies have shown that a diet rich in carbohydrates (and fiber) beneficially influences pre- and postprandial blood glucose, triglycerides, and glycohemoglobins (SIMPSON et al., 1981). Additionally, such a diet reduces the patient's insulin requirement (NEY et al., 1982). Carbohydrates thus should make up 50% to 60% of the pregnant diabetic's diet (about 215 g, 18 bread units), and over 12 g should be fiber. Carbohydrates with a low blood glycemic index are preferrable. Monomeric carbohydrates and free starch should be avoided. The antiquated concept of bread units should be dropped, since it encourages the patient to eat white bread (glycemic index 69) which, while not prohibited, is undesirable (MEHNERT, 1979).

Recent studies at the University of Graz Department of Obstetrics evaluated the absorption kinetics of common carbohydrate-supplying foodstuffs (EDLINGER, 1986). The artificial pancreas (Biostator) was used to follow the course of the blood glucose concentration continuously; serial determinations were done also of insulin, C-peptide, and cortisol after the intake of isocaloric amounts of potatoes, pasta, white bread, or rice (1 g carbohydrate/kg).

The absorption curves of the same foodstuff differed widely among the individual subjects, making a transverse comparison difficult (Fig. 15). Consequently, longitudinal studies were done on 28 healthy probands, since different carbohydrates show qualitatively very similar resorption patterns in the individual subject (Fig. 15).

Glucose was administered first (orally), and the resulting resorption profile used as the reference for the other foodstuffs in that proband. Plotting the blood glucose values after glucose intake against blood glucose values at the same time intervals after intake of other isocaloric carbohydrates produces a regression line (Fig. 16). The smaller the slope of the regression lines, the smaller the blood glucose increase produced by the carbohydrate in question, and consequently, the smaller the stimulation of insulin production (Figs. 16, 17). Potatoes are not an ideal carbohydrate source for diabetics, and should be eaten only sparingly. The blood glucose profile after the intake of potatoes resembles that after the intake of glucose, both qualitatively and quantitatively (Fig. 15). Similar findings have been reported by CRAPO et al. (1976), BANTLE et al. (1983), and COULSTEN et al. (1984). This property is the result of the loose structure of the starch granule in the potato bulb. Boiling, which causes some hydrolysis of starch to limit-dextrins and to di- or monosaccharides, transforms the carbohydrate polymer made up of 1,4-α-glycoside bonds into the sol state. This accelerates the enzymatic cleaving of the polymer and leads to a blood glucose surge (WAKHLOO et al., 1984).

Rice is by far the most suitable carbohydrate source for diabetics. Its resorption profile is protracted, and the blood glucose increase is small (Figs. 15, 16). Rice should be favored

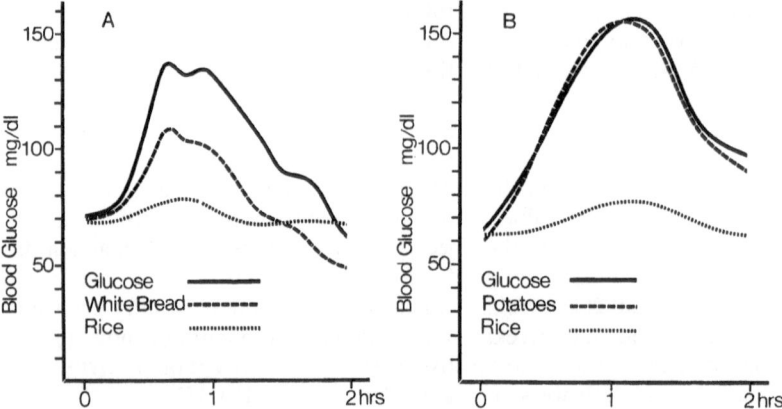

Fig. 15. Resorption patterns after the intake of various carbohydrates in two healthy probands (**A** and **B**) with continuous monitoring by the artificial pancreas (Biostator). The curves demonstrate an individual resorption pattern in different probands. The intake of rice does not lead to a surge of blood glucose while the resorption of potatoes is similar to pure glucose

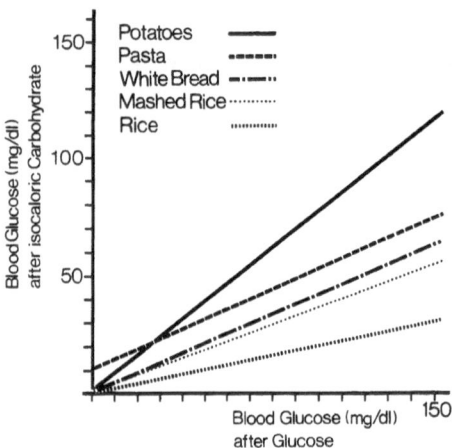

Quotient	Glucose		Insulin		C Peptide		Cortisol	
Glucose Potatoes								
Glucose Pasta								
Glucose W. Bread								
Glucose M. Rice								
Glucose Rice								
Quotient	<1	≥1	<1	≥1	<1	≥1	<1	≥1

Fig. 17. R × C (rows and columns) table of the blood-glucose-, insulin-, C-peptide-, and cortisol-quotients (< 1, > 1) after a glucose load and a load of a specific carbohydrate in the same patient (n = 28)

Fig. 16. Regression lines resulting from the plotting of blood glucose values after glucose intake (abscissa) against those after the intake of another isocaloric carbohydrate (coordinate). The smaller the slope of the regression line, the smaller the blood glucose increase produced by the carbohydrate in question

when planning a diet. Rice seeds contain large carbohydrate amounts with 1,4-α-1,6-α glycoside bonds (amylopectin), which soak up water when boiled, but which are difficult to hydrolize (MANNERS, 1974). The favorable resorptive characteristics of rice are also a

result of its granular form. After mashing, the resorptive profile becomes less ideal (Fig. 16).

Legumes have not been studied as extensively, because in our area they have only a minor role as a carbohydrate source. However, preliminary studies indicate a favorable resorptive profile.

The pregnant woman has a high protein requirement (about 25% of the total caloric requirement). The daily protein intake should be 1.5 to 2.0 g/kg, and preferably not less than 100 g per day. When choosing protein, the patient should beware of hidden fat. Lean meat and fish are preferrable to processed meats (sausages, spreads); cheeses should be low-fat.

In conventional diabetes diets, a carbohydrate-protein-fat ratio of 40 to 20 to 40 was suggested. But, as mentioned above, a carbohydrate portion of 50% to 60% and a protein portion of 25% are preferred; thus fat should make up only 15% to 25% of the caloric intake. Hidden fats must be considered when planning the diet. Fats as spreads on bread (for example butter, lard) should be omitted. In oils and fats for cooking and salads, those with polyunsaturated fatty acids (linoleic acid, arachidonic acid) are preferrable; they are supposed to have a beneficial influence on microangiopathy (MATZKIES et al., 1982).

A low-sodium and high-potassium intake is advantageous. Calcium, magnesium and phosphorous requirements are also increased in pregnancy. Comprehensive information on the protein, fat, carbohydrate, water, bulk, cholesterol, vitamin, and mineral content of about 800 foods is given in the "Big Nutritional Value Table" (CREMER et al., 1983).

Daily food intake is divided into three main meals and three snacks. Eating many small meals at unconventional times of day does not correspond to the eating habits of our culture, and labels the diabetic as a social outsider; the patients rarely accept such a regimen. Established eating habits must be respected when planning a diet. It is, for example, pointless to divert calories for metabolic reasons from breakfast to a late meal if the patient then goes hungry in the mornings. Insulin treatment must be adapted to eating habits, not eating habits to insulin treatment.

A carbohydrate-restrictive diet was usually prescribed for gestational (White Class A) diabetes (AUINGER, 1978). However, re-

stricting carbohydrates favors the formation of ketones (metabolic catabolism) (POTTER et al., 1982). Ketones also increase promptly when the daily caloric requirement (about 1,500 kcal; 6,280 kJ) is not met, since the pregnant diabetic slips into a catabolic state easier than the non-pregnant diabetic ("accelerated starvation", FREINKEL, 1964; FELIG and LYNCH, 1970; HOET, 1977). Because ketones may cause embryonal malformations, ketosis is to be strictly avoided in early pregnancy. Thus a diet according to the guidelines described above (balanced in calories and rich in carbohydrates and fiber) is important especially in early gestation. In gestational (latent) diabetes, a fiber-rich diet frequently normalizes the abnormal postprandial insulin peak (FRASER et al., 1983). There are no experimental studies indicating that ketones are harmful to the fetus after the embryonal period. The fetus can even use ketones as an energy source, thereby protecting itself from maternal starvation ketosis (RUDOLF and SHERWIN, 1983). Earlier reports of a detrimental effect of ketones on fetal brain development (CHURCHIL et al., 1969) did not consider that it is not ketosis or ketoacidosis that affect the fetus, but the metabolic dysregulation that causes the ketosis. STELDINGER and coworkers describe on p. 176 ff., a diabetic patient with continuous ketosis and normal offsprings in three pregnancies because of an Atkins diet.

If the diabetic patient is of normal weight, and her eating habits not highly imprudent, then no drastic dietary changes need be imposed because of a pregnancy – the patient would hardly accept them. Smaller corrections should be proposed to comply with the strategy described above. The patient's willing and able cooperation are essential to dietary treatment.

The daily insulin requirement is essentially an individual value, and is only slightly affected by dietary adjustments. Recent studies with the artifical pancreas (Biostator, Live Science Instruments) show that the insulin requirement with isocaloric food intake stays practically constant when the patient varies her carbohydrate intake. The same is true when she eats constant carbohydrate amounts within meals with varying caloric quantity (NAJEMNIC et al., 1981). The interval between meals and insulin administration depends on the onset and peak of insulin action. and on the route of administration (subcutaneous, intravenous, intraperitoneal). Since the peak effect of regular insulin occurs about 1.5 hours after subcutaneous administration, while carbohydrate absorption peaks about 1 hour after meals, regular insulin should be injected approximately 30 minutes before a meal. It must also be considered that human short-acting insulin is absorbed somewhat faster than regular insulin derived from animals, and that carbohydrate absorption peaks later as the gestation advances. Consequently the appropriate insulin-meal interval must be tailored to the individual.

Hypoglycemic episodes should not be treated with sugar (sugar water, dextrose, sweetened tea). The prompt absorption of sugar is potentiated by counter regulatory mechanisms resulting in erratic fluctuations and subsequent instability of blood glucose. In case of subjective hypoglycemia and a blood glucose level under 70 mg/dl (3.9 mmol/l) (self tested), the patient should drink a glass of milk (250 ml) and recheck her blood glucose after 15 minutes. If it is still under 70 mg/dl the procedure should be repeated (JOVANOVIC et al., 1982).

SPELLACY et al. (1977) report a relative deficiency of vitamin B_6 (pyridoxine) in gestational diabetes. They describe a therapeutic dose of 100 mg pyridoxine daily as causing a decrease of plasma insulin and glucose concentrations, as well as a significant improvement of glucose tolerance. GILLMER and MAZIBUKO (1979), however, were unable to reproduce these findings.

Insulin treatment

Insulin treatment in gestational diabetes is primarily for the benefit of the intrauterine patient, who demands much stricter metabolic control than the gravida herself. Undisturbed embryonal and fetal development requires normoglycemia, e.g. a mean maternal blood glucose of 70 to 100 mg/dl (3.9 to 5.5 mmol/l) with the smallest possible fluctuations (IRSIGLER, 1978; TAMAS et al., 1981; SERVICE et al., 1970). Also ideal insulin treatment should imitate the physiologic hyperinsulinism and the decrease of mean blood glucose that occur in normal pregnancy. Insulin treatment must be tailored to the individual patient considering the features mentioned above.

Insulin is indicated when, despite diet, fasting blood glucose values repeatedly exceed 110 mg/dl (6.1 mmol/l), and in patients with a mean blood glucose exceeding 130 mg/dl even when their fasting blood glucose is below 110 mg/dl. These are patients who first manifest insulin-requiring diabetes in pregnancy (Bo according to the Graz classification). More often insulin administration is indicated by amniotic-fluid insulin levels that exceed the 97th percentile of normal (Class AB diabetes) (see p. 6 and Fig. 1).

Class AB diabetic patients are usually Type-II diabetics with elevated plasma insulin levels and peripheral insulin resistance (MUCK and HOMMEL, 1977; TURNER et al., 1977; DEMPE et al., 1978; HOMMEL and MUCK, 1977). As a rule, these gravidae have an especially high insulin requirement (HOFMANN and WEISS, 1986), which is due to the above mentioned insulin resistance and to the anti-insulin placental hormones (Table 10) that dominate mainly the last trimester of pregnancy (JOVANOVIC and PETERSON, 1982). Especially human placental lactogen (HPL) seems to be elevated in gestational diabetic patients (FUHRMANN et al., 1980).

Insulin treatment of Class AB diabetes requires some reconsideration of conventional medical wisdom. At first glance, it seems odd to give very high insulin doses to a patient with "*normal*" blood glucose values (Fig. 19). The first effect of therapeutic insulin, next to down-regulation of insulin receptors, is the inhibition of the patient's own insulin (hyper-)secretion; a distinct therapeutic effect is at first absent. Only above a

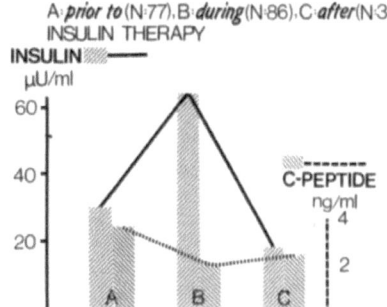

Fig. 18. Serum insulin (solid line) and C-peptide (broken line) prior to (n = 77), during (n = 86), and after (n = 31) insulin treatment (HOFMANN and WEISS, 1987)

certain, individually different dose does the mean blood glucose level begin to decline. This mechanism can be demonstrated by sequentially monitoring serum insulin and C-peptide. Appropriate insulin treatment more than doubles the serum insulin level, while the C-peptide level declines by half (Fig. 18). Hence, endogenous 24-hour insulin production declines by about 25 to 30 units; a therapeutic effect of exogenous insulin can thus be expected only when the insulin dose exceeds this amount. During insulin treatment, the insulin/C-peptide ratio (μU/ml/ng/ml) rises on average from 8 to over 100. If the ratio is under 30, then the insulin dose is probably too low.

The insulin requirement can be determined easily with an artificial pancreas (Biostator). Fig. 19a shows the metabolic adjustment with the aid of an artifical pancreas in a 33-year-old Class AB diabetic patient in the 27th gestational week. While the Biostator continually monitors blood glucose, we administer 8 U of regular insulin every hour subcutaneously. The insulin requirement (or tolerance) is 144 U/24 hours. With this device, the blood glucose profile is normoglycemic, and shows no exaggerated counterregulation, fluctuation, or hypoglycemia. Since endogenous insulin production must first be inhibited, and the deficit caused by pregnancy covered, the insulin requirement at the beginning of treatment is higher than that in the further course. Accordingly the

expected daily insulin requirement can be estimated by subtracting the insulin amount of the first 6 hours from the total dose (Table 17: Patient 1). The distribution within a day can be estimated by adding up 6-Biostator-hour blocks, or by the treatment schedules in Table 18 and 19.

A similar biostator procedure is used when instituting insulin treatment in a patient previously treated with sulfonylureas. These are also Type-II diabetic patients with a high insulin requirement because of peripheral insulin resistance. Oral antidiabetic agents must be discontinued long enough before beginning insulin administration to avoid an overlapping effect. Figure 19b shows the course of a 36-year-old Class-Bo diabetic in the 14th gestational week who was switched from oral medication to insulin therapy. Despite a high initial subcutaneous insulin dose (236 U/24 hours), no hypoglycemia occurs. The dose is then reduced, as in Class-AB diabetics. After a few days, the dose was set at 180 U/24 hours in 6 doses every 4 hours (Table 17: Patient 2).

Diabetes that first appears during pregnancy may be MOD(Y) Type desease (*anticipated Type-II Diabetes*) (see p. 4) with a high insulin requirement, or JOD Type desease (*delayed Type-I Diabetes*) with a low insulin requirement. Type-I diabetic patients are mostly normal- or underweight women without peripheral insulin resistance. They are sensitive to insulin, so their relatively low insulin requirement must be determined carefully. The metabolic management is otherwise unproblematic, as in all diabetic patients with residual beta-cell function. The metabolic condition remains stable, and they do not tend to larger glycemic excursions or to hypoglycemia. Figure 19c shows the initial insulin adjustment of a 27-year-old Class-Bo Type-I diabetic women in the 21st gestational week. Because of the high initial blood sugar value, treatment is begun with high hourly insulin doses, leading to a rapid drop of the blood glucose. Thereafter, subcutaneous insulin doses are administered when blood glucose exceeds 90 mg/dl

Table 17. Metabolic control prior to and after biostator adjustment

Age	Classification	Biostator-adjustment	Gestational week	Insulin-dosage	Insulin U/24 h	6h	8h	9h	12h	15h	18h	21h	24h	3h	6h	Mean blood glucose	HbA$_1$ prior to-, 1 month after biostator adjustment
Patient 1 33 years	AB	prior to	27	0	0	80	92	148	67	91	104	103	71	74	80	91	7.6%
		after	28	Actrapid 36/28/24/8 U	96	66	104	56	104	81	75	94	82	86	68	82	7.0%
Patient 2 36 years	B$_0$	prior to	14	Glutril 1/—/1	0	95	100	157	113	102	116	110	73	63	93	102	10.6%
		after	15	Actrapid 40/20/40/ 20/40/20 U	180	89	123	84	96	112	100	65	61	82	70	88	6.8%
Patient 3 28 years	B$_0$	prior to	21	0	0	196	—	345	231	153	137	143	287	176	205	208	9.5%
		after	22	Actrapid 16/8/4 U	36	124	—	111	108	78	192	186	163	62	101	125	5.8%

Glucose-profile (mg/dl) (in hours)

BIOSTATOR

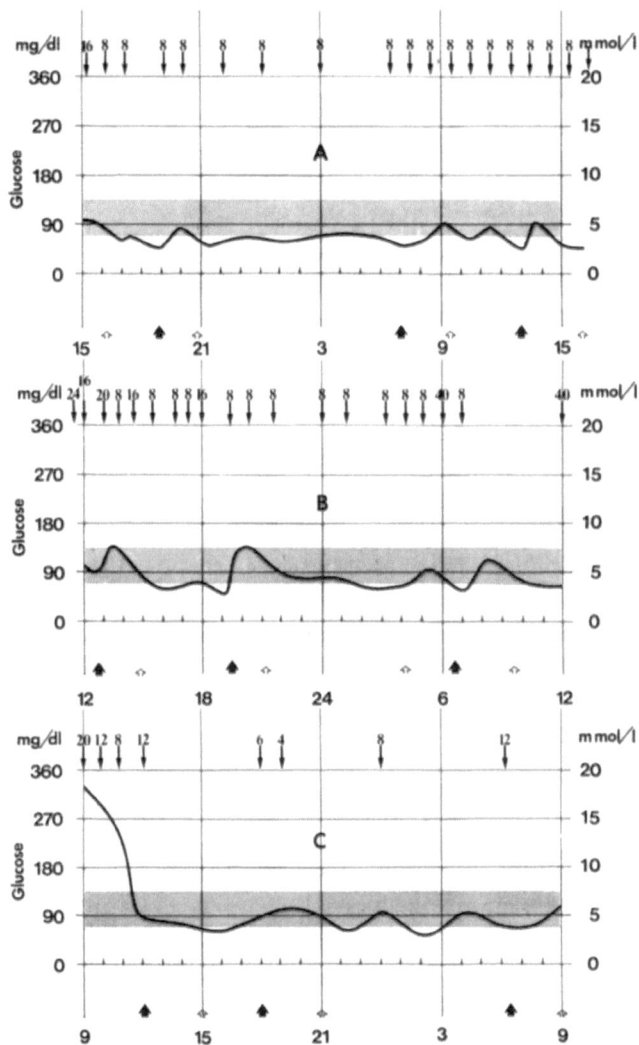

Fig. 19. Metabolic adjustment by subcutaneous insulin injections during blood glucose monitoring by an "artificial pancreas" (Biostator) (WEISS and HOFMANN 1985). **A** a Class AB diabetic; **B** a Class Bo diabetic after oral antidiabetic therapy, and **C** a Class Bo patient with first manifestation of diabetes in pregnancy. Downward pointing arrows: insulin administration; solid line: course of blood glucose, gray zone: range of normoglycemia; upward pointing arrows: meals and snacks; horizontal: time scale. Detailed data are provided in the text

(5.0 mmol/l) and at the main meals. The appropriate insulin requirement for subsequent treatment can be calculated by subtracting the first 3 doses (that were necessary to cover the insulin deficit and to bring blood glucose into the normal range) from the total in the first 24 hours (Table 17: Patient 3).

A Biostator is seldom available, and so the insulin requirement must usually be determined empirically. This may be done according to the principle of maximum insulin tolerance (ROVERSI et al., 1979). Insulin doses are increased stepwise until slight signs of hypoglycemia appear. Then the corresponding dose is reduced by 4 U. As a rule of thumb approximately 0.7 to 1.5 U/kg per 24 hours are tolerated. To avoid insulin-antibody induction, human insulin is preferable in women undergoing insulin treatment for the first time.

We consider a single dose of depot insulin inappropriate for treatment of gestational diabetes because it further elevates the already abnormal high basal insulin level (HOLLINGWORTH, 1983; PERSSON and LUNELL, 1975). Apart from insulin resistance gestational diabetes is primarily a result of the delayed insulin response to nutritional stimuli (PERSSON and LUNELL, 1975; CHENEY et al., 1985; KÜHL, 1986) (Fig. 20). Thus we consider it important to cover the main meals with fast-acting, short-lasting regular insulin. Once the individual insulin requirement has been determined, the further metabolic management of Type-AB diabetic patients is straightforward. Since a residual function of the islet organ persists, and since treatment usually starts only late in pregnancy, the insulin requirement of Class AB diabetics does not increase as gestation advances.

For the distribution of the insulin dose over 24 hours we have developed three therapeutic schedules (WEISS and HOFMANN, 1984 b, 1985). They use regular insulin alone, or a combination of short-acting insulin with intermediate or long-acting insulin, respectively (Fig. 21). The distribution and combination of the insulin types is the logical

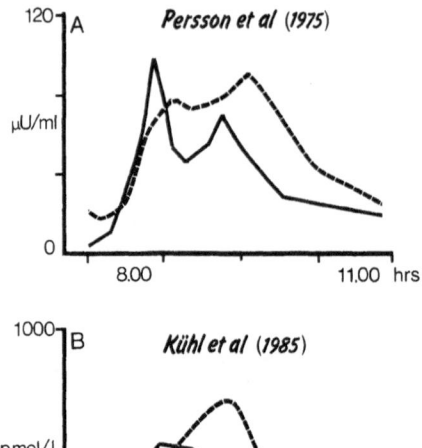

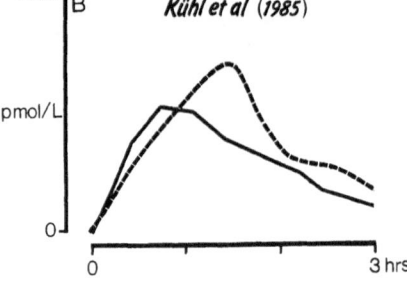

Fig. 20. Plasma insulin response in healthy (solid line) and gestational diabetic (broken line) pregnant women after normal food intake (top) and after a glucose load (bottom)

consequence of their pharmacokinetics, specifically their absorption kinetics and the duration of their activity (SCHLICHTKRULL, 1977). The insulin requirement is highest in the morning hours, independently of food intake. This is probably related to the circadian peak of cortisol secretion in the morning hours. We mainly used Aktrapid MC and Monotard MC (mostly as Humanactrapid and Humanmonotard), but with small variations these schedules can also use other appropriate insulin types. Figure 21 b–d shows the absorption patterns of the insulin combinations used in the respective schedules. They add up to summation curves that almost perfectly imitate the insulin-secretion profile of metabolically healthy gravidae (LEWIS et al., 1976) (Fig. 21 a).

When planning treatment, one must consider that the duration of depot insulin activity is about one third shorter in pregnancy than outside of pregnancy (BARANYI et al., 1981). Thus in pregnancy Monotard for ex-

Table 18. Distribution of four doses of short-acting insulin (Aktrapid) over 24 hours (WEISS and HOFMANN, 1984)

Insulin requirement U/24 h	Distribution in units Aktrapid				Insulin requirement U/24 h	Distribution in units Aktrapid			
	6^{00}	12^{00}	18^{00}	24^{00}		6^{00}	12^{00}	18^{00}	24^{00}
100%	39%	24%	25%	12%	100%	39%	24%	25%	12%
28	10.92	6.72	7.00	3.36	68	26.52	16.32	17.00	8.16
30	11.70	7.20	7.50	3.60	70	27.30	16.80	17.50	8.40
32	12.48	7.68	8.00	3.84	72	28.08	17.28	18.00	8.64
34	13.26	8.16	8.50	4.08	74	28.86	17.76	18.50	8.88
36	14.04	8.64	9.00	4.32	76	29.64	18.24	19.00	2.12
38	14.82	9.12	9.50	4.56	78	30.42	18.72	19.50	9.36
40	15.60	9.60	10.00	4.80	80	31.20	19.20	20.00	9.60
42	16.38	10.08	10.50	5.04	82	31.98	19.68	20.50	9.84
44	17.16	10.56	11.00	5.28	84	32.76	20.16	21.00	10.08
46	17.94	11.04	11.50	5.52	86	33.54	20.64	21.50	10.32
48	18.72	11.52	12.00	5.76	88	34.32	21.12	22.00	10.56
50	19.50	12.00	12.50	6.00	90	35.10	21.60	22.50	10.80
52	20.28	12.48	13.00	6.24	92	35.88	22.08	23.00	11.04
54	21.06	12.96	13.50	6.48	94	36.66	22.56	23.50	11.28
56	21.84	13.44	14.00	6.72	96	37.44	23.04	24.00	11.52
58	22.62	13.92	14.50	6.96	98	38.22	23.52	24.50	11.76
60	23.40	14.40	15.00	7.20	100	39.00	24.00	25.00	12.00
62	24.18	14.88	15.50	7.44	102	39.78	24.48	25.50	12.24
64	24.96	15.36	16.00	7.68	104	40.56	24.96	26.00	12.48
66	25.74	15.84	16.50	7.92	106	41.34	25.44	26.50	12.72

ample must be considered an intermediate, not a long-acting insulin.

Schedule 1. Treatment with four doses of regular insulin (Fig. 21 b)

The distribution over 24 hours is 39%, 24%, 25%, and 12% at about 6 a.m., 12 noon, 6 p.m., and midnight, respectively. This distribution of insulin stays constant, even when the total insulin requirement rises during pregnancy. If the actual insulin requirement is known, the individual doses can be found in Table 18. This schedule is suited primarily to patients who have residual basal insulin secretion. It is thus used mainly in Class AB, Bo and White-Class B diabetics. Due to the duration of regular insulin activity, it must be administered at least every 6 hours.

Schedule 2. Treatment with three doses of regular insulin and one of intermediate insulin (Fig. 21 c)

This schedule is a modification of the first. 33%, 21%, and 20% of the daily insulin dose is given at 6 a.m., 12 noon, and 6 p.m. respectively; 26% is given as intermediate insulin also at 6 p.m. The ratio of the single doses stays constant even when the total requirement rises. If the daily requirement is known the single doses can be found in Table 19. This schedule is indicated in the same patients as in Schedule 1. Its advantage is that no insulin is administered at midnight.

Schedule 3. Treatment with three doses of regular insulin and one of long-acting insulin (Fig. 21 d)

We are increasingly using a plan with three doses of Actrapid prior to the main meals and additionally one dose of Ultratard, a

Table 19. Distribution of 3 doses of short acting insulin (Actrapid) and one dose intermediate acting insulin over 24 hours (Weiss and Hofmann, 1984)

Insulin require-ment U/24 h	Distribution in units				Insulin require-ment U/24 h	Distribution in units			
	AR 6^{00}	AR 12^{00}	AR 18^{00}	MT 18^{00}		AR 6^{00}	AR 12^{00}	AR 18^{00}	MT 18^{00}
100%	33%	21%	20%	26%	100%	33%	21%	20%	26%
28	9.24	5.88	5.60	7.28	68	22.44	14.28	13.60	17.68
30	9.90	6.30	6.00	7.80	70	23.10	14.70	14.00	18.20
32	10.56	6.72	6.40	8.32	72	23.76	15.12	14.40	18.72
34	11.22	7.14	6.80	8.84	74	24.42	15.54	14.80	19.24
36	11.88	7.56	7.20	9.36	76	25.08	15.96	15.20	19.76
38	12.54	7.98	7.60	9.88	78	25.74	16.38	15.60	20.28
40	13.20	8.40	8.00	10.40	80	26.40	16.80	16.00	20.80
42	13.86	8.82	8.40	10.92	82	27.06	17.22	16.40	21.32
44	14.52	9.24	8.80	11.44	84	27.72	17.64	16.80	21.84
46	15.18	9.66	9.20	11.96	86	28.38	18.06	17.20	22.36
48	15.84	10.08	9.60	12.48	88	29.04	18.48	17.60	22.88
50	16.50	10.50	10.00	13.00	90	29.70	18.90	18.00	23.40
52	17.16	10.92	10.40	13.52	92	30.36	19.32	18.40	23.92
54	17.82	11.34	10.80	14.04	94	31.02	19.74	18.80	24.44
56	18.48	11.76	11.20	14.56	96	31.68	20.16	19.20	24.96
58	19.14	12.18	11.60	15.08	98	32.34	20.58	19.60	25.48
60	19.80	12.60	12.00	15.60	100	33.00	21.00	20.00	26.00
62	20.46	13.02	12.40	16.12	102	33.66	21.42	20.40	26.52
64	21.12	13.44	12.80	16.64	104	43.32	21.84	20.80	27.04
66	21.78	13.86	13.20	17.16	106	34.98	22.26	21.20	27.56

long-acting depot human insulin, at 10 p.m. (Hofmann and Weiss, 1986). The distribution is approximately 30%, 20%, and 20% of the daily insulin dose as Humanactrapid and 30% as Ultratard. Actrapid is administered with the Novopen (Novo, Copenhagen) a new device that makes dosage easy and application almost painless. Thus the treatment is more acceptable to the patient. However, the Ultratard dose must be carefully adjusted; a too large dose may lead to protracted hypoglycemia, especially in the early morning hours. Detailed data are given in the contribution of Hofmann et al. on page 142 ff.

These three schedules meet the needs of most patients. Only rarely must individual plans be developed. Sometimes, especially if the insulin requirement is very high, the patient can be managed only with 5 or 6 scattered doses of regular insulin, since a single dose of more than 40 U is of no value (Table 17).

As to the insulin pump, we consider it inappropriate for the treatment of gestational diabetes.

Some reports in the literature are not encouraging regarding insulin treatment of gestational diabetes (Kalkhoff, 1985). The reason may be the lack of fetal parameters to distinguish patients requiring insulin from those adequately managed by diet alone. Beyond that, the reported daily insulin doses of 10 to 30 units (O'Sullivan et al., 1966; O'Sullivan et al., 1974) are too low. Treatment directed towards fetal needs (as determined by repeated amniotic-fluid insulin evaluations) requires a mean daily insulin dose of about 65 units in Class AB-patients and about 90 units in Class Bo-patients (Weiss et al., 1986; Hofmann and Weiss, 1986). These doses cause the mean blood glucose level to sink by 15 to 20 mg/dl (0.8 to 1.1 mmol/l) on average (Table 20).

Should amniocentesis and amniotic fluid in-

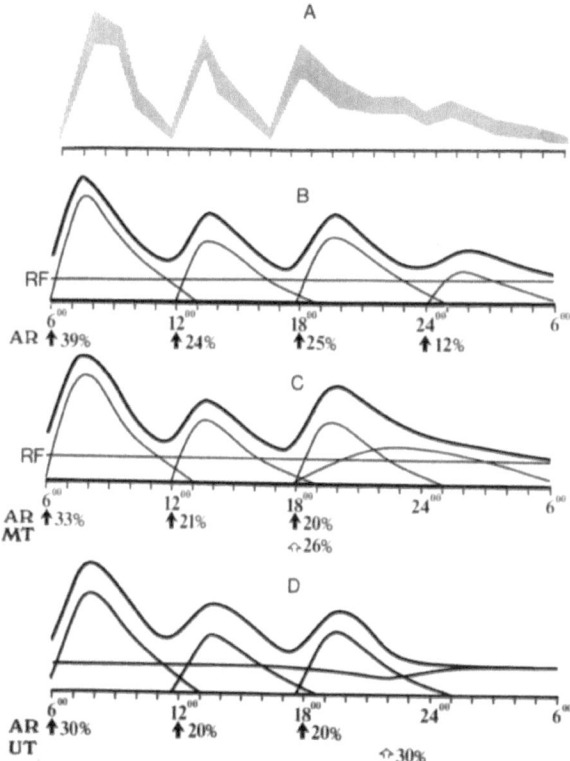

Fig. 21. Patterns of insulin absorption in Schedules 1, 2 and 3. The total insulin requirement is regarded as 100%. The individual doses are expressed in percentages of the total insulin requirement. The single peaks correspond to the percentile distribution of insulin. They add up to a summation curve corresponding approximately to the resorption pattern of the individual schedule. Scale: time (h), arrows: time of insulin injection; black arrows: regular insulin; white arrows: long-acting insulin; *AR* Actrapid; *MT* Monotard; and *UT* Ultratard. **A** Pattern of insulin secretion in six pregnant nondiabetics (Lewis et al., 1976). *RF* residual function of the islet organ

sulin determination not be feasible for legal or technical reasons, or if there is no laboratory available for amniotic fluid insulin determination, we would suggest prophylactic insulin therapy in gestational diabetes as the best alternative described by Coustan in his contribution later on.

Metabolic control during insulin treatment

Quality of metabolic control

Both the mother's mean blood glucose level and the stability of her metabolism influen-ce fetal outcome (Artal et al., 1983). The mean blood glucose (MBG) value, the mean amplitude of glycemic excursions (MAGE) (Service et al., 1970), the M value (Schlicht-krull et al., 1965), and the mean of daily differences (MODD) (Molnar et al., 1972) are widely used parameters to define metabolic stability; the methods and formulae for their determination are summarized by Nelson and Service (1980).

The calculation of these values requires continuous or hourly blood glucose checks, which is practically impossible. However, Molnar et al. (1974) report that the differ-

Table 20. Gestational diabetes. Evaluation of 139 women with potential diabetes (WEISS et al., 1986)

	Group 1: metabolically normal (n = 51)	Group 2: Class A (n = 69)	Group 3: Class AB (n = 19)	Difference between group 2 and group 3
Age	28.0 ± 5.7	28.4 ± 6.1	31.4 ± 6.0	p = 0.03
Parity	1.6 ± 1.9	1.6 ± 2.0	2.4 ± 2.0	NS
Fasting blood glucose mg/dl	82.4 ± 7.3	93.0 ± 15.4	91.8 ± 16.7	NS
HbA$_1$ in %	5.3 ± 1.8	5.8 ± 0.9	6.2 ± 0.9	NS
Amniotic fluid insulin µU/ml 27th–32nd week	6.3 ± 3.1	7.1 ± 2.0	20.7 ± 10.9	p < 0.01
Therapy	none	diet	diet/insulin	
MBG prior to insulin therapy mg/dl		87.0 ± 19.0[a]	93.0 ± 18.0	NS
MBG during insulin therapy mg/dl			82.0 ± 10.2 ⟶ p < 0.01	
Insulin requirements U/24 hours			64.6 ± 29.5	
Mean birth weight, g	3355 ± 545	3505 ± 593	3385 ± 566	NS
Mean birth length, cm	50.0 ± 2.2	50.2 ± 2.2	49.4 ± 3.2	NS
Mean gestational age, weeks	38.7 ± 2.5	39.7 ± 1.9	38.9 ± 1.3[b]	NS
Insulin in umbilical cord blood, µU/ml	10.0 ± 3.6	12.4 ± 6.8	11.1 ± 4.6	NS

[a] MBG from diagnosis of GD to delivery.
[b] 1 × birth of twins in the 36th week.

ence between the blood glucose values before and 80 minutes after breakfast correlates well with the MAGE value, and the MBG correlates with the mean of the FBG value and the blood glucose value 80 minutes after breakfast, while the MODD agrees with the difference between 2 FBG values on consecutive days. In gestational diabetes, the blood sugar values between midnight and 6 a.m. reflect the mean blood glucose. The average of the blood glucose values before breakfast, before lunch, and 2 hours after dinner correlates especially closely with the mean blood glucose (RIZVI et al., 1980). We find that the metabolic stability can be assessed adequately by the standard deviation of the blood glucose values (at least ten) over 1 week (HOFMANN et al., 1986). A standard deviation under 30 to 40 mg/dl (1.6 to 2.2 mmol/l) reflects stable metabolism.

Apart from the quality of metabolic management, the individual disposition, the residual beta-cell function, and the stage of the pregnancy influence metabolic stability. Residual beta-cell function stabilizes metabolism (LUTTERMAN et al., 1981; SERVICE and NELSON, 1980).

Monitoring blood glucose

The overwhelming majority of gestational diabetic patients are stable, and do not tend towards hypoglycemia. Their insulin requirement stays constant as the pregnancy progresses. Thus, gestational diabetics need not have their blood glucose checked as often as insulin-dependent diabetics. Gestational diabetics should have a 24-hour blood glucose profile with 9 or 10 single values drawn every 10 to 14 days, depending on their individual metabolic stability. We determine the blood glucose level at 6 a.m., 8 a.m., 9 a.m., noon, 3 p.m., 6 p.m., 9 p.m., midnight, and 3 a.m. This profile shows both preprandial and postprandial glycemia around the clock. Treatment aims to keep preprandial blood glucose values under 90 mg/dl, postprandial values under 140 mg/dl, and the MBG under 100 mg/dl.

Between the 24-hour profiles, the patient's FBG should be checked daily or every second day, depending on her individual metabolic stability. Additional checks should be done in case of suspected hyper- or hypoglycemia, dietary mistakes, exceptional psychological or physical stress, intercurrent illness, etc. Such close-meshed control of blood glucose requires self-monitoring. This is done with blood glucose test strips. These strips are evaluated with a reflectometer, which provides results very close to those obtained in the laboratory with enzymatic methods (IRSIGLER et al., 1982). The reflectometer is preferred to evaluation by eyesight alone, since the diabetic's color perception can be altered long before retinopathy appears, especially in Type I diabetes (LaCROIX et al., 1982).

To check the accuracy of self monitoring, the patient should, before an outpatient examination, draw a 24-hour blood glucose profile at home using the test strips, and simultaneously draw a microcapillary of blood (10 µl). These capillaries are transferred to numbered, sealed vials containing hemolysing reagent (Gluco-quant, Boehringer Mannheim; Gluc-DH, E. Merck, Darmstadt). The hemolysate can be stored in a refrigerator for at least 1 week. In the hospital, the blood glucose level of these samples is determined and compared with the test-strip profile results.

Glycosylated hemoglobin

The glycosylated hemoglobins can be used to retrospectively evaluate glycemia in the previous 8 weeks (O'SHAUGHNESSY et al., 1979; SCHERNTHANER et al., 1980; NATHAN et al., 1984).

Either total glycosylated hemoglobins (HbA_1) or HbA_{1c} (the main fraction of the minor components) is determined. These 2 parameters seem to be of equal clinical value. An HbA_1 value of 7.0% (or an HbA_{1c} value of 5%) corresponds to a MBG of 87 mg/dl (4.8 mmol/l); on average an HbA_1 value of 11% (HbA_{1c} value of about 8%) corresponds to a MBG of about 200 mg/dl (11.0 mmol/l) (BERGER and SONNENBERG, 1980).

In metabolically healthy women, both a decline of the glycosylated hemoglobins with advancing pregnancy (WIDNESS et al., 1980) and a bipolar course with the nadir in the 24th gestational week (PHELPS et al., 1983) have been described. A significant correlation with age is reflected in a 0.053% annual increase of glycosylated hemoglobin (GRÄFENSTEIN and DUCHNA, 1981). Elevated HbA_1 values are found after tocolytic treatment with beta-mimetic agents (NIEDERAU et al., 1981).

Glycosylated hemoglobin is solely a maternal parameter and cannot be used to predict diabetogenic fetopathy (O'SHAUGHNESSY et al., 1979; GRÄFENSTEIN and DUCHNA, 1981; FADEL et al., 1979; BURKART, 1984). But monthly determinations of glycohemoglobin levels in the diabetic gravida give a good line on the trend of long-term metabolic regulation (KJAERGAARD and DITZEL, 1979). Comprehensive studies on this field are presented in the sections by PETERSON and JOVANOVIC and by CZEKELIUS and ROLLMANN.

Patient education

Normoglycemic metabolic management of gestational diabetes requires the patient's full cooperation. She can be educated to manage the disorder to a large extent on her own (BERGER et al., 1983; JOVANOVIC et al., 1980; SKYLER et al., 1980). This drastically reduces the time she spends as an in-patient, and saves money (GOLDSTEIN et al., 1983).

Table 21. Gestational diabetes. Complications in pregnancy (1980–1985)

	Control	Class A	Class AB	Class Bo
Number	28, 126	302	39	18
Preeclampsia	3.4%	11.0%[a]	28.2%[a]	27.7%[a]
Bacteriuria (> 10^5)	8.6	12.7%[a]	35.8%[a]	27.7%[a]

[a] < 0.1.

Patients are best educated using previously published guidelines (DANKMEIJER, 1981; TRAVIS and HÜRTER, 1980; MEHNERT and STANDL, 1975; ROBBERS and TRAUMANN, 1980; BERGER et al., 1981).

Monitoring pregnancy, labor, and the puerperium

Apart from the blood glucose checks and the determinations of amniotic-fluid insulin discussed above, the gestational diabetic is followed as any high-risk patient. Depending on the individual risk, she is examined more often than the healthy gravida. To recognize excessive intrauterine growth early, the fetus must be observed especially closely. The gestational diabetic is at an above normal risk of toxemia and of urinary tract infection (Table 21), so her urine should be examined regularly.

Arrest of preterm labor with β-mimetic tocolytic agents

The incidence of premature labor and delivery is markedly elevated in patients with gestational diabetes (see Fig. 12). We observed an incidence of approximately 17% in Women with Class AB or Bo disease. Beta-mimetic agents influence carbohydrate metabolism by reducing glucose tolerance (URBAN et al., 1972). After their administration, metabolically healthy women show an increase of blood glucose and free fatty acids as well as increased glucagon and insulin levels (LIPSHITZ and VINIK, 1978; KAUPILLA et al., 1978); the mean blood glucose rises by about 40 mg/dl (2.2 mmol/l) (SMYTHE and SAKUKINI, 1981). Especially at the beginning

of treatment with β-mimetic agents, the insulin requirement of diabetes can rise by a factor of up to four (ROST et al., 1982). Frequent blood glucose checks and appropriate adjustments of the insulin dose are necessary to avoid metabolic derangements. ELLIOT (1983) reported promising results in a study of 355 patients after using magnesium sulfate for tocolysis. Since this compound does not affect carbohydrate metabolism it appears to be a good alternative for the arrest of preterm labor.

Induction of fetal lung maturation

Glucocorticoids increase blood glucose and consequently the insulin requirement. Ambroxol (Mucosolvan®), which has no effect on carbohydrate metabolism, is an alternative. It is administered intravenously (1 g/d) (LORENZ et al., 1974; ZAHN et al., 1978; MÜLLER-TYL and SALZER, 1978). This treatment can reduce the incidence of respiratory distress syndrome (RDS) in preterm infants by up to 75% (LÖWENBERG et al., 1981).

At the **onset of labor**, insulin administration should be discontinued, since the insulin requirement decreases sharply with labor and delivery. As in every high-risk gravida, the fetus should be monitored continuously during labor. A protracted duration of labor, an increased incidence of shoulder dystocia, and a higher rate of cesarean section have been associated with infants weighing more than 4,000 g (MODANLOU et al., 1980; MODANLOU et al., 1982; PARKS and ZIEL, 1978), but other authors (CALANDRA et al., 1981; RUGE and ANDERSON, 1985) did not find an increased risk of obstetric complications or

Table 22. Gestational diabetes. Data on newborns

	Mean (standard deviation)		
	General deliveries	Group A	Group AB + Bo
Number	28, 126	302	57
Week of pregnancy	39.7 (2.1)	39.7 (1.9)	38.6 (2.0)[a]
Mean birthweight	3390 (520)	3346 (529)	3170 (586)[a]
> 90. percentile	9.8%	8.9%	12.3%
Hypoglycemia (< 30 mg%)	no data	1.7%	12.3%
Cord blood insulin μU/ml (N = 764)	6.9 (5.0)	9.9 (6.4)[a]	13.4 (10.5)[a]
Hyperbilirubinemia (> 12 mg%)	8.3%	7.9%	17.5%[a]
Perinatal mortality	1.0%	1.0%	1.7%
Cesarean section rate	11.0%	8.9%	28.0%[a]

[a] $p < 0.1$.

cesarean section. We found an average birth weight of 2,910 g (under the 50th percentile) in the infants of our Class AB and Bo patients who had to be delivered by cesarean section; only 1 infant weighed over 4,000 g (Table 22).

Significant *postpartal bleeding* is described in 15.6% of mothers with a high-birth-weight infant, as compared to 0.07% of those with a normal-weight infant (PARKS and ZIEL, 1978). Thus uterotonic agents or prostaglandins are recommended after the delivery of a high-birth-weight infant.

5.2% of infants weighing over 4,000 g develop *postpartal hypoglycemia* (GOLDITCH and KIRMAN, 1978).

After delivery, women with gestational diabetes who were treated with insulin during pregnancy (especially depot insulin) can develop hypoglycemia. This is a result of a sudden imbalance between the antiinsulin placental hormones that decline precipitately at delivery and the exogenous insulin that continues to be absorbed from the subcutaneous tissue. Hence insulin treated gestational diabetics must be observed for signs of hypoglycemia on the first postpartal day. True gestational diabetic women do not require further insulin after delivery. However, WARD and his colleagues (1985) found a significantly elevated basal insulin level, a decreased insulin response to glucose stimulation, and decreased sensitivity to insulin in normoglycemic women with a history of gestational diabetes. They consider it as a preliminary stage of manifest diabetes (see also p. 50, postpartal screening for diabetes, and the sections by OATS et al. and MESTMAN later on).

Birth weight

The birth weight is especially important in the diagnosis of carbohydrate metabolism disorders, since a large-for-date newborn is often the first sign of maternal diabetes.

Valid standards are necessary to evaluate the birth weight. The Denver standards (LUBCHENCO et al., 1963) are the most widely used, but they are not suited for Europe because of differences in race, socioeconomic environment, and altitude. The Denver weight percentiles are markedly under those of the new German standards (HOHENAUER, 1980), with differences of up to 230 g. Other American, British, Scottish, and Swedish standards are also higher than the Denver standards (TANNER, 1970).

In addition to regional differences, a number of other factors must be considered before

Table 23. Factors influencing birth weight

	Number of deliveries evaluated	Factors evaluated	Additional information and aids
Thomson et al., 1968; Newcastle	52,000	Maternal: Weight, height, parity External: Social status Fetal: Sex	Tables of percentiles, table for correction according to maternal height and weight
Brenner et al., 1976; North Carolina	30,772	Maternal: Parity, age, race External: Social status Fetal: Sex	Tables of percentiles, curves of percentiles, diagrams for correction of weight according to parity, race, and sex
Wilcox, 1981; North Carolina	Not provided	Maternal: Race External: Social status, smoking Fetal: Twins, triplets, quadruplets	Growth curves for multiple pregnancies
Hohenauer, 1980; Linz	7,180	Maternal: Weight, parity External: — Fetal: Sex	Tables of percentiles, comparison of 8 different published standards

a diagnosis of abnormally high or low birth weight can be made (Table 23). Maternal height and weight is associated with a ± 400 g variation of birth weight. A variation of ± 100 g is associated both with parity and the infant's sex. Thus the weights of two healthy, mature newborns can differ by up to 1,000 g. Table 23 shows the factors that should be considered in addition to the gestational age before classifying a newborn as overweight (hypertrophic, large-for-date) or underweight (hypotrophic, dystrophic, small-for-date).

The over-4000 g newborn

Newborns over 4,000 g have always been noted. This empirical upper limit of normal birth weight for term newborns stems from the time before valid standards were available and continues to be widely applied. It corresponds to about the 90th weight percentile of the new German standards (Hohenauer, 1980), when male and female infants are considered together.

If a woman with recognized gestational diabetes is delivered of an infant weighing over 4,000 g, inadequate treatment must be suspected first. It can be excluded only by demonstrating normal levels of insulin production in the newborn (see p. 54).

If an infant of over 4,000 g is born of a seemingly uncomplicated gestation, an unrecognized disorder of carbohydrate metabolism should be sought.

Between 6% and 10% of all newborns weigh more than 4,000 g (Table 24). Overly high birth weight is associated with maternal age over 35 years, multiparity, previous heavy children, post term delivery, maternal obesity at the beginning of pregnancy, excessive weight gain during pregnancy, gestational diabetes, and diabetes mellitus (Abell et al., 1976; Parks and Ziel, 1978; Schindler and Meyfort, 1979; Modanlou et al., 1980; Gross et al., 1980; Calandra et al., 1981). Thus, genetic, uteroplacental, fetal, and nutritional factors are responsible for high birth weight (Vorher, 1982). Among the genetic factors are primary racial traits. The predominance of male infants is also genetically determined; it rises with increasing birth weight (Schindler and Meyfort, 1979). We found 70% of infants over 4,000 g, 76% of those over 4,500 g, and 82% of those over 5,000 g to be males (Table 25).

A genetic cause is also likely when metabolically healthy women bear large-for-date

Table 24. The incidence of newborns weighing more than 4,000 g

Number of newborns > 4000 g evaluated	4,000 g	4,500 g	5,000 g	Proportion of males	References
1099	9.3%	1.2%	0.1%	66.9–94.1%	SCHINDLER and MEYFORT, 1979; Tübingen
~5000	7.0–10.7%	1.3%		69.7%	MODANLOU et al., 1980; 1982; Long Beach
137		0.86%[a]		71.6%	OATS et al., 1980; Melbourne
2909	7.1%	0.83%	0.1%	n. d.	AUINGER, 1978; Wien
398	5.2%	0.3%	0.02%	n. d.	FISCHL et al., 1981; Wien
801	8.1%	1.3%	0.14%	62.0%	GOLDITCH and KIRKMAN, 1978; San Francisco
2394	6.3%	0.7%	0.04%	70.0–82.0%	Own data, Graz (see Table 25)

n. d. no data.
[a] > 4540 g.

infants in successive pregnancies. However, with a disorder of maternal carbohydrate metabolism, each further neonate typically is heavier than the last.

Since a previous pregnancy increases the capacity of the uterine vessels, the heavier birth weight of children born to multiparae may result from uteroplacental factors.

Different interpretations of the association between high birth weight and increased maternal age are possible. Since advanced age is associated both with increased parity and with decreased glucose tolerance.

Maternal over-nutrition increases fuel supply to the fetus, and thus stimulates its insulin production. Insulin is an important fetal growth factor, as has been shown also in animal experiments (PICON, 1967). Insulin's effect depends on the number of insulin receptors on the cell and on their insulin affinity. Fetal monocytes have more than 6 times as many insulin receptors, and these a twice as high affinity, than monocytes from the adult (THORSSON et al., 1977). Also the birth weight is positively correlated with the cord-blood insulin concentration, indicating the importance of insulin for fetal growth

(SPELLACY et al., 1973; WEISS et al., 1984 a). Distinct fetal hyperinsulinism does not yet occur in gravidae with normal glucose tolerance through over eating, but fetal insulin biosynthesis is in the upper range of normal. During food rationing in the postwar years 1949–1952, the incidence at our hospital of newborns weighing more than 4,000 g was constant at 4.3%. After rationing was discontinued in 1952 the rate jumped to 6.3% (Table 25). Since the 1960s, wide spread over-nutrition has become a public health concern. Between 1979–1981, the incidence of newborns over 4,000 g was 7.4%.

Impaired carbohydrate tolerance during pregnancy can induce fetal hyperinsulinemia – the insulin concentration in fetal blood can be 20 to 30 times the norm. Insulin measurements in the cord blood of newborns over 4,000 g showed that 24.5% were hyperinsulinemic (> 20 μU/ml), indicating a disorder of maternal carbohydrate metabolism (WEISS et al., 1984 a). Assuming that the incidence of carbohydrate tolerance disorders is at least halved in times of food shortages, Table 25 shows that excessive food intake contributes to the high birth

Table 25. The impact of nutrition on the incidence of newborns above 4000 g

Year	Number of deliveries	Newborns ≥ 4000 g number (%)	Males number (%)	Newborns ≥ 4500 g number (%)	Males number (%)	Newborns ≥ 5000 g number (%)	Males number (%)
			A Group 1 (Food shortages and rationing)				
1949	1842	66 (3.6%)	49 (74%)	5 (0.3%)	4	0	0
1950	1776	87 (4.8%)	58 (67%)	9 (0.5%)	6	0	0
1951	1715	77 (4.5%)	60 (78%)	16 (0.9%)	16	1	1
1952	1750	77 (4.4%)	52 (68%)	11 (0.6%)	8	2	2
Total	7063	307 (4.3%)	219 (71%)	41 (0.6%)	34 (83%)	3 (0.04%)	3 (100%)
			B Group 2 (Termination of rationing, normal food supply)				
1953	2054	133 (6.7%)	95 (71%)	14 (0.7%)	8	1	0
1954	2288	141 (6.2%)	104 (74%)	17 (0.7%)	14	0	0
1955	2591	160 (6.2%)	107 (67%)	19 (0.7%)	11	2	1
1956	2958	175 (6.0%)	119 (68%)	19 (0.6%)	12	0	0
1957	3154	203 (6.4%)	142 (70%)	24 (0.8%)	19	2	2
1958	3382	215 (6.4%)	160 (74%)	26 (0.8%)	24	3	3
Total	16427	1027 (6.3%)	727 (71%)	119 (0.7%)	88 (74%)	9 (0.05%)	6 (67%)
			C Group 3 (Nutritional over-supply, abundant overnutrition)				
1979	4560	339 (7.4%)	238 (70%)	20 (0.4%)	15	0	0
1980	4770	353 (7.4%)	232 (66%)	47 (1.0%)	38	3	3
1981	4974	368 (7.4%)	267 (73%)	32 (0.6%)	22	2	2
Total	14304	1060 (7.4%)	737 (70%)	99 (0.7%)	75 (76%)	5 (0.03%)	5 (100%)
Mean values A + B + C	37814	2394 (6.33%)	1683 (70%)	259 (0.7%)	197 (76%)	17 (0.04%)	14 (82%)

Table 26. Formulas for the calculation of the fetal weight from ultrasonic measurements

Formulas	References
Log 10 (birth weight) = $-$ 1.7492 + 0.166 (BPD) + 0.046 (AC) $-$ 2.646 (AC × BPD)/1,000	SHEPPARD et al., 1982
Birth weight = 378 BPD + 416 Th Tr $-$ 4.450 or Birth weight = 349 BPD + 384 Th Tr + 22 TR $-$ 4,260	MILLER[a], 1980
Birth weight = $-$ 1.05775 (BPD) + 0.649145 (Th Tr) + 0.0930707 BPD^2 $-$ 0.0205620 Th Tr^2 + 0.515263 or Birth weight = $-$ 1.33450 BPD + 0.798429 Th AP + 0.103458 BPD^2 $-$ 0.0254788 Th AP^2 + 1.35470	HANSMANN[a], 1976

BPD Biparietal diameter.
AC Abdominal circumference.
Th Tr Transverse thoracic diameter.
TR Truncus-length.
Th AP Anterior-posterior thoracic diameter.
[a] Nomograms within the publications.

weight in 27.2% of the newborns over 4,000 g. This would mean that 1.8% of the newborns at our hospital weighing over 4,000 g did so because of a maternal glucose tolerance disorder, 2.0% because of maternal over-eating, and 3.6% because of genetic factors — a total of 7.4%.

Sonographic estimation of fetal weight and ultrasound diagnosis of diabetogenic fetopathy

Biometric parameters correlate with birth-weight percentiles (HANSMANN, 1976; WARSOF et al., 1977; MILLER, 1980; OTT and DOYLE, 1982; SHEPARD et al., 1982), but a single parameter such as the biparietal diameter allows only a rough estimation of fetal weight. With two or three parameters, computer analysis or empirically obtained coefficients were used to calculate formulae or tables (nomograms) with which fetal weight can be approximated (Table 26). While the genetically predetermined large fetus develops in normal proportions, the fetuses of women with excessive food intake and of gestational diabetics develop a disparity between the size of the head and of the thorax. WLADIMIROFF and colleagues

(1978) found that the cranial measurements of overly large fetuses did not exceed the 97th percentile more often than those of a normal collective, but that the thoracic measurements did so in 47%. Thus in suspected gestational diabetes, the thoracic measurements are of more diagnostic value than the biparietal diameter. The excess weight of these fetuses is mostly a result of increased subcutaneous fat. The skin-fold thickness of newborns with diabetogenic fetopathy is linearly related to the maternal fasting blood glucose level (WHITELAW, 1977). If a large fetus is suspected at an ultrasound examination, the mother should be screened for a disorder of carbohydrate metabolism.

The fetus of an insulin-dependent diabetic mother develops differently from that of a metabolically healthy gravida, a result of disturbed trophoblast development in the first half of the gestation. Embryonic growth can be retarded as early as in the 7th–14th gestational week (PEDERSEN and MOLSTED PEDERSEN, 1981). Between the 20th and the 32nd week the average biparietal diameter is significantly under the expected value (AANTAA, 1980, 1277 measurements in diabetics). Accelerated growth occurs only when the fetus becomes hyperinsulinemic – between the 26th and 30th week, causing fetal weight to surpass the norm only late in pregnancy (36–40 week). Thus, in some cases diabetogenic fetopathy must be diagnosed by ultrasound while the fetal weight is still appropriate-for-dates: when the transverse abdominal diameter lies above the 95th percentile as a result of visceromegaly (Grandjean et al.,

1980; Holländer, 1970), or when disproportion exists between the size of the skull and the abdomen, the thorax and the biparietal diameter, or the transverse abdominal diameter and the femur length. A thorax-skull ratio of 1.4 or more (Elliot et al., 1982), and an abdomen femur ratio of 1.385 or more (Bracero et al., 1985) suggest diabetogenic macrosomia.

Postpartal screening for diabetes

After being delivered of an infant weighing more than 4,000 g, the mother should be screened for a disorder of carbohydrate tolerance. However, the oral glucose-tolerance test is known to be unreliable in the puerperium (Benjamin, 1968). The abrupt cessation of the diabetogenic stress of pregnancy results in an improvement of glucose tolerance. While during pregnancy 30% to 40% of potential diabetics have abnormally high blood glucose levels after glucose loading (Fuhrmann, 1982; Weiss et al., 1984), only 2,5% have them in the puerperium (Salzberger et al., 1975 a). We found repeated normal oGTT results in documented gestational diabetics even after delivery of a fetopathic infant.

The glycosylated hemoglobins (GH, HA_1, HbA_1, HbA_{1c}) are also of little value in screening for impaired glucose tolerance in the puerperium (O'Shaughnessy et al., 1979; Fadel et al., 1979). Fasting blood glucose levels of 100 and 140 mg/dl (5.5 and 7.7 mmol/l) can not be distinguished with accuracy. While Pollak and Brehm (1981) described significantly elevated glycosylated hemoglobin values in mothers of heavy newborns, others (Gräfenstein and Duchna, 1981; Coen et al., 1980; Fadel et al., 1979;

O'Shaughnessy et al., 1979) found no such correlation.

Since glucose oversupply stimulates fetal insulin production (Obenshain et al., 1970; Weiss et al., 1984, 1984 a), the cord blood insulin level is the most specific indicator of impaired maternal glucose tolerance severe enough to affect the fetus. For the management of a further pregnancy, this is what matters most to the obstetrician. Thus, we recommend determining cord-blood insulin in all newborns over 4,000 g, or when the birth weight exceeds the 90th percentile, and when the mother shows signs of potential diabetes. Normal values can be found on page 54.

Women who have been delivered of overly high birth weight infants should be screened for a disorder of carbohydrate metabolism at least every two years, especially if the cord-blood insulin level was abnormally high. Mickal and coworkers (1966) found that 60% of mothers of infants over 4,500 g developed diabetes during a 12-year follow-up. The further course of patients with proven gestational diabetes is described in detail in the contributions by Mestmann, and by Oates and coworkers in separate sections.

Newborns of diabetic mothers

About two thirds of newborns of overt diabetics have some form of disorders caused by maternal diabetes (Gabbe et al., 1977).

The incidence is even higher in unrecognized gestational diabetes. Fetal pathology is mainly a result of an over or undersupply

Table 27. Fetal implications of maternal hyperglycemia (patients without vascular disease—white classes A, B, C and D) (SARLES and ADAMSONS, 1978)

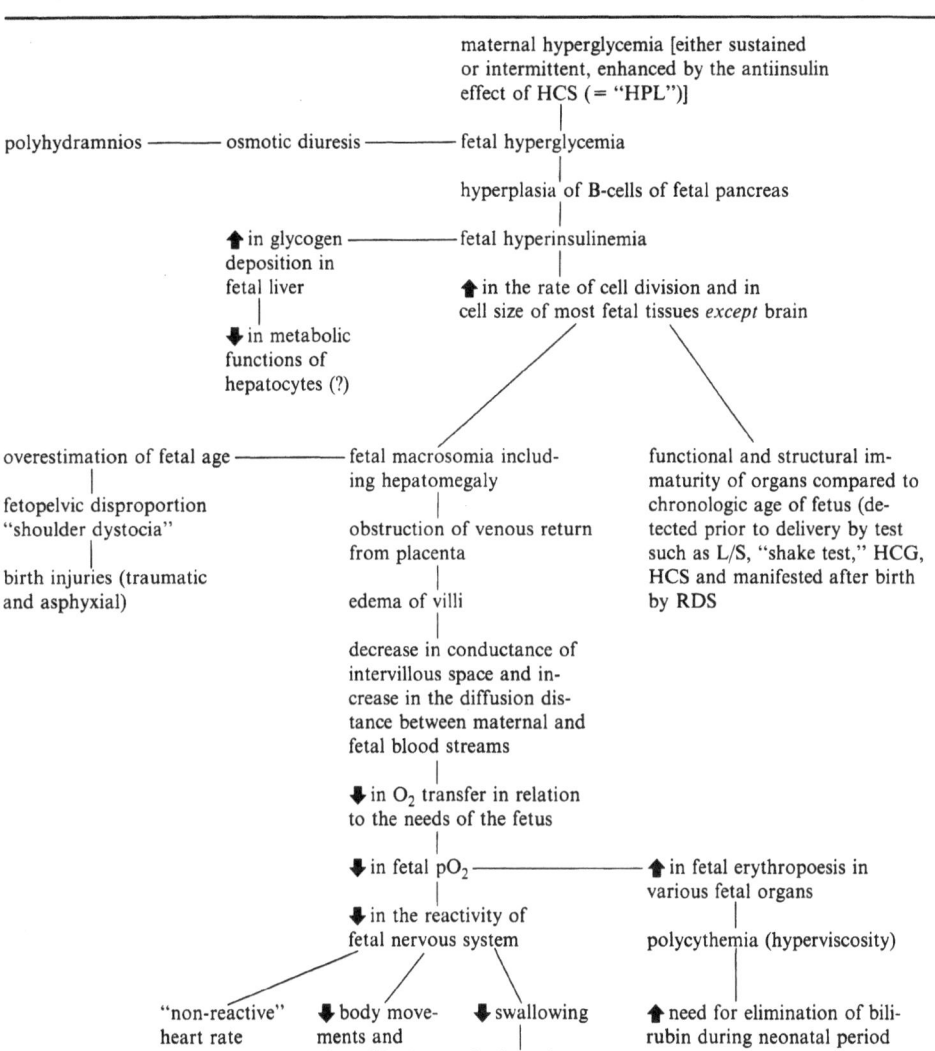

Risks to fetus or newborn
increased frequency of congenital malformations
impaired intellectual development
death during episodes of diabetic coma
"silent" death prior to term
intrapartum asphyxia
obstetric trauma due to macrosomia
pulmonary and hepatic immaturity when delivered prior (or even at) term
hypoglycemia

of glucose to the fetus and its ensuing insulin response (Gabbe and Quilligan, 1977; Bellmann, 1978; Shelley et al., 1975; Sarles and Adamson, 1978).
Diabetogenic fetopathy, the specific syndrome, is a direct result of fetal hyperinsulinism and its consequences (Table 27) But in diabetics with systemic vascular disease or gestational diabetics with preeclampsia, placental insufficiency can lead to a fetal glucose deficit, hypoinsulinism, and prenatal dystrophy (Table 28). These contrary disturbances can also overlap, *i.e.* prenatal dystrophy can coincide with hyperinsulinism. Such a seemingly paradoxical situation may be a result of an unhindered transport of glucose across the placenta while the supply of other essential nutritional elements, such as amino acids or fatty acids, is impaired (Saintonge et al., 1983; Hull, 1975).

Table 28. Fetal implications of maternal hyperglycemia in patients with vascular disease or toxemia (Sarles and Adamsons, 1978)

vascular disease of mother
|
⬇ uterine blood flow
|
underdevelopment of placenta
|
⬇ perfusion of intervillous space
|
presentation of *normal* or even less than normal *quantity* of glucose to the fetus even in the presence of *maternal hyperglycemia*
|
normoglycemia of fetus
|
absence of hyperplasia of B-cells of pancreas
|
"normo-insulinemia" of fetus
|
absence of hypersomia (more likely fetal hyposomia [= "SGA"]) immaturity of cells and organ systems

Risks to fetus
premature onset of labor, premature separation of placenta, intrapartum asphyxia

Assessment of the quality of metabolic management during pregnancy by the condition of the newborn

The extent of pathophysiologic changes in the newborn (Table 27) reflects the quality of metabolic management during the pregnancy. The demonstration of fetal hyperinsulinism is the most specific criterion. Thalme and Edström (1974) have suggested a score for the assessment of the newborn, to which we have added the insulin content of the cord blood as a further parameter (Table 29).
Since a variety of factors influence the *weight of the newborn* (p. 46 f.), it alone should not be used as a diagnostic index. A child weighing over 4,000 g can be perfectly healthy. The birth weight must be assessed in light of regional standards. By definition, 10% of newborns of metabolically healthy women exceed the 90th and 3% the 97th percentile, respectively. A similar incidence can be expected in newborns of diabetic mothers with

good metabolic control during the pregnancy. Thus the birth weight is not a specific criterion of diabetes.
The *clinical aspect* of the newborn with diabetogenic fetopathy (somatic fetopathy) often is characteristic. It consists of a cushingoid appearance; plethora; tomato-red skin; short neck; well developed fat tissue; thick, dark hair; hypertrichosis of the ear lobes; hypotonic musculature; the so-called frog position; as well as hepato-, spleno-, and cardiomegaly a hypertrophic umbilical cord is frequently found. The separation of "mild" and "severe" forms is subjective, but newborns with mild fetopathy should present only a few features mentioned above. Pastor and his colleagues describe the significance of objective somatometric parameters of the newborn (p. 171 ff.). The marked *weight loss* of fetopathic children in

Table 29. Evaluation of the newborn to asses the quality of metabolic management during pregnancy

Points	0	1	2	4
Birth weight[a]/gestational age	< 90%	≥ 90%	≥ 97%	
Diabetogenic fetopathy (clinical appearance)	none	mild	severe	
Weight loss	< 9%	9–11%	> 11%	
Respiratory distress syndrome clinically[b] or	0–2	3–5	6–10	
radiologically[c]	I	II–III	IV	
Hyperexcitability	none	mild	severe	
Hypoglycemia in the first 3[h] of life	> 30 mg%	20–29%	< 20 mg%	
Hyperbilirubinemia	< 12 mg%	≥ 12 mg%		
Polycythemia. Hb or	Hb < 20 g%	Hb ≥ 20 g%		
Venous hematocrit	Hkt < 70	Hkt ≥ 70		
Congenital abnormalities	none	present		
Hypocalcemia			≤ 4 m Eq/l	
Cord blood insulin[d] or	< 20 μE/ml	20–39 μE/ml	40–59 μE/ml	≥ 60 μE/ml
C-peptide[e]	< 1,7 ng/ml	1,7–3,6 ng/ml	3,7–5,6 ng/ml	≥ 5,7 ng/ml

[a] Weight percentiles according to HOHENAUER.
[b] Silverman Index (SILVERMAN and ANDERSON, 1956).
[c] GIEDION et al., 1973.
[d] Phadebas® Insulin Test.
[e] RIA-gnost® hC-Peptid.
(Minimum: 0, maximum: 21).

the first postpartal days is caused by water and electrolyte imbalances, that cause an increased urinary output.

The incidence of the *respiratory distress syndrome (RDS)* is increased in infants of diabetic mothers, but the condition is not pathognomic. RDS can be evaluated clinically by the SILVERMAN index (SILVERMAN and ANDERSON, 1956) or radiographically (GIEDION et al., 1973).

Hyperexcitability of the newborn is also associated with maternal diabetes and may occur without hypocalcemia or hypoglycemia. Gross angulations of the extremities, coarse tremor of the hands and a fine tremor of the fingertips are signs of mild hyperexcitability. Seizures suggest a severe disorder but usually occur with hypocalcemia.

Hypoglycemia of the fetopathic newborn (< 30 mg/dl or 1.7 mmol/l in mature infants; < 20 mg/dl or 1.1 mmol/l in preterm infants) is due primarily to the infant's hyperinsulinism. Diminished glucose production (KALHAN et al., 1977) and/or glucagon deficiency (WILLIAMS et al., 1979) probably also contribute to hypoglycemia. Furthermore tocolytic treatment of premature labor with β-mimetic agents can induce hypoglycemia in the newborn (WEIDINGER et al., 1976; ROSANELLI et al., 1982). Four fifths of newborns with hypoglycemia are asymptomatic (GABBE and QUILLIGAN, 1977). The blood glucose level of newborns of diabetic mothers should thus be checked routinely, especially in the first few hours of life. Test strips are imprecise at low glucose concentrations and provide only a rough estimate. The incidence of *hyperbilirubinemia, poly-*

Table 30. Range of naval cord serum insulin level in offspring born to mothers with normal glucose tolerance during pregnancy

Gender	♂ + ♀	♂	♀
Number of cases	764	433	331
Birth weight[a] (SD)	3334 (503)	3406 (526)	3239 (456)
Week of pregnancy[a] (SD)	40.0 (1.2)	39.8 (1.5)	40.2 (0.9)
Cord serum glucose[a] (SD)	65 (29) mg/dl	66 (32) mg/dl	64 (25) mg/dl
Cord serum insulin[a] (SD)	6.9 (5.0) µU/ml	6.7 (4.7) µU/ml	7.1 (5.4) µU/ml
Centiles (µU/ml)			
3rd	2.0	1.9	2.0
10th	2.8	2.8	3.0
50th	5.7	5.7	5.8
90th	12.1	11.4	12.5
97th	17.4	17.5	16.1

[a] Mean and standard deviation (SD).

Table 31. Insulin content in umbilical cord blood and urine in eight metabolically healthy neonates of metabolically healthy mothers and in ten neonates with diabetogenic fetopathy in gestational diabetes (Weiss et al., 1984)

	Metabolically healthy		Diabetogenic fetopathy	
	A cord blood	A₁ urine	B cord blood	B₁ urine
	10.6	3.2	32.0	78.0
	12.6	15.3	130.0	150.0
	15.5	17.6	73.6	55.0
	14.8	14.4	46.0	39.3
	1.6	3.6	29.0	39.1
	6.8	3.9	23.1	50.0
	13.4	2.5	53.6	31.3
	20.0	27.9	47.4	36.4
			58.1[a]	296.0[a]
			21.3	22.1
Mean values	11.9	11.1	50.7	55.7

Data in µU/mL.
r A : A₁ = 0.775 (P < 0.05); r B : B₁ = 0.813 (P < 0.01). Both the regression lines are tested for parallelity (P < 0.001). Their average Y compared by analysis of variance is different with a significance level of P < 0.001.
[a] Not included in the calculation of the mean values, because the large discrepancy is likely to be due to glucose infusion in the newborn.
Reprinted with permission from The American College of Obstetricians and Gynecologists. (Weiss et al. [1984] Obstetrics and Gynecology 63: 776).

cythemia, and *congenital malformations* is also increased in newborns of diabetic mothers without being specific for the disease. Hypocalcemia in fetopathic newborns is probably due to a parathyroid hormone deficit (Cruikshank et al., 1983). The *insulin and C-peptide content* of the cord blood reflect the rate of insulin synthesis in the newborn (Weiss et al., 1984 a; Sosenko et al., 1979). It will be higher the poorer the metabolic control of the gravida, especially in the third trimester. While the incidence of the above mentioned disorders is higher in pregnancies complicated by maternal diabetes (without being specific), hyperinsulinism is specific for diabetes and thus a "hard" parameter of the quality of maternal metabolic management during pregnancy. The increased fetal insulin synthesis need not be associated with macrosomia (SOSENKO et al., 1982). Since hyperinsulinism precedes macrosomia (somatic fetopathy), and since placental and fetal factors can prevent macrosomia despite fetal hyperinsulinism, normal fetal weight does not rule out hyperinsulinism (biochemical fetopathy). In fact, hypoglycemia is three to four times as common as macrosomia in newborns of diabetic mothers (ADASHI et al., 1979). The determination of insulin or C-peptide in the cord blood should thus be carried out routinely in newborns of diabetic mothers. Table 30 shows the normal values of the cord blood insulin concentration. The concentration of C-peptide in the cord blood is significantly correlated with that of insulin, both in meta-bolically healthy and in gestational diabetic mothers (K: 0.758, A: −0.2895) (Table 29). Insulin excretion in the urine of the newborn also reflects fetal insulin production. The insulin concentration of the urine on the first postpartal day is practically identical to that of the amniotic fluid at the end of the gestation (WEISS et al., 1984, 1985) (Table 31). If the newborn receives glucose solutions intravenously, the additional stimulation of the fetal β-islet cells can lead to a surge of urinary insulin values.

More recent studies by SOSENKO et al. (1982) suggest that the glycosylated hemoglobin level in the cord blood may also reflect the quality of maternal metabolic control during the pregnancy. The section by CZEKELIUS and ROLLMANN (p. 115 ff.) and that by DOMENECH et al. (p. 167 ff.) describe in detail the importance of glycosylated hemoglobin and also of fetal hemoglobin.

A. The Significance of Gestational Diabetes

2

The World Scope of Gestational Diabetes

J. J. Hoet

Department of Medicine, Catholic University of Louvain, Brussels, Belgium

Introduction

The objective of this chapter is to discuss the adverse effects of diabetes upon maternal and neonatal health in circumstances which together with diabetes could be responsible for pathological events during childbearing especially in the tropical and subtropical zones. The maternal mortality rate related to childbearing and delivery, as well as the perinatal (one week after birth) or neonatal (28 days after birth) mortality, remain distressingly high in these areas.

Biological and environmental factors characteristic for the developing world may influence obstetrical events in diabetic women. They may produce a clinical pattern which is different than in the developed world. Circumstances which are likely to alter the general health status in the tropical and subtropical zones may also favour metabolic derangements, possibly by inducing pathologic changes in the endocrine pancreas as well. This global interaction may lead to maladaptation of insulin secretion during pregnancy, and to metabolic alterations which may affect the outcome of pregnancy. These considerations may strengthen an overall problem solving philosophy which should lead to a better assessment of the metabolic status of adolescent or adult women with childbearing potential.

The latter should be of the highest priority in the delivery of health care where planning and management should envisage an holistic approach. Ninety percent of the births throughout the world take place in the tropical and subtropical zones where diabetes is highly prevalent. The rate of low birth weight infants may reach 80% of the births, with a perinatal mortality as high as 70% (EBRAHIM, 1979; DUNN, 1979). The surviving infant may be afflicted by permanent cerebral dysfunction, major vision defects or other longterm handicap.

Determinant biological factors of pathologic childbearing

This discussion will not consider the influence of cultural beliefs and practices, nor of education, upon the women's destiny. It emphasizes rather some biological factors which may alter the homeostasis during pregnancy. In tropical and subtropical zones undernutrition, which is also favoured by heavy manual labour, results in the birth of small for dates and infants with low weights.

Concomitant insufficiencies such as iron deficiencies, often starting during puberty, may be maintained or initiated during childbearing. Protein supplementation to the mother and increase in maternal weight add 200 to 300 gm to the low neonatal weight. However, in underprivileged conditions infants also may be born weighing more than 4 000 gm. Their rate may reach 2 to 3% of the births (HABICHT 1974). While under-nutrition as well as frequently associated anaemia and infection are conductive to impaired fetal growth, maternal metabolic disturbances could provide increased endogenous nutrients to the fetus which may explain the fetal overgrowth. The infants born in the developing and developed world share, up to 26–28 weeks of gestation, a comparable body weight notwithstanding the food shortage, but they differ qualitatively because of protein deficit, iron insufficiencies and low vitamin stores (A, B 2 and folate) in the latter. Food shortage in the mother during the second part of pregnancy induces the greatest reduction in neonatal size, but it leaves the possibility of catch-up growth. Factors such as infection and anaemia early in pregnancy, frequently associated with undernutrition, also influence fetal growth and determine the size at birth with a possible reduction of cell numbers and an uncertain catch-up growth. The most important infections on a worldwide scale are malaria, tuberculosis, syphilis, rubella, urinary tract infections, intestinal infestations and amniotic cavity infections. Virus diseases such as hepatitis, cytomegalic virus and also slow virusses are determinants of maternal and neonatal health. Infections may affect the metabolic status of the mother and enhance the catabolic tendency of pregnancy as well as induce an acidotic-ketotic state (EBRAHIM, 1979). Comparable immune complexes are found in placentas of diabetics and patients with malaria (FAULK). Congenital syphilis is associated with major pathologic changes in the pancreas including its endocrine part (NAEYE, 1979). Cytomegalic virus may also affect the fetal endocrine pancreas (CRAIGHEAD, 1985) and congential rubella may provoke a deferred diabetic syndrome (FORREST, 1971).

Determinants of diabetes

Diabetes is a heterogenic disease in its clinical features and outcome. Type I diabetes may be initiated by an immune or auto-immune process which results in an insulin dependent syndrome. This is rather specific in young subjects of the caucasian race with specific histocompatible antigens. Diabetes in the young may have other etiologies than immune-dependent disorders and occurs in subjects of different races such as black, indian and caucasian. The heterogeneity of diabetes becomes further apparent in the adult onset (Type II) syndrome, which seems to affect subjects of all races but may have a higher prevalence in certain communities than in others (FAJANS, 1987). Malnutrition diabetes is being described and found prevalent in certain areas of the developing world. Toxic substances such as cyanogenic agents associated with protein deficiency appear to initiate the disease (BAJAJ, 1986). These different diabetes states, which are heterogeneous in their origins, feature different clinical patterns. The pathologic findings in the endocrine pancreas highlight this heterogeneity and emphasize the various origins and the initiation processes of the disease (HOET et al.; a, b, 1987). The global trend in the diabetic pancreas is characterized by a progressive reduction of the B cell mass which is the result of ongoing aggressions. Two further considerations should be made. First, the endocrine pancreata of close to 50% of adult onset diabetics in the developed world show an increased deposit of amyloid bodies which are in conjunction with B cells. This pathologic lesion, when extensive, may be specific for adult onset diabetics. Possibly slow viruses may be implicated in the origin of this lesion. Amyloid deposits, being described in several races, indicate the ubiquity of the aggressive agents which may explain the high prevalence of diabetes in certain communities where it may add to a genetic predisposition. Viral agents or particles may be transmitted horizontally but also vertically from mother to foetus (FUJITA 1986). This would occur more frequently in zones where maternal infections and their transmission to the fetus are endemic (PARENT, 1979). Cytomegalic virus infections (CMV) do affect young infants at a high rate in developing countries. The 70% prevalence of antibodies to CMV in children under one year in a large community of Saudi Arabia, the typical inclusions of CMV in pancreatic islet cell reported in human infants and confirmed in experimental animals (SCRAIGHED, 1985), indicate that infection with this virus is widespread. In many developing countries it is acquired early in life, and even in utero (BAKIR, 1987). The endemic infection of childbearing women, often themselves in a young age group, is responsible for the continuous transmission of this disease which may attack the fetal islets. This state of events may contribute, along with other infectious diseases (rubella), to the prevalence of diabetes in certain communities of the developing world.

A second comment deals with the reduced number of B cells in low birth weight infants. The inheritance of a reduced B cell mass because of infection, malnutrition or other factors in utero should be considered when different diabetic syndromes are encountered in the young. Neonatal diabetes is reported mainly in small for dates infants, and its incidence is high in the developing countries (HOET, 1985). Reduced fetal B cell mass as a consequence of severe maternal diabetes or malnutrition has been confirmed by experimental evidence whereby the neonatal B cell replication rate analysed, in these instances, by *in vitro* testing is reduced (HOET, 1985, SNOECK 1986). Maternal diabetes may induce hyperplasia and B cell hypertrophy of the islets. Experimentally, a higher than normal rate of replication of the B cells has been described in pups born to a moderately severe diabetic mother dog (REUSSENS, 1984). In conclusion, the determinants for inducing diabetes are multifold during adulthood, puberty, infancy and in utero. These conditions are concentrated in the tropical and subtropical zones. They all may contribute to the endemic state of diabetes in certain constituencies.

Diabetic patterns during pregnancy

The heterogeneity of diabetes becomes also apparent during pregnancy. Insulin dependent (Type I) diabetes may appear suddenly and possibly be associated with an auto-immune process (BUSCHARD, et al., 1987). The adult onset (Type II) diabetes, which may be present before pregnancy, may be exaggerated by it. Gestational diabetes may manifest itself during pregnancy and disappear afterwards.

The obstetrical pathology as a consequence of diabetes during pregnancy may be different in various clinical settings. Large for dates babies with dystocia, hydramnios, and excessive maternal weight gain have been reported during pregnancies also in tropical zones. In Guam, with eighty thousand inhabitants, the mean age of the population is 18.9 years. The diabetes prevalence rises rapidly after the age of 25 and the total cost of diabetes reaches three million dollars per year. A complete screening of the pregnant women on the island indicates a 42% incidence of abnormal glucose tolerance and 29% of confirmed gestational diabetes. 10% had permanent diabetes after pregnancy. These abnormalities were associated with poor obstetrical outcomes (KUBERSKI, 1980; YEN 1964). In a large survey in three tertiary health care centres respectively in Tunisia, in Algeria and in Morocco, the clincial outcome for diabetic pregnancies was significantly abnormal: four times more spontaneous abortions than normal, prematurity three times higher, fetal death rates and neonatal death rates respectively eleven and five times higher. Macrosomia occurred six times and toxemia seven times more frequently than in a control population. The malformation rate was also increased (BEN KHALIFA, 1986).

A recent Asian experience revealed the association of vaginal moniliasis, edema, urinary tract infection, increased blood pressure, hydramnios, intrauterine death, and congenital malformation with impaired glucose tolerance during pregnancy as well as gestational diabetes. Spontaneous labor and normal delivery were reduced while induced labor, forceps delivey, and low segment cacsarean section were increased when the maternal glucose was abnormal (MADAN and MOHAN, 1986).

In the developing world where undernutrition and infection with its catabolic effect may possibly lower the basal levels of blood glucose, the normal blood glucose levels have to be assessed and the effect of its variation upon the obstetrical outcome to be verified. Each clinical situation such as infection, malnutrition or denutrition and their associated deficiencies should be analyzed because of their cumulative effect upon the fetal or neonatal survival. The high levels of stillbirth in populations of four of five Pacific islands was related to the high prevalence of abnormal glucose tolerance and diabetes (SICRÉE, 1986). This observation indicated that in the remaining island other factors, possibly still associated with abnormal glucose levels, are responsible for the stillbirths.

Epidemiologic surveys should report the prevalence and incidence of diabetes in order to assess the magnitude of the problem. In clinical practice, causal relationships have to be explored in order to implement a rewarding therapeutic program.

Epidemiologic prospective studies on impaired glucose tolerance or diabetes during pregnancy have been carried out, for comparison with the known background risk of their incidence and prevalence in the Malta population. This prospective study indicated that glucose tolerance is abnormal in 14% of the pregnant women, whereas the abnormality appears in only 1.7% of the non pregnant female subjects of the same age group. The prevalence of diabetes and impaired glucose tolerance in the Malta population is 14%. The prevalence of abnormal glucose values in the pregnant population is close to that observed in the total population, including the aged group (SCHRANZ, 1987). This concentration of abnormal metabolic events in a young age group should stimulate awareness programs to target selective efforts in high risk subjects (FREINKEL, 1986). This will result in a high yield and a better cost benefit ratio. During pregnancy it will also benefit the unborn.

Diabetes health care in pregnancy in tropical and subtropical zones

The previous considerations highlight the need for delineating the associated factors which together with an abnormal glucose value affect the pregnancy outcome. The major goals of any program should specify in a holistic approach the diagnosis and the management of diabetes and impaired glucose tolerance during pregnancy. This challenge is more delicate and greater than ever and is facing health care professionals at all levels in the developing countries.

Programs are now implemented in areas of India, North Africa and South America where primary health care workers are the first contacts with the populations and ed-

ucated to select the high risk subjects for diabetes (IDF Bulletin 1986). Their task should be to establish the diagnosis with the available means and to educate the pregnant women in acquiring skills for self evaluation of their health status, and for possibly insulin injection. They should refer to secondary and tertiary health care centers where a team consisting of a well trained internist, obstetrician and pediatrician may evaluate and guide the pregnant women with the aid of primary health care workers. This awareness program at the primary health care level including where needed the medical and obstetrical interventions in an holistic approach, has improved the perinatal mortality and some complications of diabetic pregnancies where it had remained in jeopardy over the years. A thorough educated primary health care team will prevent congestion at the secondary or tertiary level and reinforce the rewards of high technical interventions on high risk groups by specialized teams.

The neonatal morbidity, however, remains usually high. Health education should be further promoted througth basic education as it influences infant mortality and morbidity. World science should permeate health care everywhere, and investments in primary care as well as specialized care should be well-balanced. The programming of health sciences and of their application needs a global perception of the cultural assets and the biological intricacies of early life. The sharing of the advancement of applied health is the challenge for the different worlds. It has to be achieved to balance out the benefit of health for everyone.

Perceiving the influence of environmental and biological agents especially during pregnancy in the tropical and subtropical zones, may give further insight into disease processes such as diabetes which are not well understood in the developed world either. This reciprocal assessment may lead to further acquisitions for the benefit of many and especially of the unborn.

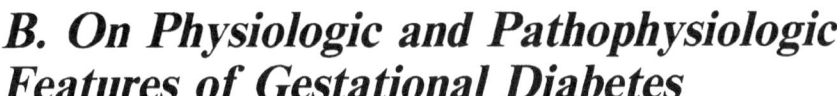

B. On Physiologic and Pathophysiologic Features of Gestational Diabetes

3

Pathophysiological Background for Gestational Diabetes

C. Kühl[1,2] and O. Andersen[2]

[1] Hvidøre Hospital, Klampenborg, and [2] Diabetes Center, Department of Obstetrics and Gynaecology Y, Rigshospitalet, University of Copenhagen, Copenhagen, Denmark

Introduction

In normal pregnancy, glucose homeostasis is changed in the direction of diabetes. Thus, glucose tolerance gradually deteriorates (LIND et al., 1973; KÜHL, 1975; HORNNES and KÜHL, 1984), for which reason pregnancy is often called "diabetogenic". In most women glucose tolerance stays within the normal range (LIND et al., 1973), but in about 1% of all pregnancies it deteriorates sufficiently for the diagnostic criteria of diabetes mellitus to be fulfilled. This condition is termed "gestational diabetes mellitus" (KÜHL, 1977).

Women with gestational diabetes have a considerably increased risk of developing manifest diabetes later on (O'SULLIVAN, 1978) and, furthermore, the perinatal mortality and morbidity of their infants appear to be increased (COUSTAN and LEWIS, 1978; ADASHI et al., 1979). For these reasons it is important to recognize and treat the disorder.

Why pregnancy is capable of inducing a temporary diabetic state is still obscure (KÜHL, 1977; KÜHL and HORNNES, 1984). Among the possible explanations are reduced insulin secretion, increased concentrations of proinsulin, increased insulin degradation, increased secretion of hormones with an anti-insulin effect, reduced tissue sensitivity to insulin, or a combination of two or more of these mechanisms. In the present article we will focus upon these etiological possibilities as explanations for the pathophysiological background for gestational diabetes.

Plasma insulin in pregnancy

Fasting plasma insulin increases gradually in pregnancy (KÜHL, 1975) and in late pregnancy it is almost twice as high as postpartum both in normal pregnant women and gestational diabetics (KÜHL, 1975; KÜHL and HOLST, 1976; HORNNES et al., 1981). Normal pregnant women and normal weight gestational diabetics have comparable fasting insulin levels (KÜHL and HOLST, 1976; KÜHL, 1976), whereas the levels in obese gestational diabetics are considerably higher (HORNNES et al., 1981).

Following an oral glucose load to gestational diabetics, significantly higher insulin levels are reached in pregnancy than postpartum (KÜHL and HOLST, 1976; HORNNES et al., 1981). This finding compares well with that obtained in normal pregnant women (KÜHL and HOLST, 1976). However, although the insulin reponses in absolute terms are nearly identical in normal women and gestational diabetics, the insulin response per unit of glycaemia is significantly higher in normal pregnant women than in gestational diabetics (KÜHL and HOLST, 1976). Moreover, when glucose is infused intravenously at rates which result in identical physiological elevations of plasma glucose in normal and gestational diabetic women in late pregnancy and postpartum, the insulin response to glucose is four times increased in normal late pregnancy (HORNNES and KÜHL, 1980) as opposed to approximately three times in late gestational diabetic pregnancy (HORNNES and KÜHL, 1980 a) when compared to postpartum. Thus, there is evidence for a slightly reduced insulin response to glucose in gestational diabetics when compared with normal pregnant women.

The insulin response to a protein-rich meal or oral administration of single amino acids is also considerably increased in late gesta-tional diabetic (HORNNES et al., 1982) and normal (METZGER et al., 1977; KITZMILLER et al., 1980; HORNNES et al., 1981) pregnancy. In response to identical protein-rich meals, normal pregnant women exhibit greater in-crements in insulin responses as compared to postpartum than those found in gesta-tional diabetics (HORNNES et al., 1982; HORNNES et al., 1981). Following similar physiological elevations of plasma amino acids by means of an intravenous amino acid infusion, the insulin response is approxi-mately three times increased both in normal and gestational diabetic pregnancy (KÜHL and HORNNES, 1982).

These results clearly demonstrate that the B-cell sensitivity to glucose and amino acids is significantly enhanced in gestation. It also appears that the insulin secretory capacity of normal weight gestational diabetics is gen-erally less than that found in healthy preg-nant controls.

Serum proinsulin in pregnancy

Most insulin radioimmunoassays also de-termine proinsulin which, however, is much less biologically active than insulin (LAZARUS et al., 1970). The enhancement of the mea-sured plasma immunoreactive insulin levels in pregnancy might reflect an increased con-centration of proinsulin thereby explaining the apparent inefficacy of the raised plasma insulin levels in maintaining normal glucose tolerance in gestation.

The absolute concentration of proinsulin is increased in late gestational diabetic and normal pregnancy, but the fraction of proin-sulin (i.e. the percentage of total insulin im-munoreactivity accounted for by proinsulin) is unchanged (KÜHL, 1976; PHELPS et al., 1975). There is therefore no evidence that glucose tolerance deteriorates in normal and gestational diabetic pregnancy because of an increased secretion of proinsulin. Likewise, the quantiatively different impairment of glucose tolerance in healthy pregnant women and gestational diabetics is not due to different proportions of serum proinsulin in these patients (KÜHL, 1976).

Insulin degradation in pregnancy

Human investigations have demonstrated that insulin degradation is unaffected by pregnancy (BELLMANN and HARTMANN, 1975; KÜHL et al., 1981). The decrease in glu-cose tolerance in gestation is therefore not ascribable to an increased degradation of insulin.

Glucagon secretion in pregnancy

Glucagon secretion is often abnormal in diabetic patients for which reason it has been proposed that glucagon plays an essential role in the pathogenesis of diabetes.

Fasting plasma glucagon is slightly but significantly increased in late normal pregnancy (KÜHL and HOLST, 1976; HORNNES et al., 1981; METZGER et al., 1977). In late gestational diabetic pregnancy, fasting plasma glucagon has been reported to be either unchanged (KÜHL and HOLST, 1976; KITZMILLER et al., 1980) or enhanced (HORNNES et al., 1981). However, the fasting molar insulin: glucagon ratio is increased in late normal and in gestational diabetic pregnancy (KÜHL and HOLST, 1976) whereas the opposite finding is a characteristic of insulin-dependent diabetes.

After oral glucose, glucagon suppression below fasting levels is normally exaggerated (KÜHL and HOLST, 1976; HORNNES et al., 1981) in late normal and gestational diabetic pregnancy. These findings contrast the lack of suppressability of glucagon during hyperglycaemia found in non-pregnant insulin-dependent diabetics. The increased suppressability of glucagon during gestation seems to be due to the higher plasma glucose levels reached after glucose administration in pregnancy. Thus, if plasma glucose is similarly increased by graded intravenous glucose infusions in the same women in late pregnancy and postpartum, identical suppressions of plasma glucagon below the fasting levels are seen in normal women (HORNNES and KÜHL, 1980) and gestational diabetics (HORNNES and KÜHL, 1980a).

Oral intake of alanine elicited a greater rise in plasma glucagon in normal women investigated in late pregnancy as compared with postpartum (KITZMILLER et al., 1980). Despite the rise in plasma glucagon, blood glucose remained unchanged at both occasions (KITZMILLER et al., 1980). This is probably due to the concomitant increase in plasma insulin which was much higher in pregnancy than postpartum.

The glucagon response to a protein-rich meal has been reported to be unaffected (METZGER et al., 1977; HORNNES et al., 1981) by both normal and gestational diabetic pregnancy (HORNNES et al., 1982). Graded intravenous amino acid infusions which led to almost identical plasma amino acid levels in pregnancy and postpartum elicited glucagon responses which were similar in pregnancy and postpartum in normal pregnant women whereas, in gestational diabetics, an enhanced glucagon response to amino acids was found in pregnancy both as compared to postpartum and to the normal pregnant women (KÜHL and HORNNES, 1982). The physiologic implications of the higher glucagon levels reached in the gestational diabetics after intravenous amino acids are unknown as is the reason why these differences in glucagon levels were not seen after the ingestion of a protein-rich meal (HORNNES et al., 1982; HORNNES et al., 1981).

In conclusion, the available data show that abnormalities in glucagon secretion in pregnancy are not involved in the pathogenesis of gestational diabetes.

Insulin resistance in pregnancy

In pregnancy glucose tolerance deteriorates in spite of steadily increasing levels of insulin in plasma. This points to pregnancy as a state of insulin resistance.

The explanation for the insulin resistance in pregnancy is unknown. The available data on insulin receptor binding in normal pregnant women are conflicting. Increased insulin binding to monocytes (GRATACOS et al., 1981; PUAVILAI et al., 1982), unchanged binding to monocytes (TSIBRIS et al., 1980) and decreased binding to monocytes (BECK-NIELSEN et al., 1979) and adipocytes (PAGANO et al., 1980) have been reported. Increased insulin binding to monocytes was recently found in non-insulin treated gestational diabetics compared to healthy pregnant controls (GRATACOS et al., 1981) whereas in insulin-treated gestational diabetics, the insulin binding to adipocytes was decreased (PAGANO et al., 1980).

In Copenhagen, a serial study of the insulin receptor binding to monocytes from normal pregnant women and untreated gestational diabetics has been carried out (ANDERSEN et al., 1986). In normal pregnant women a significant increase in insulin receptor binding to monocytes was found in mid pregnancy. This was followed by a significant decrease in late pregnancy and probably again an increase postpartum. The insulin concentration necessary to reduce tracer insulin binding by 50% (ID_{50}) remained unchanged in the normal pregnant women. No differences in insulin binding at tracer insulin concentration to monocytes from gesta-

tional diabetics and normal pregnant women were found, but the ID_{50} was significantly lower in gestational diabetics diagnosed in late pregnancy than in healthy pregnant women. Since the insulin binding at tracer insulin concentration was similar in the two groups, it seems that the number of insulin receptors on monocytes is decreased in women with gestational diabetes late in pregnancy compared to normal pregnant controls. Thus, gestational diabetes could, at least partly, be the result of a decreased insulin receptor binding to target cells at higher concentrations of insulin combined with a relative deficiency of circulating insulin. The existence or postreceptor defect(s) is also possible, but studies on insulin receptor binding and biological effect in adipocytes show that in normal pregnant women the tissue resistance to insulin is mainly located outside the adipose tissue, e.g. in the hepatocyte and/or muscle cell (ANDERSEN and KÜHL, 1985). Whether this is also true in gestational diabetics remains to be investigated.

Concluding remarks

The data available concerning insulin secretion, insulin degradation, and insulin receptor binding to monocytes in pregnancy clearly point to pregnancy as a state of insulin resistance. This is also reflected by a diminished decline in plasma glucose following intraveneously administered insulin (LIND et al., 1977; BURT and DAVIDSON,

1974). As reported here, most pregnant women are able to counteract the insulin resistance of pregnancy by increasing their insulin secretion. However, when the capacity of insulin secretion is not sufficiently large to meet the resistance, glucose intolerance develops and the woman becomes a gestational diabetic.

Summary

Even though glucose tolerance normally deteriorates in human pregnancy, 99% of all pregnant women retain a normal glucose tolerance whereas the remaining 1% develop abnormal glucose tolerance (gestational diabetes). The possibility that glucose toler-

ance deteriorates in pregnancy because of diabetes-like changes in the pancreatic A- and B-cell function, insulin degradation and/ or changes in tissue sensitivity to insulin has been investigated. Even though the insulin responses to oral glucose and mixed meals are equally large, in absolute terms, in gestational diabetics and normal pregnant women, they differ in two pertinent ways. First, the insulin response is delayed, and second, the insulin response per unit of glycaemic stimulus is usually significantly lower in gestational diabetics that that seen in normal pregnant women. The fasting molar insulin : glucagon ratio is increased and the suppression of glucagon by hyperglycaemia is exaggerated both in normal pregnant women and in gestational diabetics. The glucagon response to mixed meals is unaffected by pregnancy in both groups. Glucagon is thus not involved in the pathogenesis of gestational diabetes. Insulin degradation is unaffected by pregnancy and the proinsulin share of total plasma insulin immunoreactivity does not increase in pregnancy. Insulin receptor binding to monocytes from normal women is unchanged or slightly decreased in late pregnancy compared to postpartum. No difference in insulin receptor binding to monocytes, at tracer insulin concentration, is seen between normal pregnant women and gestational diabetics. The insulin concentration necessary to reduce tracer insulin binding by 50% is lower in gestational diabetics diagnosed in late pregnancy than in pregnant controls. These findings indicate that the number of insulin receptors on monocytes is decreased in gestational diabetics diagnosed late in pregnancy. The main reason for the diabetogenicity of pregnancy is thus insulin resistance combined with a relative lack of circulating insulin. Most pregnant women are able to increase their insulin secretion and thereby counteract the resistance. Some pregnant women do, however, have a more limited insulin secretory capacity which eventually may lead to gestational diabetes.

4

The Human Placenta in Gestational Diabetes

G. Desoye

Department of Obstetrics and Gynecology, University of Graz, Graz, Austria

Introduction

The placenta is a complex and so far poorly understood organ which plays the central metabolic role in pregnancy. In addition to synthesizing various hormones it regulates the transport of maternal fuels to the fetus and facilitates maternal metabolic adaptations to different stages of pregnancy (MUNRO et al., 1983). The placenta of the diabetic woman has attracted much interest largely because it is thought that placental damage may be partially responsible for the unduly high incidence of fetal complications that occur in pregnancies complicated by diabetes mellitus (PERSSON et al., 1978). The overwhelming number of reports involve structural analysis by light and electron microscopy as excellently reviewed by FOX (1978). Furthermore, for obvious reasons, biochemical studies have focused on alterations in placental carbohydrate metabolism, mainly glycogen content.

However, in recent years a growing body of data from different fields of interest has emerged, and thus the time appears ripe for a brief review. No attempts will be made to compile the extraordinary extensive literature on placental histology and morphology in pregnancies complicated by diabetes mellitus. Instead, selective reviews will be cited for the interested reader.

Many findings in the literature are controversial. These discrepancies seem to arise because studies variously include mothers with mild, moderate or severe forms of diabetes mellitus, with and without superimposed vascular disease or toxemia of pregnancy, and without regard to the clinical control of maternal diabetes. In this summary discussion will be confined to gestational diabetes whenever pertinent data are available. Otherwise, alterations observed in overt diabetes mellitus or in a mixed population will be described. Gestational diabetes as a term is used interchangeably with diabetes mellitus class A according to WHITE (WHITE, 1978) throughout the article.

The majority of results shows that the changes found in gestational diabetes are identical in character to those found in placentae from women with well established overt diabetes mellitus although the (pathological) alterations tend to be less pronounced and to occur with decreased frequency than in placentae from overt diabetic women.

Histopathology

In contrast to the enormous amout of literature dealing with morphologic and histologic placental changes which may occur in women with overt diabetes mellitus (FOX, 1978; HAUST, 1981; SINGER, 1984) the changes found in gestational diabetes have received scant attention.

Gross examination of placentae from ges-

Table 32. Changes in placentae from woman with gestational diabetes (taken from Jones CJP and Fox H, 1976)

Patchy syncytial necrosis.
Increased number of syncytial secretory droplets.
Dilatation of the syncytial rough endoplasmic reticulum.
Increased syncytial lysosomal activity.
Cytotrophoblastic hyperplasia.
Increased number of active cytotrophoblastic cells.
Degenerative changes in occasional cytotrophoblastic cells.
Focal thickening of the trophoblast basement membrane.
Narrowing of the lumen of already small fetal vessels by enlarged endothelial cells.

---

tational diabetic mothers did not reveal discernible alterations in placental weight or the number of small infarcts and calcified areas (Aladjem, 1967). In overt diabetes mellitus placentae, however, there is some agreement that placental size is increased in most cases and infarcts are about five times as numerous as in control specimens (Fox, 1978; Haust, 1981; Singer, 1984). Generally, most authors find that placental changes occur in pregnancies complicated by overt diabetes mellitus.

Using phase contrast microscopy three types of placental pathology were found (Aladjem, 1967):

1. syncytial pathology including detachment, rupture or disappearance of the syncytium; subsyncytial edema and hyperplasia of the syncytium;

2. stromal pathology displaying edema and intravillous hemorrhage;

3. mixed syncytial and stromal pathology.

Attempts to correlate placental pathology with fetal outcome revealed that all neonatal complications and perinatal deaths occured with mixed syncytial and stromal pathology or stromal pathology alone, whereas syncytial pathology alone did not appear to adversly affect the infant. Syncytial pathology resembled findings in the normal placenta at term but was exaggerated. Early aging of the placenta is also suggested in another study (Emmerich et al., 1978). There is now general agreement as to the presence of premature senescence or persistant immaturity, of varying degrees, particularly of the terminal villi (Jones and Fox, 1976).

The placental lesions associated with gestional diabetes, apparent with the electron microscope, are summarized in Table 32. These abnormalities were not all present in every placenta, but were completely absent in none. Each of these changes might also occur in normal pregnancies but to a much more limited extent. The majority of the placentae in gestational diabetes showed changes identical in character to those found in placentae from overt diabetes mellitus. Thus, there is a certain degree of overlap, although these changes were less marked and occured with less frequency than in the overt diabetes group. Very interestingly, treatment of gestational diabetes with insulin did not alter the changes observed compared to placentae from patients whose maternal serum glucose level was controlled by diet alone (Jones and Fox, 1976). This contrasts with our findings concerning the affinity and number of insulin receptors, which differ significantly in insulin treated gestational diabetic mothers (Desoye et al., 1986). This discrepancy might be due to different states of maternal glycaemic control. In most centers the daily insulin dosage does not exceed 10 to 30 units (Kalkhoff, 1985) whereas at our institution 60 to 90 units of insulin are administered on average.

Distribution and quantity of the enzymes acid phosphatase, aryl sulfatase and alkaline phosphatase did not differ from placentae from uncomplicated full term pregnancies as assessed by ultrahistochemical means (Jones and Fox, 1976). From the ultrastructural studies it is suggested that elimination of hyperglycemia does not in itself prevent the development of placental histologic abnormalities and these must therefore be due to some still unknown components of the disorder which are only partially affected by diet or insulin.

Histomorphometry

In recent years attempts have been made to quantify the morphologic changes in placentae associated with deterioration of maternal carbohydrate metabolism by morphometric analysis in order to define precisely the morphological lesions. One of the studies reported was confined to mothers with diabetes mellitus class A, and data were compared with a carefully matched control group. Thus it was assured that neither different degrees of diabetic severity nor superimposed complications or different gestational ages could confuse the picture and obscure the significance of the results (TEASDALE, 1981).

The placentae from gestational diabetic mothers were somewhat larger in total weight compared to the control group, although this difference was not statistically significant. The increase in placental weight was mainly due to a significant increase of the parenchymal tissue weight, and was accompanied by significantly higher surface areas and a higher cellular content. The parenchymal tissue comprised the trophoblast layer, the fetal capillaries of both peripheral and stem villi, and the intervillous space, and thus contained all structures or compartments which are concerned with the metabolic exchange between mother and fetus.

The finding of a cytotrophoblastic hyperplasia (JONES and Fox, 1976) was not supported by a selective increase in the number of cytotrophoblast nuclei in the placentae from gestational diabetic mothers. The number of highly specialized syncytiovascular membranes was equal in placentae from both groups, contrary to another report where a decreased number of these membranes was found in diabetic placentae (LIEBHART, 1971).

The increased area of exchange quantified in terms of peripheral and villous capillary surface areas and intervillous space volume, suggests that in gestational diabetes placental function is adapted to compensate for dysfunctional metabolic processes, and thus can adequately sustain fetal growth.

Immunpathology

Given the high frequency of diabetes and the apparent involvement of immunological phenomena in its causation it comes as no surprise that immunpathological investigations have been applied to the placenta in maternal diabetes mellitus. These studies were also justified by results suggesting that immunological mechanisms have to be considered in certain aspects of the outcome of diabetic pregnancy (reviewed by GALBRAITH, 1979).

In placentae from overt diabetic mothers the amount and distribution of immunoglobulin, complement factors, certain clotting factors and protease inhibitors were studied. Severe and generalized differences were found in amount, although not necessarily in distribution, when compared with placentae from normal pregnancies. Particularly large quantities of the complement components C 3 and C 9 were found on the trophoblast basement membrane which strongly suggested immunpathological events in the placentae of mothers with longstanding diabetes mellitus (GALBRAITH and FAULK, 1979).

Placentae from gestational diabetic mothers were investigated for the presence and distribution of certain immunoglobulins (IgG and complement components C4, C3, and C9), clotting factors (plasminogen and fibrinogen) and protease inhibitors (alpha-1-antitrypsin and alpha-2-macroglobulin). The most frequently detected differences compared to placentae from healthy mothers were in alpha-2-macroglobulin, complement factors C4 and C3 in the intervillous space and clotting factors related to areas of fibrinoid necrosis. Less commonly found differences were increased depositions of alpha-1-antitrypsin, IgG in stroma and of C9 on trophoblast basement membranes (GALBRAITH et al., 1981). These differences are parallel to those found in placentae from overt diabetic mothers, although they appeared to be less severe and suggest that deterioration of carbohydrate metabolism in pregnancy, even of minor degree, is frequently associated with immunpathological processes reflected in the placenta.

However, the increased deposition of complement components on the trophoblast basement membrane was also detected under other pathological conditions like preeclamptic toxemia (SINHA et al. 1984) and malaria (GALBRAITH, 1980) indicating the nonspecificity of these alterations (immunpathological findings). The discrete location of C4 in the intervillous spaces has been interpreted to indicate that activation of complement within the villi may occur via the alternate pathway (FAULK et al., 1980). The increased amount of fibrinogen, plasminogen and alpha-2-macroglobulin found is in line with the increase in areas of fibrinoid necrosis frequently observed in diabetic placentae (FOX, 1978). It can be speculated that this reflects an abnormality of clotting mechanisms which might possibly affect the fetus. Some support is added by the observations of a diminished release of prostacyclin PGI_2 from trophoblast cells in tissue culture by elevated medium glucose concentrations (RAKOCZI et al., 1986) since PGI_2 is a potent antiaggregating factor, and prevents platelet aggregation and thrombus formation (SAMUELSON et al., 1978). Further-

more, evidence was provided for a decreased release of endogenous heparin or heparin-like substances from its complexes with placental protein 5 (PP 5) upon the addition of protamine in placentae from gestational diabetics. This indicates lower levels of heparin-like substances in the placental circulation, and that PP 5 may be involved in the coagulation system (SALEM et al., 1981).

Apart from the above characterized quantitative alterations in products deposited in diabetic placentae which seem to be very nonspecific, observations have been reported which are more specifically related to the particular metabolic events associated with disturbances of carbohydrate metabolism. The placental transport of maternal immunoglobulins serves as a protective mechanism to the fetus (BRAMBELL, 1970). Attention has focused on the transfer of antibodies to insulin, to insulin receptors, and to islet cells. In fact, the presence of maternal islet cell antibodies in the fetal circulation was detected in an offspring of a gestational diabetic mother (GAMLEN et al., 1977). Furthermore, binding of fluorescein-labeled insulin to placental tissue was observed in the majority of diabetic pregnancies, whereas no similar binding could be demonstrated in placentae from healthy mothers (BURSTEIN et al., 1963), which was interpreted as an indication of the presence of insulin antibodies in the diabetic placenta. However, on the basis of current knowledge on placental insulin receptors this notion seems to be doubtful.

The above considerations indicate that the placenta is immunologically more active during diabetic pregnancies. The finding of increased deposition of antibodies and immune complexes in the placenta might result in the tissue damage well known in both gestational and insulin dependent diabetes (FOX, 1978). On the basis of current literature it seems that the findings discussed above do not appear to correlate in extent with the severity of the maternal metabolic disturbance. Thus one reasonably can speculate that maternal diabetes might superimpose an immunological stress upon normal pregnancy, leading finally to placental damage with potentially severe pathological consequences for the fetus (FAULK et al., 1980).

Carbohydrate metabolism

In addition to morphologic and histologic changes in placentae from diabetic mothers, carbohydrate metabolism has received enormous interest. Whereas early reports were centered around the production and breakdown of glycogen, in recent years other aspects such as variations of enzyme activities or insulin receptors have been studied.

The concentration of D-glucose is higher in maternal than in fetal plasma (MORRISS et al., 1975; KALHAN et al., 1979) and glucose in the fetal compartment stems entirely from the mother (KALHAN et al., 1979). Furthermore, glucose not only crosses the placenta from mother to fetus, but also serves as the major substrate for aerobic metabolism in the placenta itself (MESCHIA et al., 1980). About 50–60% of the glucose leaving the maternal compartment is utilized by the placenta and only the remaining glucose is taken up by the fetus (MESCHIA et al., 1980; SCHNEIDER et al., 1981).

The metabolic fate of glucose in the placenta was studied by using radioactively labelled glucose. The fraction of glucose metabolized by different pathways was calculated from the distribution of label in different metabolites. In term placentae 90% of label was recovered in the Embden-Meyerhof pathway; the pentose phosphate and the non-triose phosphate pathways (including glycogen synthesis) each accounted for about

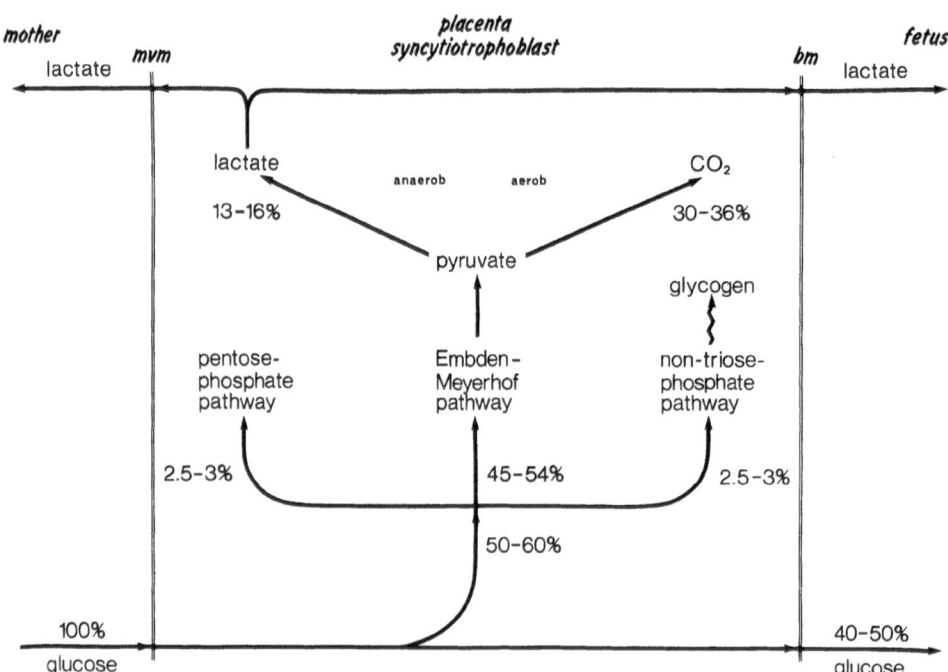

Fig. 22. Fate of maternal glucose in the placenta. Percentage values refer to 100% maternal glucose before entering the placenta. *mvm* microvillous membrane, *bm* basal membrane

5% of metabolized glucose (SAKURAI et al., 1969). The largest portion of glucose is utilized for providing energy to support the different transport activities and processes of protein synthesis unique to the placenta (Fig. 22).

About 30% of the glucose taken up by the placenta is converted to lactate (MESCHIA et al., 1980; SCHNEIDER et al., 1981) which then is excreted primarily to the maternal side (SCHNEIDER et al., 1981), very likely by a non-carrier-mediated diffusion (IILLSLEY et al., 1986). In sheep the small amount of lactate which is transfered to the fetus might serve as substrate for oxygen consumption but similar information is lacking for the human (FABER and THORNBURG, 1983).

Uptake and transport of glucose to the fetus by the placenta is a saturable, sodium-independent process which is inhibited by phloretin, suggesting a carrier mediated mechanism. The placental structures concerned with glucose uptake and transport are located in the trophoblast and syncytiotrophoblast which has been shown to be polarized in terms of specialization of the distinct plasma membranes (WHITSETT, 1980; TRUMAN and FORD, 1984). The brush border membrane is rich in insulin receptors and in alkaline phosphatase and contains a Na^+ dependent amino acid transporter. The non-brush border membrane is rich in hormone sensitive adenylate cyclase activity, beta-adrenergic receptors and an ATP-dependent Ca^{++}-transport system. This polarization allows different interactions at the maternal-placental and fetal-placental interfaces.

Both membranes have been shown to contain glucose transport molecules (BISSON-NETTE et al., 1981; JOHNSON and SMITH, 1985) with similar properties and similar capacity for transport in terms of unit membrane area. However, the basal membrane surface area in vivo is only one-fifth that of the microvillous membrane surface area (LAGA et al., 1973). Thus the basal membrane has less total capacity and might be the limiting factor for glucose transport through the placenta.

High concentrations of estriol and progesterone inhibit the glucose transport system in both microvillous and basal membrane (JOHNSON and SMITH, 1985). The level of steroid hormones is continuously increasing throughout pregnancy in maternal blood (DESOYE et al., 1987). In addition the local concentrations in placenta in vivo are reportedly high enough to inhibit glucose transport (SMITH and BRUSH, 1978). Whether this form of control over glucose flow occurs in vivo is still an open question but would be unique with regard to the mechanisms involved. Another possible regulating hormone for glucose transport is insulin which is known to stimulate the recycling of glucose transport molecules into the plasma membrane in fat cells (KONO, 1983). However, in perfusion studies neither uptake and transfer nor lactate release were significantly affected by insulin (CHALLIER et al., 1986). Placental glucose transport is apparently not regulated by maternal hyper- or hypoinsulinemia and seems to be unaffected by the saturation of the glucose transporter indicating the lack of feedback mechanisms (NESBITT et al., 1973). Thus glucose uptake and transfer by the human placenta seem to be autonomous from mother and fetus and only be limited by the transport capacity of the trophoblast basal membrane. In the event of maternal diabetes or gestational diabetes the placenta is faced with large amounts of glucose. Some of the glucose is believed to be diverted to the synthesis of polysaccharides including glycosaminoglycans, as manifested by the accumulation of these substances in placental tissue (ABRAMOVICI et al., 1978). This could reflect a mechanism of local homeostasis generating enhanced glucose utilization which leads to higher polymers such as polysaccharides, glycoproteins or glycogen.

The human placenta contains appreciable quantities of glycogen which are higher in the first trimester compared to term. This glycogen is located in the main stem chorionic villi and in larger fetal blood vessels (ROBB and HYTTEN, 1976), in the amniotic epithelium (WANG, 1980) and also in small clusters and individual particles as well as in the cytoplasm and in the syncytial sprouts in the syncytium (BOYD, 1958; FOX, 1978 a). It was first thought that placental glycogen might provide an energy source for the fetus but the fetal liver is much better equipped to deal with emergency fetal demands since it stores ten-fold more glycogen than the placenta. Thus it is more likely that glycogen is stored for local use and probably plays an important role as an emergency energy reserve for vasomotor function supporting the placental vessels to regulate fetal levels of essential substances such as oxygen and carbon dioxide. This notion is substantiated by increased glycogen breakdown in cases of hypoxia (DEMERS et al., 1972; GABBE et al., 1972 a). Furthermore, glycogen utilization might yield the energy for the transport of amino acids into the placenta anaerobically against a concentration gradient (LONGO et al., 1973).

Decreased glycogen contents in placentae are the net result of both a decreased rate of synthesis caused by a shift from glucose-6-phosphate independent to dependent glycogen synthetase (enzyme catalogue number, EC 2.4.1.11) and a concommitant degradation by activation of glycogen phosphorylase (EC 2.4.1.1) (DEMERS et al., 1972). However, the likehood of an energy providing role of glycogen is limited by reports of very low concentrations (DI PIETRO et al., 1967) or the absence from the placenta (WALKER et al., 1967) of one of the rate limiting enzymes for the conversion of glycogen to glucose – glucose-6-phosphatase (EC 3.1.3.9). Perhaps glucose is formed by the terminal action of a non-specific placental alkaline phosphatase (EC 3.1.3.1) which was found to hydrolyze glucose-6-phosphate in vitro (HAFEZ, 1964). This enzyme is only present in third trimester placenta when the glycogen content is decreased (JONES and FOX, 1976). Its localization in the syncytiotrophoblast and at the apposition of cytotrophoblastic and syncytiotrophoblastic membranes (HULSTAERT et al., 1973; JONES

and FOX, 1976) renders the involvement of such a phosphatase in the process of glycogen utilization topologically possible.

Since glycogen synthetase as well as glycogen phosphorylase are insulin dependent enzymes as shown in liver (BISHOP and LARNER, 1967) and muscle (VILLAR-PALASI and LARNER, 1961), it is reasonable to assume a regulatory role of insulin in placental carbohydrate metabolism, too. The stimulation of placental glucose uptake by insulin is documented (VILLES, 1953) as well as a glycogenic response of placental tissue culture when insulin is added to the culture medium (DEMERS et al., 1972). The latter effect was evoked by a shift from a dependent to an independent form of glycogen synthetase resulting in an increased glycogen synthesis regardless of the medium glucose concentration. However, another study failed to show an effect on placental glycogen content of maternal diabetes mellitus or maternal or fetal insulin (ROBB and HYTTEN, 1976). Moreover, no effects of insulin on glucose uptake nor on glucose oxidation were detected (SZABO and GRIMALDI, 1970). Based on correlations of glycogen content with the concentrations of insulin receptors in the placenta and the insulin-insulin receptor complexes in placentae from normal and diet treated diabetic women it was inferred in a very recent study that insulin by its receptor regulates the placental glycogen concentration (DESOYE et al., 1986).

To the authors knowledge no attempts have been made as yet to unravel conceivable autocrine effects of human placental lactogen on placental glycogen although the diabetogenic properties of the hormone (BECK and DAUGHADAY, 1967) should stimulate such studies.

Generally, an increased glycogen content in placentae from insulin-dependent diabetic women is unequivocally documented (HEIJKENSKHOLD and GEMZELL, 1957; FISCHER and HORKY, 1966; GABBE et al., 1972; DIAMANT et al., 1982; DIAMANT and KISSILEVITZ, 1983) as well as an increased lactate production (GINSBURG and JEACOCK,

Table 33. Alterations of some enzyme activities with regulatory functions in different pathways of carbohydrate metabolism in gestational (GD) and overt, insulin-dependent (ODM) diabetes mellitus compared to groups of normal pregnant women. Enzymes are related to glycogen metabolism (1, 2), to glycolysis (3), to pentose phosphate pathway (NADP generation) (4, 5) and to gluconeogenesis (6–8)

Enzyme			GD	ODM	Reference
1	Glycogensynthetase EC 2.4.1.11				GABBE et al., 1972
	Gluc-6-P	indep.	=	↓	
		dep.	↑	=	
2	Glycogenphosphorylase EC 2.4.1.1				GABBE et al., 1972
	5'-AMP	indep.	=	↑	
		dep.	↑	↑	
3	Pyruvatkinase EC 2.7.1.40		↑	↓	DIAMANT et al., 1984
4	Glucose-6-phosphate-dehydrogenase EC 1.1.1.49		↑	↑	DIAMANT et al., 1984
5	6-Phospho-gluconat-dehydrogenase EC 1.1.1.43		↑	↑	DIAMANT et al., 1984
6	PEP-carboxylase EC 4.1.1.31		=	=	DIAMANT et al., 1984
7	Aspartate amino transferase EC 2.6.1.1		=	=	DIAMANT et al., 1984
8	NADP-malate dehydrogenase EC 1.1.1.40				DIAMANT et al., 1984
	per g prot, g DNA		=	↑	
	per total placenta		↑	=	

1966). Assessment of such data for the placentae of gestational diabetics requires a more refined approach, since glycogen content has been shown to be increased if expressed as g glycogen per total placenta, whereas it was unaltered if related to mg DNA or g placental tissue (GABBE et al., 1972; DIAMANT and KISSILEVITZ, 1983). An even more complicated picture emerged when diet treated were compared with insulin treated gestational diabetics. In the former case there was decreased, but in the latter case no change in, placental glycogen per g tissue (DESOYE et al., 1986). These results can be explained by the altered placental enzyme activities in gestational and overt diabetes mellitus (Table 33).

The affinity and concentration of insulin receptors are additional determinants of insulin effects in target tissue. Human placenta has been shown to contain high concentrations of insulin receptors (POSNER, 1973; HAOUR and BERTRAND, 1974; MARSHALL et al., 1974; TAKANO et al., 1975; STEEL et al., 1979; ARMSTRONG et al., 1982) and is the tissue of origin for the receptor isolation (WILLIAMS and TURTLE, 1979; HARRISON and ITIN, 1980). Autoradiography with ^{125}I-insulin and electron microscopy using insulin-ferritin conjugates demonstrated that the insulin binding sites are located on the microvillous membrane of the syncytiotrophoblast (NELSON et al., 1978; WHITSETT and LESSARD, 1978). The physiological significance of the placental insulin receptor still awaits an unambigous clarification. As in other cells, insulin has been shown to stimulate tyrosine phosphorylation of its placental receptor (AVRUCH et al., 1982), a process which is thought to represent one of the key steps in the signalling mechanism for intra-cellular insulin action. Furthermore, a protein kinase activity is associated with the placental insulin receptor, which *in vitro* phosphorylates rabbit muscle actin (MACH-ICAO et al., 1983) and endogenous (membrane) as well as exogenous phosphatidyli-nositol (MACHICAO and WIELAND, 1984). Indirect evidence suggests that the insulin receptor exerts its function on placental glycogen and N-acetyl-neuraminic acid – a carbohydrate molecule present in every membrane glycoprotein (DESOYE et al., 1986).

The status of placental insulin receptors in diabetes mellitus is debated. One report demonstrated that insulin binding to the receptor was unaffected by maternal diabetes without specifying the severity of the disorder (POSNER, 1973). A significant decrease in the receptor concentration to about half the normal was observed in a microsomal membrane fraction from placentae of insulin dependent mothers (HARRISON et al., 1977). The number of receptors reportedly decreased in gestational diabetic mothers (DURAN-GARCIA et al., 1979). The importance of considering the severity of maternal diabetes is stressed by the finding of unaltered receptor-affinity and -concentration for a mixed group of woman with gestational and longstanding diabetes (DESOYE and WEISS, 1987 a). The discrimination of gestational diabetics who were either treated with diet alone or with diet plus exogenous insulin according to a recently developed scheme (WEISS et al., 1986) revealed a significant increase in affinity and a significant lower receptor concentration only in the diet treated group. Women receiving exogenous insulin showed unaltered insulin binding

with respect to the receptor status (DESOYE et al., 1986).

In mechanistic terms the decrease in the receptor concentration is the net result of a lower rate of receptor recycling into the membrane and/or lower synthesis and faster degradation of receptors. The increase in the receptor affinity may be generated by an altered lipid composition of the microvillous membrane which is one of the determinants of receptor affinity. Alterations of the membrane lipid environment might provide a means for regulating receptor affinity (DE-SOYE et al., 1986 a; DESOYE and WEISS, 1987 a) and more indirectly intracellular insulin effects.

The current picture of placental carbohydrate metabolism in gestational diabetes can be summarized as follows:

1) decreased glycogen content per functional unit of the placenta;

2) glycogen synthetase is less activated by insulin (decreased fraction of the dependent form) as a consequence of diminished intracellular insulin effects due to a receptor defect (decreased receptor concentration);

3) gluconeogenesis is unaltered if expressed per functional unit;

4) glycolysis is increased, perhaps resulting in increased lactate production in cases of hypoxia;

5) generation of NADP is increased, perhaps to meet placental demands for energy since the glycogen content is decreased.

Unfortunately, to the authors knowledge, no data are available on the effects of gestational diabetes on glucose and lactate transport through the placenta. Such studies would help in unravelling the complex processes of placental carbohydrate metabolism associated with diabetes.

Lipid metabolism

The quantity of lipids in the placenta is small (ROBERTSON and SPRECHER, 1968; YOU-NOSZAI and HAWORTH, 1969), indicating a minimal placental involvement in lipid me-

tabolism. However, free fatty acids can be transported from the maternal circulation to the fetus. A considerable contribution to the transfer of maternal serum free fatty acids is assigned to the placenta. Furthermore, in view of the pronounced changes which maternal plasma lipid and lipoprotein levels undergo throughout pregnancy (DESOYE et al., 1987) which differ in gestational diabetic from normal pregnancies, lipid metabolism of the placenta has to be discussed in the context of that of the mother.

Significantly higher than normal fasting levels of plasma triglycerides in the second and third trimester of gestational diabetic pregnancies are consistently reported (KNOPP et al., 1972; WARTH et al., 1977; KNOPP et al., 1980; HOLLINGSWORTH and GRUNDY, 1982). Cholesterol levels are either unaltered compared to normal (WARTH et al., 1977; KNOPP et al., 1980) or decreased (KNOPP et al., 1977; HOLLINGSWORTH and GRUNDY, 1982). The decrease of cholesterol is due to a reduction in the cholesterol content of high density lipoproteins, probably as a result of the exchange of triglycerides for cholesterol esters in the core lipids of high density lipoprotein (HOLLINGSWORTH and GRUNDY, 1982).

These variations of maternal lipid concentrations from normal are paralleled by higher triglyceride and phospholipid-phosphorus contents in the placentae (DIAMANT et al., 1982; DIAMANT and KISSILEVITZ, 1983). Total cholesterol is unchanged which might be due to unaltered or only minimally elevated levels of cholesterol rich low density lipoproteins in diabetic pregnancies compared to very low density lipoproteins (KNOPP et al., 1978). The human placenta is able to extract cholesterol in esterified form from the maternal circulation by means of low density lipoprotein receptors located on the surface of the syncytial microvillous membrane (WINKEL et al., 1981; ALSAT et al., 1982; ALSAT et al., 1984; MALASSINE et al., 1984). The cholesterol taken up from the mother serves as the principal precursor for human placental progesterone biosynthesis, since trophoblastic tissue has only a limited capacity for synthesizing cholesterol de novo (VAN LEUSDEN and VILLEE, 1965). Triglycerides are part of chylomicrons and very low

density lipoproteins, both of which are too large to cross the placental barrier. Instead they appear to be formed in the placenta by esterification of fatty acids supplied either from the maternal circulation (YOSHIOKA and ROUX, 1972; SZABO et al., 1973; HULL and ELPHICK, 1979) or synthesized de novo (HOSOYA et al., 1960; KLEINE, 1967; DIAMANT et al., 1975). However, the key enzyme of fatty acid synthesis − acetyl-CoA-carboxylase (EC 6.4.1.2.) − is present only in negligible amounts in both early and term placenta (DIAMANT et al., 1975), and it thus seems unlikely that de novo synthesized fatty acids are a major source for triglyceride synthesis in vivo.

Gestational diabetes is accompanied by elevated levels of placental lactogen in maternal circulation (SAMAAN et al., 1985) which accounts for (TURTLE and KIPNIS, 1967) a higher concentration of free fatty acids (MOELSTED-PEDERSEN et al., 1972). This might give rise to increased free fatty acid and in turn to increased triglyceride contents in the placenta. Apart from fatty acids in the placenta, triglyceride synthesis depends on continued availability of the precursor glycerophosphate, which in turn is linked to glucose breakdown. Since glycolysis seems to be activated in gestational diabetics (see previous section) one can presume that sufficient amounts of the precursor are provided.

About 75% of placental lipids are in the form of phospholipids (HEIJKENSKJOLD, 1958) which is likely the result of placental synthesis rather than uptake from maternal blood, since intact phospholipids do not cross the placenta to the fetus (POPJACK and BEECKMANS, 1950). The placenta contains all the necessary enzymes for phospholipid synthesis (KARP et al., 1973) reaching maximum levels at term (ROUX and GREEN, 1968). Higher concentration in placentae from gestational diabetics (DIAMANT et al., 1982) may be attributed to higher placental pentose shunt activity (see previous section) and presumably associated higher NADPH-levels. Whether insulin regulates the activities of enzymes involved in placental lipid production, and/or accumulation as in adipose tissue, has not been investigated so far.

The described differences in the placental lipid pool may result in alterations in metabolic fuels that are available to the fetus.

Fetal adiposity which is frequently seen in poorly controlled gestational diabetics might be attributed to increased levels of maternal fatty acids and their subsequent transfer, esterification and incorporation into the fetal triglyceride pool.

Nucleic acids, amino acids and proteins

Placental proteins are of major importance for the progress and successful outcome of gestation. Growth of the placenta is dependent on protein metabolism. The protection of the fetus depends on proteins – on those enzymes of the endoplasmic reticulum which modify or degrade environmental pollutants. Furthermore, a considerable number and quantity of proteins is synthesized and excreted by the placenta, including peptide hormones and some proteins whose biological and physiological significance is obscure so far. Protein synthesis requires nucleic acids for the generation of the template for the encoded genetic information as well as amino acids as final precursors of proteins.

Nucleic acids and nucleotides should be present in sufficient quantities at any time of gestation since nucleotides undergo a continous turnover. Furthermore, degradation of DNA and RNA by exo- and endonucleases should guarantee a continuous level of nucleotides in the intracellular nucleotide pool. Studies on placental transfer of nucleic acids and nucleotides are lacking for the human, but in the guinea pig it appears that only oxypurines and guanines are transported from mother to fetus (VAN KREEL et al., 1982). Amino acids required for protein synthesis are mainly derived from the mother and transfered to the placenta. The mechanism of amino acid transport through the trophoblast membranes have been identified (YUDILEVICH and SWEIRY, 1985) and have been found to be coupled to an inward sodium gradient, demonstrating its energy dependence associated with an electrical potential across the membrane. The energy for amino acid uptake and transfer is generally provided by aerobic mechanisms (SMITH et al., 1973; LONGO et al., 1973; SCHNEIDER and DANCIS, 1974; ENDERS et al., 1976; RUZYCKI et al., 1977; MONTGOMERY and YOUNG, 1982). No data are reported so far as to whether gestational diabetes affects transport and uptake of amino acids. However, uptake of amino acids by placental villous tissue has been shown to be insulin independent (STEEL et al., 1979). Furthermore, despite an elevated total protein content of placentae from diabetic mothers, the increase in protein content is proportional to the increase in DNA, demonstrating that the amount of protein per cell has not increased (WINICK and NOBLE, 1967). Thus any impact of placental amino acid uptake seems very unlikely.

The key rate-limiting factor in protein synthesis is presumably the availability of pentose sugars for nucleic acid production, since pharmacologically induced changes in pentose shunt activity were accompanied by corresponding changes in nucleic acid and protein synthesis (BEACONSFIELD et al., 1965). In view of the enhanced activity of the pentose shunt pathway in gestational diabetics (see previous section) one might assume that pentose sugars should be available in sufficient amounts for the production of increased amounts of total DNA (WINICK and NOBLE, 1976), total RNA (MILSS et al., 1985) and of total protein (WINICK and NOBLE, 1967).

The regulation of placental protein synthesis at the translational level has recently been demonstrated as another part of significant determinant of the magnitude of the response to physiological or pathophysiological stimuli. Ribosomes from explants of term placenta from diabetic mothers translate the messenger RNA at a much higher speed compared to normal tissue (ILAN et al., 1984). In addition to this increased average ribosome half-transit time the average polypeptide molecular weight was lower than those of normal pregnancies (ILAN et al., 1984).

The major protein with respect to weight synthesized in human term placentae is placental lactogen; its synthesis represents as much as 25% of total protein synthesis from isolated messenger RNA (SHERWOOD et al., 1979). Higher levels in serum of gestational diabetics compared to healthy mothers (SAMAAN et al., 1985) reflect increased placental synthesis (HUBERT et al., 1981). Attempts to demonstrate increased levels of translatable messenger RNA coding for placental lactogen (hPL-mRNA) showed similar portions of HPL-mRNA from total RNA in normal, gestational diabetic and class C diabetic placentae. However, the amount of total RNA per g tissue is greater in gestational diabetics and accordingly these placentae produce more proteins including hPL (MILLS et al., 1985). In the case of severe maternal diabetes mellitus – class R – translatable hPL-mRNA was reduced relative to total mRNA. Unfortunately, in this study maternal hPL levels were not recorded (MILLS et al., 1985).

Insulin has been shown to have positive (PEAVY et al., 1978) and negative (CIMBALA et al., 1982; GRAUNER et al., 1983) effects on some specific mRNAs and it is appealing to postulate that the expression of placental proteins is regulated by insulin. This could be effectuated via an altered phosphorylation of the ribosomal protein S 6 subsequent to insulin binding (SMITH et al., 1979; THOMAS et al., 1982) giving rise to altered mean residue times of mRNAs on ribosomes. Such a mechanism could be compatible with the finding of altered levels of some other proteins synthesized by the placenta, such as placental protein 5 (PP 5) (SALEM et al., 1981), placental protein 10 (PP 10) (TIITINEN and YLINEN, 1986) and pregnancy-associated plasma protein B (PAPP-B) (LIN et al., 1978), whereas the concentration of pregnancy specific $beta_1$-glycoprotein (SP 1) was unaltered (SINGH et al., 1979). Such a regulation by insulin could represent a protective mechanism by and possibly for the placenta to compensate for high levels of placental lactogen by the counteracting hormone insulin.

However, the detailed mechanisms of regulation of placental protein synthesis are still unknown and await further investigations. Without such information it is impossible to discern whether maternal diabetes mellitus or gestational diabetes intimately affects protein synthesis or if the increased levels of placental proteins – reflected in elevated maternal serum levels – are only due to increases in trophoblast mass. In addition, the processes of protein excretion as well as the metabolism of the distinct proteins need further study.

The placenta as affector of fetal growth

Fetal requirements have usually been considered the dominant driving force for placental growth, mainly because placental size generally correlates with fetal size and number. However, continued growth of the placenta after fetectomy and continued production of hormones even in the absence of a fetus (PETROPOULOS, 1973) suggests that control of fetal growth is exerted by the placenta. Offspring of gestational diabetic mothers tend to have significantly higher birth weight than those from control mothers (FREINKEL et al., 1985). Therefore, an altered regulation of fetal growth by the placenta can reasonably be assumed in gestational diabetes. Maternal malnutrition can frequently entail fetal growth retardation (NAEYE et al., 1973; LECHTING et al., 1975;

MUNRO, 1980; ROSSO, 1980) supporting this concept as does the observed association between rapid fetal growth and elevated maternal plasma concentrations of amino acids (CHRISTENSEN and STREICHER, 1948).

Basically, two mechanisms for placental effects on fetal development may be hypothesized. Firstly, the placenta serves as buffer, transport vehicle and selective barrier for transfer of maternal nutrients and precursors. Thereby the placenta would play a more or less passive role as mediator of fluctuations and alterations of maternal metabolism to the fetus. In the classical 'Pedersen' hypothesis (PEDERSEN et al., 1977) such a role is assumed. Secondly, the placenta itself can synthesize distinct molecules generating control and regulation of fetal growth. Insulin has been regarded as likely prime candidate for growth-promoting actions in the fetus (MILNER and HILL, 1984) because of its manifold anabolic effects during postnatal life and because of its well known growth-stimulatory actions (STRAUS, 1984). However, there is no evidence for transplacental passage of considerable amounts of insulin (CHALLIER et al., 1986). Furthermore, insulin secretion by fetal pancreatic islets of Langerhans normally displays only limited adaptions and responses to acute changes in cord blood glucose until approximately week 28 of gestation (ESPINOSA DE LOS MONTEROS et al., 1970; SHELLEY et al., 1975). Thus a direct trophic action of insulin on the fetus seems to be unlikely.

Several studies in mammalian species have shown that intrauterine insulin administration could produce increased levels of insulin-like growth factor-I (IGF-I, Somatomedin C) in the fetus, suggesting, if any, an indirect mechanism by which insulin could augment fetal growth (HILL and MILNER, 1980; SPENCER et al., 1983).

Insulin-like growth factors-I and -II (IGF-I, IGF-II) are peptide hormones similar to insulin with respect to structure and biological activities such as acute anabolic and long-term growth promoting effects (RINDERKNECHT and HUMBEL, 1976). There are several lines of evidence which suggest that these peptides are probably involved in stimulating fetal growth (Table 34) (UNDERWOOD and D'ERCOLE, 1984).

The failure to detect significant production of IGF-I in mouse placenta and the observation that no growth factor was transfered into the fetal circulation of dogs, sheep and rats (D'ERCOLE et al., 1980) has necessitated the search for a placental mediator effecting fetal levels of IGF. Human placental lactogen has been proposed to constitute the primary regulator of IGF in the fetus (UNDERWOOD and D'ERCOLE, 1984). However, several facts argue against this concept (HOUGHTON et al., 1984) and it is still unclear what factors regulate fetal growth factor concentrations and whether placental lactogen is involved.

Very recent results might render the assumption of such mediators unnecessary. By hybridization techniques insulin-related sequences were detected in human placentae at various stages of development and they were estimated to represent about 0.03–0.1% of the total poly(A)$^+$ RNA in placenta (LIU et al., 1985). This quantity could code for substantial amounts of protein which may well be a regulatory hormone for fetal development. In view of the high insulin degrading activity of human placentae (POSNER, 1973) it remains to be established if this protein constitutes a nondegradable modi-

---

Table 34. Evidence that IGF-I has a role in fetal growth (from UNDERWOOD LE and D'ERCOLE AJ, 1984)

Somatomedin stimulates the proliferation of fetal cells in vitro.
Multiple fetal tissues possess somatomedin receptors.
Multiple fetal tissues synthesize somatomedin.
Somatomedin in cord blood correlates with birth size.

---

fied insulin. Moreover, mRNA coding for IGF-II was detected in the placenta as early as in week 11 of gestation (SHEN et al., 1986). Apart from the autocrine and/or paracrine mechanisms by which way it might act locally in placentae it may play a role in fetal growth and development as well.

In gestational diabetic placentae an increased relative abundance of the respective messages (mRNA) has been detected. It was hypothesized that the increased levels of hPL-mRNA (MILES et al., 1985) might bring about higher levels of IGF-II mRNA which in turn might give rise to enhanced and macrosomic fetal growth (SHEN et al., 1986).

Irrespective of the detailed function of these growth factors and their involvement in the regulation of fetal development the data which have been emerging and accumulating during recent years entail the assumption that the well known 'Pedersen' hypothesis (PEDERSEN, 1977) or its modified version (FREINKEL, 1980) might not suffice to explain the mechanisms determining fetal development. In fact, it seems that particularly with the detection of various growth promoting factors new concepts are required which have to be reconciled with the rapidly growing information. It may well be that the two mechanisms discussed above have to be combined in that the primary stimulus stems from the placenta – represented by growth factors – which also serves as supplier of maternal nutrients and precursors to sustain fetal growth. Thus 'Pedersen's' hypothesis would represent the metabolic part of an expanded concept. The placenta as determining factor of fetal growth, and overgrowth in diabetic placentae, will likely be better understood in the future.

Miscellaneous

Several data which do not fit in any of the previous sections will be summarized herein. The glycosaminoglycan (GAG) composition of connective tissue components of placental syncytiovascular membrane was characterized as a step towards understanding the involvement of the membrane in transfer processes. Total GAG content was increased in diabetic placentae (WASSERMAN et al., 1980) as well as in placentae of rats with chemically induced diabetes (ABRAMOVICI et al., 1978). Furthermore, among the GAG-subfractions measured, hyaluronic acid was markedly (LAURETI et al., 1982; WASSERMAN et al., 1980) and heparan sulphate was slightly increased (WASSERMAN et al., 1980). Heparan sulphate as well as dermatan sulphate, whose portion was considerably decreased, are involved in the blood coagulation process. Pregnancy in general, and diabetic pregnancy in particular, is accompanied by deteriorated regulation of blood clotting leading to a hypercoagulable state and greater risk of blood clotting. This is reflected in thrombi and deposition of fibrin found histologically in diabetic placenta (FOX, 1978). Studies in rat placentae have revealed a direct correlation between the degree of maternal hyperglycemia and the amount of GAG (ABRAMOVICI and SVEJCAR, 1982). Similar studies have not included the human but would be of importance to clarify whether the alterations observed in GAG are etiological for the histopathologic changes in diabetic placentae, or if they merely reflect the impaired physiological conditions.

The thickening of placental basement membranes observed in mothers with manifest diabetes mellitus (OKUDAIRA et al., 1966) might be attributed to increased amounts of collagen – the major connective tissue protein – reflected in about 30% higher concentrations of 4-hydroxyproline in these placentae (LAURETI et al., 1982).

In view of the large quantity of steroid hormones produced by the placenta, particularly in the second half of gestation, steroid synthesis was assessed in pregnancies complicated by maternal diabetes. The aromatizing capacity, i.e. the formation of estrogens from androgens, was measured in placental tissue homogenates (SYBULSKI, 1969) as well as in perfused placentae (Pajszczyk-Kieszkiewicz, 1972). Diminished levels of estrone account for the decreased amounts of total estrogens found in diabetic placentae since levels of estradiol were unaffected. Furthermore, the response to exogenous NADPH — a cofactor for enzymes involved in the synthesis of estrone and 17-beta-estradiol — was lower in the diabetics, which suggests diminished activity of aromatizing enzymes. It should be recalled that in gestational diabetes the increased generation of NADP (see previous section) should provide an amount sufficient for the synthetic processes. In addition, a direct negative correlation of estradiol secretion by cultured trophoblast cells with intracellular glucose was demonstrated as well as a decrease of estradiol secretion dependent on the insulin dose (Hochberg et al., 1982).

It has been hypothesized that estradiol produces insulin resistance which in turn, by lowering intracellular glucose concentration and utilization, stimulates estradiol secretion. Although one has to keep in mind that data obtained in isolated cells need not necessarily reflect the *in vivo* situation this notion seems attractive, particularly because it might help explain the low values of plasma estriol found in some diabetic women during pregnancy (Ratanasopa et al., 1967). Increased concentrations of insulin during insulin treatment of diabetics give rise to enhanced glucose transport and metabolism and may lead to a decreased secretion of estradiol and conceivably of estriol, too.

5

Prevention of Placental Insufficiency Syndrome in Diabetic Pregnancies

W. Burkart, U. Cirkel, J. P. Hanker, and *H. P. G. Schneider*

University Clinic of Obstetrics and Gynecology, Münster, Federal Republic of Germany

Introduction

Diabetes mellitus represents a serious complication of pregnancy. Besides maternal jeopardy by metabolic derangements (HOLLINGSWORTH, 1984) increased perinatal and neonatal mortality are well documented (WEISS and HOFMANN, 1985). About 50% of the perinatal mortality is caused by severe congenital malformations which are found about two to three times more frequently in diabetic pregnancies (BURKART et al., 1987; FUHRMANN, 1982). The stillbirth rate due to acute placental insufficiency is still increased in comparison to a population of nondiabetic mothers (WEISS and HOFMANN, 1985). In prior years premterm delivery was recommended to avoid stillbirth; the risks of an immature newborn seemed to be of less concern.

In most cases placental insufficiency does not develop suddenly but rather slowly. It is seen in macrosomic as well as in small for gestational age fetuses (SAINTONGE and CÔTÉ, 1983; HEISIG 1975). The only clinically feasible way to prevent placental insufficiency, and in consequence intrauterine demise, in diabetic pregnancies seems to be the maintenance of normoglycemia throughout pregnancy (BURKART et al., 1987; GABBE, 1979). In order to detect metabolic alterations of the placenta caused by maternal diabetes we first studied the placental glycogen and protein content as well as the activity of placental enzymes in alloxan-diabetic rats (CIRKEL et al., 1986). The results were verified in the human placenta and we tried to find out which components of placental insufficiency could be influenced by good metabolic control during pregnancy.

Materials and methods

21 placentas which were obtained by vaginal delivery or by cesarean section after gestational week 37 were studied. Immediately after birth the material was cleansed from ovum membranes, and umbilical cord residues and frozen ($-18°C$). A control group consisted of 9 placentas from healthy mothers in which diabetes was excluded by history and by a single HbA1 determination. 12 placentas originated from insulin dependent mothers (n = 11) and one patient in whom diabetes was not known before pregnancy. White's classification modified according to WEISS and HOFMANN (1985) allows to subdivide the diabetics as Bo (n = 1), B (n = 5), C (n = 2), D (n = 2), and RF (n = 1). Three

Table 35. Mean birth weight and mean placental wet weight

	Birth weight [g]	Gestational age (weeks)	Placental wet weight	n
HbAl > 8.5%	4233 ± 660	40.3 ± 0.6	704 ± 156	3
HbAl < 7.5%	3380 ± 656	39.5 ± 1.2	502 ± 125	9
Controls	3465 ± 534	40.3 ± 1.5	537 ± 104	9
	n.s.	n.s.	p < 0,05	

of the patients were not normoglycemic as estimated by a mean HbA1 value of > 8.5%, the remaining 9 patients maintained near-normoglycemia from the beginning of pregnancy or at least since the first trimester. The mean concentration of glycosylated hemoglobins in these latter patients was below 7.5%.

In the diabetic patients the mean HbA1 value was calculated from the single values measured biweekly between gestational week 30 and delivery. In case of late booking we tried to have at least two single values before and postpartum. For the determination of the amount of glycosylated hemoglobin we used the method of PanChem (Kleinwallstadt, FRG). The analysis of all other parameters was previously described (CIRKEL et al., 1986). For statistical evaluation we used the Mann-Whitney-Test.

Results

Table 35 gives the mean birth weight, the mean placental wet weight and the gestational age at delivery. In the group of poorly controlled diabetics the mean birthweight was 4233 ± 660 g which differed, but not significantly from the mean birthweight in the normoglycemic diabetics (3380 ± 656) or from the control group, respectively (3465 ± 534 g). The three groups did not differ significantly in gestational age either, but we found a difference in placental weight which was significant at the 5% level, between the poorly controlled diabetics and the normoglycemic group as well as the control group. Here, the respective data were: 704 ± 156 g in the poorly controlled diabetics, 502 ± 125 g in the normoglycemic diabetics and 537 ± 104 g in the control group.

Table 36. Protein and glycogen content of the placenta

	Protein*	Glycogen*
HbAl > 8.5	69.7 ± 13.0	2.20 ± 0.89
HbAl < 7.5	48.6 ± 13.0	1.53 ± 0.66
Controls	46.0 ± 10.3	1.43 ± 0.73
	p < 0.05	n.s.

* mg/g placental wet weight.

Table 37. Enzyme activities in the placenta

	GOT*	GPT*	γGT*	LDH*
HbAl > 8.5	1.2 ± 0.6	2.5 ± 3.4	2.9 ± 0.5	1.7 ± 0.5
HbAl < 7.5	1.3 ± 1.2	5.0 ± 7.0	9.2 ± 7.0 [+]	2.0 ± 0.7
Controls	1.1 ± 1.1	5.1 ± 5.2	12.3 ± 5.6 [++]	1.5 ± 0.9
	n.s.	n.s.	[+] p < 0,05	n.s.
			[++] p < 0,01	

* U/g protein.

The mean protein content of the placentas of poorly controlled mothers (69.7 ± 13.0 mg/g wet weight) was significantly increased (p   0.05) compared to both other groups. In the group with sufficient metabolic control protein content was 48.6 ± 13.0 mg/g, and in the control group 46.0 ± 10.3 mg/g. As Table 36 shows, a similar but statistically not significant difference was found in the glycogen content.

Enzyme activity in relation to the total protein content of the placenta did not exhibit a clear cut trend (Table 37). The activities of some enzymes, e.g. GOT and LDH, were not significantly different in the three groups, whereas the activities of GPT and gammaGT were lower in the cases of poorly controlled diabetes mellitus. The mean values of GOT activity were 5.04 ± 7.05 U/g in diabetics with good control and 5.06 ± 5.2 U/g in the controls compared to 2.48 ± 3.4 U/g in the case of poor control. The difference between the mean values mentioned was not statistically significant. In the case of gammaGT the lowest activity (2.91 ± 0.54 U/g) was found in the placentas of poorly controlled diabetics, and was significantly different from the mean value of 9.18 ± 7.04 U/g in placentas of well controlled diabetics, and from the controls (12.30 ± 5.56 U/g).

Discussion

The function of the placenta includes a materno-fetal substrate and oxygen transfer. Maternal blood is supplied by the spiral arteries and flows through the intervillous space and back to decidual veins. The diffusion of substrates takes place at the intervillous space; here a close contact between maternal blood and fetal sinusoidal capillaries is guaranteed. The diffusion is facilitated by minimal diffusional distances in the region of the epithelial plates (SCHUHMANN, 1986).

In diabetic pregnancies the placenta may be gigantic, weighing nearly twice as much as a normal placenta. There is histological evidence which clearly shows different stages of maturity in the same placenta of a diabetic (THOMSEN and LIESCHKE, 1958). Advanced maturity of the villi with an increase in peripheral villous tissue is found, as well as the lack of formation of villi or an increased number of immature villi. Often fibrous and atretic regions are seen. The capillary endothelium may be swollen, thus increasing the diffusional distance (THOMSEN and LIESCHKE, 1958). Histomorphometric findings are confusing in so far as some authors found an increase (BOYD et al., 1986, BJÖRK and PERSSON, 1984) and others described a decrease in villous surface in diabetic placentas (CLAVERO and BOTELLA LLUSIA, 1963). Placentas of different species contain glycogen (HORKY, 1965). As pregnancy proceeds the placental concentration of glycogen decreases inversly as the glycogen content of the fetal liver increases (VILLEE, 1962). As long as the high glucose support from the mother remains, the fetus does not have to fall back on its placental glycogen reserve. This means, however, that the glycogen stored in the placenta no longer has any functional significance. Our assumption is consistent with findings of VILLEE (1962), who demonstrated decreased hexokinase activity in placentas of diabetics. Also, UHER (1969) observed reduced placental utilisation of glucose in cases of diabetes mellitus.

The amount of placental glycogen is increased in diabetic pregnancies and depends on the mother's mean blood glucose. Probably the swelling of the capillary endothelium is directly linked to the glycogen content of the placenta (THOMSEN and LIESCHKE, 1958). FISCHER and HORKY (1966) were able to demonstrate that placental glycogen decreases with insulin therapy. In our study we did not find any difference in glycogen content between well controlled diabetics and normal controls. In contrast, in the cases of poor metabolic control, higher glycogen levels could be demonstrated in human placentas as well as in rat placentas (Cirkel et al., 1986).

The deposition of protein observed is also of importance in this context. Since we could not find any increase but rather a decrease in enzymatic activity we conclude that the increased protein content represents an increase in structural protein. This assumption is mainly supported by our findings in rat placental metabolism (Cirkel et al., 1986); in human placentas a fall in GOT and GPT activity could be shown.

Other placental enzyme activities are reduced in maternal diabetes. There is a reduction of mitochondrial activities of glutamate dehydrogenase and GOT which are closely related to a fall in placental lactogen (Brandau, 1974) via a derangement of protein synthesis. A decrease of fructose-6-phosphate kinase activity, which represents the total flow via glycolysis, and also a fall in the activities of malate dehydrogenase, isocitrate dehydrogenase and glucose-6-phosphat-dehydrogenase, are associated with decreased excretion of estrogen.

Wulf (1981) developed a working concept of the placental insufficiency syndrome. Three subgroups were suggested: membranous, haemodynamic and cellular-parenchymatous insufficiency. Our observations of placental morphological and metabolic changes in diabetes mellitus add data to this theoretical classification.

Membranous placental insufficiency is demonstrated by the swelling of capillary endothelium which may be caused by the increased glycogen and/or protein content. This component is further worsened in the case of diminished villous surface.

Haemodynamic insufficiency may be caused by:
1) atheromatosis of spiral arteries,
2) diabetic vascular disease (microangiopathy),
3) decreased rheological properties.

The fall in enzyme activities as well as the lowered estrogen and placental lactogen synthesis represent the cellular-parenchymatous component of placental insufficiency.

In conclusion, in the case of diabetic pregnancy one can postulate a global placental insufficiency which is manifest clinically as an increased proportion of intrauterine death and intrauterine growth retardation (Burkart et al., 1987). Some of the components of placental insufficiency may be influenced in a positive way by normoglycemic blood glucose in pregnancy. Our study shows that the glycogen and protein content of the placenta is normal after sufficient metabolic control in pregnancy. Also, the activities of placental enzymes were found equal to the controls. The rheological parameters seem to normalize after some weeks of normoglycemia.

It therefore seems as if at least the cellular-parenchymatous and the membranous component of placental insufficiency could be influenced by normoglycemia during pregnancy. Some doubt as to the benefit of normoglycemia on the immaturity of placental villi exists. In the work of Senft et al. (1986) and others (Schmid, 1983) there were still measurable differences in villous surface of diabetic placentas as compared to placentas of diabetic pregnancies after optimized metabolic control during pregnancy. Finally, vascular lesions in the sense of diabetic microangiopathy can not be influenced by normoglycemia over a few weeks or during pregnancy. Normoglycemia improves the outcome in diabetic pregnancies as was clinically demonstrated, but obviously a part of placental insufficiency still remains which classifies the diabetic pregnant woman as a patient at risk.

C. On Screening and Diagnostics in Gestational Diabetes

6

Gestational Diabetes, Significance of Risk Factors and Results of a Follow-up Study 8 Years After Delivery

K. Fuhrmann

Department of Obstetrics and Gynecology Kaulsdorf, Berlin, German Democratic Republic, GDR

Introduction

Accurate estimates of the incidence of abnormal glucose tolerance during pregnancy and the frequency and the utility of gestational diabetes (GDM) risk factors are virtually nonexistent. Screening select populations of women with risk factors for the condition, and the non-random, nonpopulation − based nature of most studies, have given rise to wide variations in the reported incidence and significance of risk factors and GDM.

Methods

We examined a randomized population − based group of 2510 pregnant women from Berlin (GDR). All women who came for the first visit during early pregnancy to a pregnancy clinic were counted, and the patients with an odd number were asked to take part in the study. In order to ensure a representative social-economic status of the group three typical districts of Berlin were choosen. A careful medical history was documented and in a follow up study all the laboratory values used generally during pregnancy were documented. During the 26th and 36th week of pregnancy a 50 g OGTT was performed under standardized conditions. To study the reproducability of a 50 g OGTT during pregnancy, a second test was performed after 14 days in 150 cases. During the GTT insulin and triglyceride levels were estimated.

Criteria for normality in the OGTT are as follows (mean plus 2 SD of the whole population):

fasting 5.55 mmol/l (100 mg/dl),
1 h 8.88 mmol/l (160 mg/dl),
2 h 7.22 mmol/l (130 mg/dl) (Table 38).

Women identified as having two or more abnormal OGTT values were called gestational diabetes (GDM). Those with one abnormal OGTT value were classified as having impaired glucose tolerance (IGT).

Result and discussion

The GDM prevalence in this closed population was 1.1%; 6% had an IGT and 92.9% had a normal glucose tolerance. GDM prevalence increases with the duration of pregnancy as does in a more pronounced way the IGT-prevalence (Table 39). There is no

Table 38. Mean blood glucose values (mmol/l) and standard deviation (SD) of 2510 50 g OGTT during the third trimester of pregnancy in a normal city population

	Mean	Mean + 1 SD	IGT	Mean + 2 SD gestational diabetes	Dimension
Fasting	3.89 ± 0.83	≤ 4.79 ≥	—	≥ 5.55	mmol/l
60 min	5.99 ± 1.44	≤ 7.49 ≥	—	≥ 8.88	mmol/l
120 min	4.66 ± 1.22	≤ 5.83 ≥	—	≥ 7.22	mmol/l

Table 39. Glucose tolerance depending on gestational age

Week of pregnancy	Normal	IGT	GDM
Mean:	92.9 %	6.0 %	1.1 %
≤ 28 weeks of pregnancy	94.6 %	4.4 %	1.0
29th – 32nd pregnancy	93.3 %	5.9 %	0.8 %
33 – 36th pregnancy	88.9 %	8.8 %	2.3
≥ 37 pregnancy	87.7 %	10.8 %	1.5 %

Table 40. Blood glucose mean (mmol/l) and SD depending on the weeks of preagnancy

Weeks of pregnancy	0 min	60 min	120 min
≥ 28th week	3.8 ± 0.8	5.8 ± 1.3	4.5 ± 1.1
29th – 32nd week	3.9 ± 0.8	5.9 ± 1.3	4.7 ± 1.1
33rd – 36th week	3.9 ± 0.9	6.2 ± 1.4	4.8 ± 1.2
≤ 37th week	3.9 ± 0.8	6.4 ± 1.3	4.7 ± 1.1

Table 41. Mean blood glucose values ± SEM in the first (I) and second (II) oGTT

Normal glucose tolerance			
0 min	4.11 ± 0.87	4.16 ± 0.89	mmol/l
60 min	7.08 ± 1.66	7.38 ± 1.84	mmol/l
120 min	5.21 ± 1.42	5.55 ± 1.70	mmol/l
IGT			
0 min	5.05 ± 1.08	4.45 ± 0.99	mmol/l
60 min	8.24 ± 1.52	7.18 ± 1.58	mmol/l
120 min	6.58 ± 1.08	5.89 ± 1.89	mmol/l
Gestational diabetes			
0 min	6.08 ± 0.92	3.91 ± 0.76	mmol/l
60 min	9.04 ± 1.54	7.97 ± 0.60	mmol/l
120 min	7.16 ± 2.47	5.62 ± 2.15	mmol/l

significant blood glucose rise during the later stages of pregnancy (Table 40). The reproducability of a 50 g OGTT is with more than 70% good in pregnant women with normal carbohydrate metabolism. In women with IGT or in GDM the reproducability is worse (Table 41 and Fig. 23).

The insulin response after a 50 g OGTT is not significantly different among the normal, impaired, and pathological glucose tolerance group (Fig. 24).

The increase of glycerol after an OGTT during the first 60 min in the gestational diabetic group (Fig. 25) as well as a significantly reduced insulin glucose index (Table 42) indicates a relative lack of circulating insulin. The historical and clinical factors traditionally employed to identify patients who are at high risk for the occurrence of gestational diabetes (risk factors) were determined and evaluated (Table 43). 85.5% of the whole population had at least one of the presumed risk factors mentioned in the literature. Only 14.5% were without risk factors.

Most of the risk factors used for GDM screening in the literature really indicate an

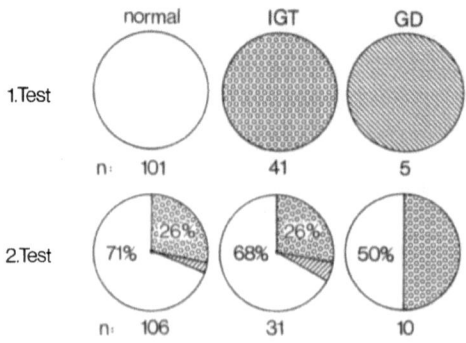

Fig. 23. Reproducability of a 50 g OGTT during pregnancy. Second OGTT after 14 days

Table 43. The frequency of risk factors in 2510 pregnant women of a city population and the number of GDM maximal discernible with

Risk factors	n	Frequency in % of population	Frequency % of gestational diabetics
1. Age ⩾ 25 years	740	29.6	65.5
2. Overweight ⩾ 20%	277	11.1	48.3
3. Glucosuria during the first half of pregnancy (24th week)	223	8.9	51.7
4. Birthweight ⩾ 4000 g in the history	79	3.2	13.8
5. Stillbirth in the history	27	1.1	0
6. Abortions in the history	259	11.8	24.1
7. Previous deliveries	967	38.7	72.4
8. Blood pressure ⩾ 140/90	598	23.9	34.5
9. Pathological weight gain during pregnancy	1596	63.4	62.0
10. Proteinuria	78	3.12	13.8
11. Diabetes heredity	600	24.0	48.3
12. No risk factors	364	14.5	0

Table 44. Combination of risk factors existing in a city population and the number of GDM maximal discernible with

1. Risk factor	2. Risk factor	3. Risk factor	% of population	% of gestational diabetes
Overweight ⩾ 20%	glucosuria		1.3	31.0
Overweight ⩾ 20%	family history		4.0	27.6
Overweight ⩾ 20%	abortions		1.5	6.9
Overweight ⩾ 20%	glucosuria	family history	0.56	17.3
Overweight ⩾ 20%	family history	abortions	0.70	3.45
Glucosuria	family history		3.3	27.6
Glucosuria	abortions		1.3	10.4
Family history	abortions		4.3	6.9
Family history	pathol. weight gain		20.4	24.2

Table 45. Sensitivity and specifity of fasting and postprandial blood glucose values for a GDM screening. Screening with fasting (FBG) and postprandial blood glucose values

Fasting	⩾ 4.44 mmol/l	⩾ 5.0 mmol/l	⩾ 5.55 mmol/l
Normal oGTT	20.8%	7.4%	3.0%
IGT	80.8%	64.0%	27.8%
Gestational diabetes	100.0%	86.3%	75.9%

60 min postprandial	⩾ 7.22 mmol/l	⩾ 7.77 mmol/l
Normal oGTT	15.0%	6.0%
IGT	74.6%	63.4%
Gestational diabetes	96.6%	96.6%

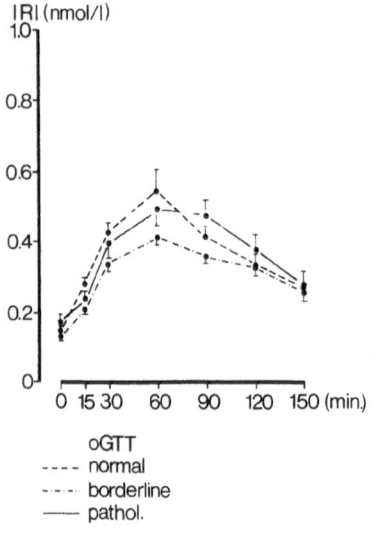

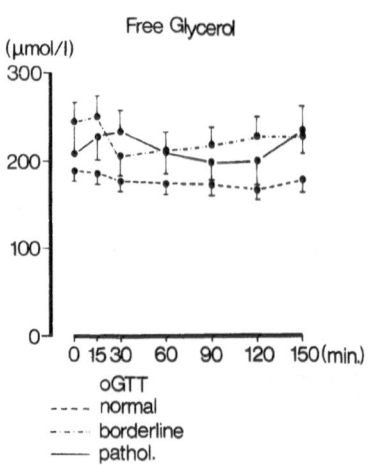

Fig. 25. Free glycerol during 50 g OGTT in pregnant women with normal-, impaired- and pathological glucose tolerance

Fig. 24. Immunoreactive insulin during a 50 g OGTT in pregnant women with normal-, impaired- and pathological glucose tolerance

increased risk of GDM. Only a past history of fetal death and pathological weight gain during pregnancy do not indicate an increased risk of having GDM. There is no specific risk factor for GDM. Therefore it is of importance to determine how often a risk factor occurs in the population. With a single risk factor it is possible to detect from 13 to 72% of all GDM (Table 43). When only patients with a combination of two risk factors are tested, a maximum of 31% of the GDM are detectible; a combination of 3 risk factors allows the identification of a maximum of 17% of all GDM (Table 44). Screening based on the presence of any of the risk factors demonstrated in this study except pathological weight gain and previous still birth detects a maximum of ~ 70% of all GDM, more or less independent of which risk factors are used. Only the number of OGTT's performed depends on the risk factors which are chosen. In our study the risk factors overweight ≥ 20%, or age ≥ 30 years, or glucosuria, or heredity first degree, or large for gestational age baby ≥ 4000 g, or ≥ 3 deliveries were found in 27% of the whole population. In this group

Table 42. Δ Insulin/Δ Glucose index after a 50 g oGTT in women with normal glucose tolerance and without risk factors, in gestational diabetics and in pregnant women with different risk factors

Δ Insulin/Δ Glucose	
Normal glucose tolerance	1.58
Gestational diabetes	0.53
Age ≥ 30 years	0.70
Relat. overweight ≥ 30%	0.79
Glucosuria	0.82
Family history	1.41

we found 72.4% of all gestational diabetics. The risk factors overweight ≥ 30%, glucosuria, age ≥ 30 years, and heredity first degree were found in 21.2% of the whole population. In this group also 72.4% of all gestational diabetics were present. The risk factors investigated are of differing importance. In women with one risk factor we found GDM two to twelve times more often when compared with women without risk factors. In women with two risk factors, GDM was found up to 20 times more often in com-

parison to women without risk factors. Unfortunately only 1.3% of the population had the combination of two risk factors (glucosuria and overweight ⩾ 30%) and therefore only 30% of the GDMs could be detected. The risk factors we investigated with the highest GDM-prevalence were:
glucosuria before the 24th week of pregnancy,

overweight ⩾ 30%,
age ⩾ 30 years
and: heredity of the mother.
Patients with these risk factors had also a significantly decreased insulin glucose index (Table 42).
They are present in 20% of the population and represent 70% of all women with GDM.

Screening using random blood glucose values

We evaluated fasting blood glucose (FBG) – and postglucose challenge 1 h and 2 h BG values for their suitability in a GDM – screening program.
Out of 2510 pregnant women 26.8% exceeded a FBG of ⩾ 4.4 mmol/l (70 mg/dl). 100% of the GDM had a FBG of ⩾ 4.4 mmol/l. 17.8% exceeded ⩾ 4.72 mmol/l (84 mg/dl). 93% of the GDM could be discovered. 12.3% of pregnant women exceeded 5.0 mmol/l (90 mg/dl), 86.4% of the GDM could be discovered, and 5.6% of women exceeded 5.5 mmol/l (98 mg/dl). In this group, 76% of all GDM could still be discovered (Table 45).
A GDM screening by a FBG has a sensitivity of 100% if the threshold value is 4.4 mmol/l, but in 26.8% of all pregnant women an OGTT would be necessary. With a FBG of 5.5 mmol/l the sensitivity is 76% and 5.6% of all pregnant women would need an OGTT. Using a postprandial blood glucose value for a GDM-screening program (1 h after a 50 g glucose load) the sensitivity is 96.6% when the borderline value is 7.22 mmol/l (129 mg/dl). 20.6% of all pregnant women exceeded 7.22 mmol/l and would need an OGTT. A threshold value of 7.77 mmol/l (138 mg/dl) included 11% of the population and still detected 96.6% of all GDM-women 2 h after a 50 g glucose load 207 out of 2510 pregnant women (8.3%) exceeded 6.38 mmol/l (114 mg/dl), and 76.2% of all GDM women were detected. A 2 h postprandial BG value of 6.94 mmol/

l (124 mg/dl) included only 3.8% of normals, but 76.2% of the GDM women. Fasting, and postglucose load blood glucose values are suitable for screening for GDM. The sensitivity and specifity is nearly similar among the 3 testing conditions. For the discovery of about 76% of all GDM women about 4 to 6% of all pregnant women would need an OGTT depending on the kind of BG-screening used.
Screening with risk factors and blood glucose values enables a less expensive diagnosis of GDM to be made. Two thirds fewer OGTTs are needed to detect the same number of GDM in comparison to a screening by risk factors only. Screening by blood glucose alone is more sensitive than screening by risk factors, but more expensive. All pregnant women need one blood glucose value and 4–6% need one OGTT to detect ~ 76% of the GDM.
For general screening in the GDR we advise:
First step: Documentation of all women (~ 20% of all pregnant women) with one of the following risk factors:
Glucosuria before the 24th week of pregnancy.
Overweight ⩾ 30%.
Age ⩾ 30 years.
Heredity first degree (especially the mother) during week 20–24 of pregnancy.
Second step: FBG value in the risk group. Women with a FBG ⩾ 4.44 mmol/l require.
Third step: an OGTT.

Table 46. 8-year follow up of glucose tolerance in women with gestational diabetes (n = 50)

	3 month pp.	6 month pp.	12 month pp.	8 years pp.
Diabetes:	4% (2)	4% (2)	18% (5)	46% (23)
IGT:	4% (2)	8% (4)	20% (10)	16% (8)
Normal:	92% (46)	88% (44)	62% (31)	38% (19)

Table 47. 8-year follow up of glucose tolerance in women with IGT during pregnancy (n = 29)

Diabetes mellitus:	3% (1)
IGT:	28% (8)
Normal:	69% (20)

Fourth step: GDM need a day (night) BG-profile to be performed everymonth.

Women with a mean blood glucose value of $\geqslant 5.0$ mmol/l are transfered to a high risk pregnancy clinic. We start insulin therapy if the MBG exceedes $\geqslant 5.5$ mmol/l.

Follow-up: To estimate the OGTT criteria we used in our study a follow-up was performed during the 8 years after delivery of all IGT and gestational diabetic women. For our follow-up program we used according to the WHO recomodations, the 75 g OGTT (WHO criteria).

75 g OGTT's were performed:
3 month,
6 month,
1 year
and 8 years after pregnancy.

Within 8 years 46% of the GDM became diabetics, mostly NIDDM (Table 46). Only 3% of the IGT-group were diabetic after 8 years (Table 47). From the risk factors investigated during pregnancy only overweight and age showed significant correlations with the diabetes incidence 1 and 8 years after pregnancy. Close correlations were also found between both the need for insulin treatment during pregnancy, and the occurrence of diabetic fetopathy, and the diabetes incidence 1 and 8 years after pregnancy.

Conclusions

− The GDM incidence in a german city population is about 1%.
− 4 out of 100 GDM have overt diabetes.
− One year after pregnancy 18%, and 8 years after pregnancy 46%, of the GDM had become diabetics, mostly NIDDM. The diabetes incidence of borderline cases was only 3% after 8 years.
− The reproducability of a 50 g OGTT is good only in pregnant women with normal carbohydrate metabolism, a pathological test is poorly reproduceable.
− Women with one pathological OGTT are at risk and should be followed up during pregnancy and afterwards.
− Screening for GDM is possible with blood glucose values fasting as well as postprandial. Screening with risk factors and fasting or postprandial blood glucose values is of less expense.

7

The Oral 100-g-Glucose-Tolerance Test—Special Criteria for Evaluation in Late Pregnancy

O. Bellmann, N. Lang, H. Schlebusch, M. Niesen, and R. Schönhardt

Department of Obstetrics and Gynecology, University of Bonn, Federal Republic of Germany

Introduction

If a disorder of carbohydrate metabolism in pregnancy is suspected, an oral glucose tolerance test is usually performed. If 100 g glucose are used, the test results are generally interpreted according to the so-called pregnancy-specific norm criteria of O'SULLIVAN and MAHAN (1964). An indication for the test most often appears in the third trimester. Even if a test was already performed in early pregnancy and found to be normal, it should be repeated at this time. This is because the chance of revealing a disorder seems greatest in the third trimester, in which a marked deterioration of oral glucose tolerance is physiologic (KÜHL, 1975; LIND et al., 1973). The question arises whether the oral 100-g-glucose-tolerance test performed in the third trimester can be adequately evaluated by the criteria of O'SULLIVAN and MAHAN (1964). These criteria are based on investigations only half of which were done in the third trimester, while the rest were done earlier in

pregnancy. The increasing alterations of carbohydrate metabolism, particularly in the second half of pregnancy make it seem necessary to evaluate the oral 100-g-glucose-tolerance test in late pregnancy according to criteria based on investigations exclusively in this period. Otherwise, we may find an unduly high rate of abnormal glucose-tolerance tests, which, in turn, might lead to an underestimation of its clinical relevance. This investigation aimed to describe the limits of the oral 100-g-glucose-tolerance test in the third trimester of normal pregnancy. The test was performed in a large unselected group of pregnant women in the third trimester at the Universitäts-Frauenklinik Bonn. The associations between the test results and special features of the mother and the newborn were studied, and a control group of pregnant women without diabetic risk factors and with clinically healthy newborns was established retrospectively.

Methods

The oral glucose tolerance test (oGTT) was performed in unselected outpatients under standardized conditions (Report of the Committee on Statistics of the American Di-

abetes Association 1969) with 100 g of an oligosaccharid mixture (Dextro oGT®, Boehringer Mannheim) dissolved in 400 ml of flavoured water, which was given within

5 minutes. Capillary blood samples were obtained at 0, 60, 120, and 180 minutes. During the study the results of the tests were evaluated according to the criteria of O'SULLIVAN and MAHAN (1964). No attention was paid to the fact that the limits of O'SULLIVAN ($\bar{x} \pm 2\,\text{S.D.}$) were derived from glucose determinations in venous blood. The glucose tolerance was considered as being impaired if two or more values exceeded the upper limits of O'SULLIVAN.

The determination of the glucose concentrations from non-hemolysed capillary blood was performed by the hexokinase/glucose-6-phosphate-dehydrogenase method (BERGMEYER et al., 1974) with commercially available test substances (Glucoquant®, Boehringer Mannheim). In contrast to the original instructions, deproteinization was omitted and replaced by running a blank for each sample. Apart from the known specificity of the hexokinase method this modification provided the benefit of easy handling. Precision and accuracy were evaluated by means of commercially available control sera. The coefficient of variation from day to day (interassay v.c.) was 2.0% for $\bar{x} = 49.3\,\text{mg/dl}$ (n = 30), 2.4% for $\bar{x} = 118.2\,\text{mg/dl}$ (n = 30) and 1.5% for $\bar{x} = 204.7\,\text{mg/dl}$ (n = 30). The coefficient of variation within this series (intraassay v.c.) was 1.9% for $\bar{x} = 47.4\,\text{mg/dl}$ (n = 20), 1.7% for $\bar{x} = 96.4\%\,\text{mg/dl}$ (n = 20) and 1.3% for $\bar{x} = 214.3\,\text{mg/dl}$ (n = 20).

644 pregnant women were studied exclusively in the third trimester. Their clinical characteristics are summarized in Table 48. The group tested between 1972 and 1974 did not differ markedly with respect to their obstetrical features from the total population of the women who delivered at the Universitäts-Frauenklinik Bonn in the years 1973 to 1977 (SCHÖNHARDT, 1981). The fact that perinatal mortality was considerably lower in the group of women tested (0.3%) than in the total population of pregnant women from 1973 to 1977 (2.6%) can be explained by having performed the tests relatively late in the third trimester. Therefore

the risks of prematurity were more or less eliminated. Furthermore, pregnant women with overt diabetes mellitus, rh-isoimmunization, or twin pregnancies were excluded from the study. Each newborn whose mother had shown an impaired glucose tolerance according to the criteria of O'SULLIVAN was routinely evaluated by a blood glucose profile starting 2 hours after birth and fed with a 25% oligosaccharid mixture (Dextroneonat®, Maizena Diät GmbH Heilbronn) after the first blood sample for glucose determination had been taken (NIESEN, 1978). The data were stored in a disc file on the IBM 370/125 computer of the Institut für Medizinische Statistik, Dokumentation and Datenverarbeitung, University of Bonn. All the data reduction was performed by computer programs written in PL/I. The area under the concentration-time-curve of the oGTT was calculated by the formula

$$F = \text{Min}\,(G_1, G_2) + \frac{(G_1 - G_2)}{2}$$

$$+ \text{Min}\,(G_2, G_3) + \frac{(G_2 - G_3)}{2}$$

$$+ \text{Min}\,(G_3, G_4) + \frac{(G_3 - G_4)}{2}$$

where G_1, G_2, G_3, and G_4 are the glucose concentrations at the times 0, 60, 120, and 180 minutes respectively and Min (G_1, G_2) means the lesser of the two quantities in brackets. One hour was chosen as the unit of time. In order to evaluate the influence of an independent variable, for example one of the diabetic risk factors of the mother, on a dependent variable, i.e. the oral glucose tolerance, the group of interest was divided into two alternative groups on the basis of the independent variable. Then the frequency of the outcome defined as the dependent variable was determined in either group. The result was tested by the chi squared-test with statistical significance ascribed at the 5% level. As the glucose concentrations at the different test times did not follow a normal distribution, percentiles were calculated to characterize the limits of oral glucose tolerance.

Results

The glucose concentrations in capillary blood of all the pregnant women tested averaged 72 mg/dl at test time 0 min (range 44–111), 156 mg/dl at 60 min (range 78–256), 130 mg/dl at 120 min (range 61–244) and 104 mg/dl at 180 min (range 43–210), the glucose area had a mean of 374 mg/dl × hour (range 221–604).

The relation between gestational age at the time of testing and oral glucose tolerance was studied. 15% of the pregnant women tested during the third trimester were tested between 27 and 33 weeks gestation. There was no difference between the results obtained before and at or after 34 weeks' gestation.

The relation between maternal age, weight, historic risk factors as to diabetes mellitus, and parity and the oral glucose tolerance test results in the third trimester on the other hand was studied. A higher frequency of impaired glucose tolerance according to the criteria of O'SULLIVAN was observed in women over 34 years of age, with more than 120% of their ideal weight (Broca-Index minus 10% before pregnancy), with family history of diabetes mellitus (relatives of 1st degree), with poor obstetrical history (birth weight > 4 000 g, unexplained perinatal death) and with more than two previous deliveries as compared to those women without these features (Table 49). 27% of the pregnant women tested revealed one or several of these known diabetic risk factors.

The relation between the sex of the newborn and the oral glucose tolerance test in the

Table 48. Characteristics of the 644 pregnant women tested for oral 100-g-glucose tolerance in the third trimester

Feature	Mean	Range	Criterion	Frequency (%)
Maternal age (years)	29	16–45	> 34	16
Maternal weight before pregnancy (% ideal[a])	104	75–214	>120	14
Parity (number of previous births)	0.87	0–11	>2	9
Family history of diabetes (1st degree relatives)	—	—	any	13
Bad obstetrical history (> 4 000 g, unexplained perinatal death)	—	—	any	9
Toxemia (moderate, severe[b])	—	—	any	6
Week of gestation at delivery	40.	32.–42.	< 38.	5
Forceps	—	—	idem	14
Cesarean section	—	—	idem	15
Birth weight (gram)	3 383	1 500–5 050	>4 000	9
Apgar score 1 min	7.9	1–10	<7	16
Apgar score 5 min	9.4	3–10	<8	2
Intrauterine growth retardation[c]	—	—	idem	2
Diabetic fetopathy	—	—	idem	1
Fetal malformation	—	—	idem	2
Perinatal mortality	—	—	idem	0.3

[a] Ideal body weight: Broca index — 10%.
[b] According to KAULHAUSEN (1980).
[c] According to LUBCHENCO et al. (1963).

Table 49. Relation between diabetic risk factors of the mother and the oral 100-g-glucose-tolerance test in the third trimester. Evaluation of the glucose concentration in capillary blood according to the criteria of O'SULLIVAN and MAHAN (1964)

Maternal risk factor	Criterion	Independent of oGTT (n)	With impaired oGTT[a]		Significant difference[b]
			(n)	(%)	
Age	≤ 34	535	127	24	yes
(years)	> 34	100	42	42	
Weight before pregnancy	≤ 120	513	130	25	yes
(% ideal[c])	> 120	80	30	38	
Family history of diabetes	yes	553	139	25	yes
(1st degree relatives)	no	82	30	37	
Bad obstetrical history	yes	576	146	25	yes
(> 4 000 g, unexplained perinatal death)	no	59	23	39	
Parity (number of previous births)	≤ 2	581	146	25	yes
	> 2	54	23	43	
One or more of these risk factors	yes	245	90	37	yes
	no	362	73	20	

[a] According to O'SULLIVAN and MAHAN (1964).
[b] Chi square test, 5% level.
[c] Ideal body weight: Broca index — 10%.

Table 50. Relation between diabetic risk factors of the infant and the oral 100-g-glucose-tolerance test in the third trimester. Evaluation of the glucose concentration in capillary blood according to the criteria of O'SULLIVAN and MAHAN (1964)

Fetal risk factor	Criterion	Independent of oGTT (n)	With impaired oGTT[a]		Significant difference[b]
			(n)	(%)	
Birth weight (gram)	≤ 4 000	580	145	25	yes
	> 4 000	55	24	44	
Apgar score 1 min	≥ 7	532	140	29	no
	< 7	103	29	28	
Apgar score 5 min	≥ 8	621	166	27	no
	< 8	14	3	21	
Week of gestation	> 38.	605	164	27	no
at delivery	< 38.	30	5	17	
Diabetic fetopathy	no	629	168	27	×
	yes	5	1	20	
Malformation	no	623	165	26	no
	yes	12	4	31	
Perinatal mortality	no	633	167	26	×
	yes	2	2	100	
One or more of these risk factors	no	468	116	25	no
	yes	167	53	32	

[a] According to O'SULLIVAN and MAHAN (1964).
[b] Chi square × too few for statistical analysis.

Table 51. Characteristics of the 278 pregnant women of the control group

Feature	Mean	Range	Criterion	Frequency (%)
Maternal age (years)	27	16–34	>34	—
Maternal weight before pregnancy (% ideal[a])	98	75–119	>120	—
Parity (number of previous births)	0.44	0–2	>2	—
Family history of diabetes (1st degree relatives)	—	—	yes	—
Bad obstetrical history (>4 000 g, unexplained perinatal death)	—	—	yes	—
Toxemia (moderate, severe[b])	—	—	yes	4
Week of gestation at delivery	40.	38.–42.	<38.	—
Forceps	—	—	yes	15
Cesarean section	—	—	yes	10
Birth weight (gram)	3 356	2 075–4 000	>4 000	—
Apgar score 1 min	8.4	7–10	7	—
Apgar score 5 min	9.6	8–10	<8	—
Intrauterine growth retardation[c]	—	—	yes	1
Diabetic fetopathy	—	—	yes	—
Fetal malformation	—	—	yes	—
Perinatal mortality	—	—	yes	—

[a] Ideal body weight, Broca index — 10%.
[b] According to KAULHAUSEN (1980).
[c] According to LUBCHENCO et al. (1963).

third trimester was studied. 55% of the newborns were male. There was no difference between the test results in mothers with male and female infants.

Furthermore, the relation between neonatal characteristics, which are frequently observed in babies of diabetic mothers, and the glucose tolerance in the third trimester, was studied. Pregnant women with a newborn infant of more than 4000 g had a higher frequency of gestational diabetes according to the criteria of O'SULLIVAN than pregnant women with a newborn infant below 4000 g of birth weight. In contrast, asphyxia, prematurity, and congenital malformations were not related to impaired glucose tolerance of the mother. The frequency of diabetic fetopathy and perinatal mortality was too low to allow an interpretation of the relations between these factors and the test results (Table 50). 27% of the newborns whose mothers had been tested in the third trimester revealed one or several of these neonatal characteristics.

In order to characterize oral glucose tolerance in normal late pregnancy a group of pregnant women with a healthy newborn infant and without diabetic risk factors was formed. 278 of the 644 women tested were assigned to this control group, which is characterized in Table 51. The limits of the oral glucose tolerance after a 100 g loading in the third trimester were described by percentiles (Table 52). If the 95th percentile was chosen as the upper limit, 52 (=8.2%) of the 638 women tested without missing values showed an impaired glucose tolerance, i.e. two or more values exceeding the upper

Table 52. The oral 100-g-glucose-tolerance test in the third trimester: calculation of percentiles for the glucose concentrations in capillary blood (mg dl) and the area under the concentration-time curve in the control group

oGTT	Glucose concentration (mg/dl) percentile						
	2nd	5th	10th	50th	90th	95th	98th
0 min	54	60	63	71	80	84	89
60 min	95	109	116	154	187	197	202
120 min	84	89	96	123	164	174	191
180 min	55	63	73	103	126	134	149
Area*	267	284	300	365	439	451	481

* Are under the concentration-time curve (mg/dl × hrs); for calculation see methods.

limit. The incidence of impaired glucose tolerance diminished from 8.2% to 3.4%, if the 98th percentile was chosen for evaluation of test results instead of the 95th percentile.

According to the criteria of O'Sullivan 27% of the women tested revealed an impaired glucose tolerance.

Discussion

The criteria for evaluating the oral 100-g-glucose tolerance in pregnancy differ considerably in the literature (Table 53). Looking for an explanation for this phenomenon one has to take into account the statistical methods applied. In this study we did not calculate the man and the standard deviation, but preferred to calculate percentiles, because in our study the glucose values of the diffent test points did not follow a normal distribution. The rather high upper limits in our control group could therefore in part be explained by the statistical method of evaluation. In addition, one has to take into account methodological details of glucose determination (Knopp et al., 1978). In this study the determination of glucose was done from capillary blood, while in other studies it was done from venous blood or from venous serum. Usually glucose concentrations obtained from capillary blood are higher than those obtained from venous blood or serum. Our investigations have shown, however, that the capillary-venous difference of glucose concentration in pregnancy is considerably lower compared to

that outside pregnancy (Bellmann, 1978). Therefore, it seems to be particularly important for the explanation of the relatively high upper limits of our control group to point out of that in this study in contrast to others the tests were performed exclusively in the third trimester. In this stage of pregnancy the rise of the glucose concentration after a glucose load is known to be very pronounced while the subsequent decrease is known to be very slow (Kühl, 1975, Lind et al., 1973). The comparatively low fasting glucose concentrations of our control group may also merely reflect the physiologic changes of carbohydrate metabolism in pregnancy being most striking at its end (Yen, 1978).

For establishing the limits of oral glucose tolerance in normal late pregnancy we formed a control group of pregnant women by including only those whose newborns appeared clinically healthy. Besides fetal overweight which was the only fetal characteristic to show a relation to the test results in this study, also asphyxia, prematurity, fetopathy, malformation and perinatal mor-

Table 53. The upper limit of the 96% tolerance area of the glucose concentrations (mg/dl) in the oral 100-g-glucose-tolerance test—a comparison between different studies

oGTT	Glucose concentration (mg/dl)				
	outside pregnancy		during pregnancy		
	U.S. Public Health Service (1964)	O'Sullivan and Mahan (1964)	Korp and Rogovitz (1973)	Mestman (1973)	this study— 3rd trimester
	venous blood	venous blood	capillary blood	venous serum	capillary blood
0 min	110	90	85	110	90
60 min	170	165	165	200	200
120 min	120	145	145	150	190
180 min	110	125	120	130	150

tality were reasons for exclusion of the mother from our control group. However, the clinical diagnosis of a healthy state of the infant at birth does not necessarily imply undisturbed metabolism after birth. A measurement of the blood glucose behaviour of the infant in the first two or three days post partum was not performed routinely in this study. It was only performed in those infants whose mothers had shown an impaired glucose tolerance according to the criteria of O'Sullivan. In a retrospective evaluation of the post partum blood glucose behaviour of these infants of mothers of our control group it turned out that none of the mature infants had a decrease of capillary blood glucose below 30 mg/dl. Only 8 (= 15%) of the 52 infants under consideration showed a transitory decline of capillary blood glucose between 30 and 40 mg/dl, inspite of early feeding, which remained asymptomatic (unpublished observations).

This study reaffirms that diabetic risk factors of the mother such as age, overweight, family history of diabetes mellitus, poor obstetrical history, and multiparity have an impact on the glucose tolerance in pregnancy (Pedersen, 1977). Therefore, a prerequisite for including a pregnant woman in our control group was that she was free of such risk factors. One would expect that such a selection should rather have contributed to a decrease than to a rise of the upper limits of oral glucose tolerance in late pregnancy. On the other hand, it is justified to suppose that

the criteria of O'Sullivan and Mahan (1964) would have been even lower, if the authors had selected their patients in a similar way.

This study offers a rational basis for the interpretation of the oral 100-g-glucose-tolerance test in the third trimester. Determination of the clinical relevance of glucose tolerance testing may also be possible by simply applying arbitrary limits. However, obstetrical relevance can only be expected if the diagnosis of impaired glucose tolerance is indeed accompanied by a true disorder of carbohydrate metabolism under the usual oral nourishment. Thus, the likelihood of such a finding seems to be greatest if the criteria for evaluation have a physiologically sound basis. At the present we are investigating the alterations of maternal carbohydrate metabolism under the usual oral nourishment and the blood glucose behaviour of the newborn in relation to the percentiles of our conrol group (Bellmann et al., 1981). Our attention is not only directed towards the upper limits, but also towards the lower limits of oral glucose tolerance. A diminished rise of blood glucose after oral glucose intake was observed to be accompanied by a higher incidence of intrauterine growth retardation (Abell and Beischer, 1976; Khouzami et al., 1981). Apparently the secretion of fetal insulin which is responsible for fetal growth becomes less stimulated (Hill, 1978; Lind et al., 1973).

Summary

During pregnancy there is a physiologic deterioration of oral glucose tolerance in man. Although glucose tolerance testing is often performed in the third trimester, criteria for the evaluation of oral glucose tolerance at this stage of pregnancy have been lacking up to now. In 644 unselected pregnant women we investigated the oral glucose tolerance test in the third trimester using 100 g of an oligosaccharid mixture. For a control group we selected all those women who were free of diabetic risk factors and gave birth to a healthy child around term. The characterization of oral glucose tolerance of the 278 normal pregnant women in the third trimester was performed by calculating percentiles. At the test points 0', 60', 120', and 180' the 98th percentile of the glucose concentration in capillary blood was 89, 202, 191, and 149 mg/dl respectively, the 2nd percentile was 54, 95, 84, and 55 mg/dl respectively. Thus, there is a considerable discrepancy between these limits evaluated exclusively in the third trimester of pregnancy and the so-called pregnancy-specific norm criteria that are usually applied not only in early, but also in late pregnancy.

8
Glucosylated Proteins in Normal and Diabetic Pregnancy

C. M. Peterson and L. Jovanovic
Sansum Medical Research Foundation, Santa Barbara, CA, U.S.A.

Introduction

As originally hypothesized (PETERSON and JONES, 1977), the measurement of glycosylated proteins has been increasingly defined as a clinical tool and the role of these types of reactions in the protean sequelae of diabetes is under intense study. One of the major developments during the past year has been the convening of an "Expert Committee" under the auspices of the National Diabetes Data Group of the National Institutes of Health whose charge it is to standardize nomenclature and investigate the possibility of references and standards for glycosylated protein measurement. The recommended nomenclature is summarized in Table 54. The term glycation has also been introduced recently to implicate all nonenzymatic adducts of sugars with proteins and has some merit in its simplicity and specificity.

The clinical utility of glycosylated hemoglobin or serum protein measurement will not achieve its full potential until appropriate references, standards, and degrees of accuracy and precision of these measurements are agreed upon. Because of the clinical implications of small changes in the assay pro-cedure (for example a 1% change in hemoglobin A_{1c} represents about a 35 mg/dl or 2 mM change in mean blood glucose), it would appear advisable that intra and interassay coefficients of variation be maintained under 5% and preferably under 2%. These types of methodological constraints are especially germain to evaluation of glucose control in pregnancy where small differences in mean blood glucose can result in appreciable changes in morbidity and mortality to infant and mother.

The present manuscript will review the biochemistry of glucosylation reactions, various methods of measurement which are now available, studies of the role of glucosylation in the complications of pregnancy complicated by diabetes mellitus, and the clinical utility of glucosylated protein measurements during pregnancy. A number of general reviews on the topic of glycosylation reactions and their relevance to diabetes mellitus are now available (MAYER and FREEDMAN, 1983; NATHAN, 1983; PETERSON, 1982; WIELAND, 1983; GABBAY, 1982; MONIER and CERAMI, 1982; KENNEDY and BAYNES, 1984).

Biochemistry of glucosylation and Browning reactions

The chemistry of the adduct formation of glucose with proteins has been well studied (MAYER and FREEDMAN, 1983). The initial reaction is the formation of a Schiff Base (labile adduct) which then may undergo an Amadori Rearrangement to a stable keto-amine. It is this latter product which is representative of glucose concentration over time as opposed to the labile adduct which reflects glucose concentration in the medium. Further chemical reaction is possible, producing what are known as nonenzymatic browning products since ultimately a brown pigment (melanoidins) will result. The chemistry of these reactions is only par-

tially characterized. One problem with the potential correlation of glucosylation reactions with the secondary problems associated with diabetes mellitus is that only the initial steps in the reaction can be measured by present methodologies.

Studies on methods of measurement

Table 54. Glossary of terms

Glucosylated Protein. Protein modified by glucose at the N-terminal residue and/or epsilon amino groups of lysine residues.

Glycosylated Protein (glyco-protein). A generic term for proteins containing glucose and or other carbohydrate at either N-terminal or lysine residues thus the sum of glycosyl adducts.

Hemoglobin A (HbA). The major adult form of hemoglobin. A tetramer consisting of two alpha and two beta chains (alpha$_2$, beta$_2$).

Hemoglobin A$_0$ (HbA$_0$). The major component of hemoglobin A identified by its chromatographic and electrophoretic properties. Post-translation modifications including glycosylation do exist, but do not significantly affect the charged properties of the protein.

Hemoglobin A$_1$ (HbA$_1$). Post-translationally modified, more negatively charged forms of hemoglobin A$_0$ (primarily glucosylation at the beta chain terminal valine residue) separable from HbA$_0$ by chromatographic and electrophoretic methods.

Hemoglobin A$_{1a}$, HbA$_{1a2}$, HbA$_{1b}$, HbA$_{1c}$. Chromatographically distinct stable components of HbA.

"Fast" Hemoglobin. The total HbA$_1$ fractions (HbA$_{1a}$, HbA$_{1a2}$, HbA$_{1b}$, HbA$_{1c}$) which, because of more negative charge, migrates toward the anode on electrophoresis and elutes earlier on cation exchange chromatography than HbA$_0$.

Hemoglobin A$_{1a}$, HbA1_{a2}, HbA$_{1b}$. "Fastest" most anionic forms of HbA consisting primarily of adducts of phosphorylated glycolytic intermediates with HbA.

Hemoglobin A$_c$. Component of HbA$_1$ which consists of 50 to 90% hemoglobin (depending on the quality of resolution of the chromatographic system) glucosylated by a ketamine linkage at the beta chain terminal valine residue.

Pre-Hemoglobin A. A labile form of glucosylated Hb containing glucose bound in aldimine linkage to the beta chain terminal valine residue. Felt to be the first step in nonenzymatic glucosylation which probably appears with all proteins. These labile forms are best eliminated through dialysis or some other procedure since they are reflective of the ambient glucose in the medium and not the mean glucose over a more extended time.

Table 55 summarizes the clinically used measurements of glycosylation and their relative advantages and disadvantages. The major problems with all methods is that as yet there are no agreed upon references and standards. The recently constituted committee of the National Diabetes Data Group may be able in part to speak to this need. The other major problem in terms of the clinical utility of these measurements lies in the problems of accuracy (difficult to approach without standards) and precision. Methods which achieve these levels of precision generally have relatively narrow ranges for "normal" populations. This observation was recently confirmed by Blouquet et al. (1983) using HPLC methodologies and bio-Rex-70 columns. The normal level was $5.4 \pm 0.4\%$.

Five new potentially useful methods of quantitating glucosylated proteins have been described: affinity chromatography, fructosamine quantification, monoclonal antibodies specific for glucosylated epsilon amino groups of lysine, a spectrophotometric assay dependent on the change in absorbance when phytic acid binds to hemoglobin A, and determination of glucitollysine by ion exchange chromatography and reverse phase liquid chromatography (Yue et al., 1982; Gould et al., 1982; Klenk et al., 1982; Vlassara et al., 1982; Little et al., 1983; Abraham et al., 1983; Garlick et al., 1983; Herold et al., 1983; Middle et al., 1983; Curtiss and Witztum, 1983; Johnson et al., 1982; Baker et al., 1983; Walinder et al., 1982; Schmid et al., 1984). All these

Table 55. Clinical methods employed for measurement of glycosylation

---

A. Physical Methods Based on Changes in pI

   1. Cation Exchange Chromatography
     PRO: Inexpensive and rapid.
     CON: Sensitive to small changes in resin packing, ionic strength, pH, temperature, column loading, and affected by the labile fraction, hemoglobin F, and hemoglobinopathies.

   2. High Performance Liquid Chromatography
     PRO: Dedicated instrument avoid many problems in 1.
     CON: Relatively expensive and still affected by the labile fraction, hemoglobin F, and hemoglobinopathies.

   3. Agarose Gel Electrophoresis
     PRO: Inexpensive, low technician time, standardized plates and conditions in kits, less sensitive to pH, triglyceride concentrations, and temperature.
     CON: Precision problems induced by scanner and loading variation. Sensitive to labile fraction, hemoglobin F and hemoglobinopathies.

   4. Isoelectric Focusing
     PRO: Separates most minor hemoglobin variants.
     CON: Precision over time dependent on use of same batch of ampholines on standardized plates. Scanning effects precision.

B. Methods Based on Chemical Principles

   1. Thiobarbituric Acid/Colorimetric Assay
     PRO: Minimally effected by storage conditions, fructose or 5 hydroxy methyl furtural standards may be incorporated.
     CON: Difficult to establish, large amount of technical time required, and affected by labile fraction and ambient glucose concentration.

   2. Affinity Chromatography with Immobilized m-phenyl-boronate
     PRO: Rapid, inexpensive, minimally effected by chromatographic conditions, eliminates labile adduct, not influenced by hemoglobin F or hemoglobinopathies.
     CON. Resins vary within and between manufactures.

   3. Fructosamine Determination by Nitroblue Tetrazolium Reduction
     PRO: Inexpensive, standards incorporated, may be automated, not effected by labile adduct or minor hemoglobins
     CON: Only for serum, lipids may interfere, reducing substances (eg ascorbate) interfere.

   4. Radioimmunoassay
     PRO: Inexpensive, rapid, sensitive, specific, not effected by labile adduct or minor hemoglobins.
     CON: Antibodies difficult to raise and not commercially available

---

methods are now in the process of being tested in clinical situations.

Affinity chromatography has received the most attention. The resins bind cis-diols and therefore bind glucosylated amino acids, peptides, and proteins (DUNCAN and GILHAM, 1975; WEITH et al., 1980; MALIA et al., 1981). Therefore both serum protein and hemoglobin glucosylation can be measured by this method in human as well as animal specimens. Hemoglobinopathies do not effect the measurement unless accompanied by a shortened erythrocyte survival (YUE et al., 1982; ABRAHAM et al., 1983). The measurement is not affected by the labile fraction of hemoglobin or hemoglobin F (KLENK et al., 1982; MIDDLE et al., 1983). Samples may be stored for as long as 21 days at room temperature without affecting the value (LITTLE et al., 1983) although storage properties may change if the cells are washed and the hemoglobin refrigerated or frozen (PETERSON et al., 1984). The precision of these methods in general appears good although it may vary greatly depending on the source of the resin or prepacked columns (LITTLE et al.,

1983; Herold et al., 1983; Peterson et al., 1984). The utility of the method during pregnancy appears promising but to date few studies have been performed with this methodology.

The other methodologies have been less well evaluated. Murine monoclonal antibodies have been isolated (Curtiss and Witztum, 1983) with the dominant epitope recognized being glucitollysine, the reduced hexose alcohol form of glucose conjugated to the epsilon amino group of lysine. The antibody recognized glucitollysine epitopes in reduced high density lipoprotein, albumin, hemoglobin, transferrin, and plasma proteins. Whether this methodology will have the desired precision for clinical use remains to be determined. Furthermore, reducing agents may be selective in their relative reactivity with various glucosylated epsilon amino groups of lysine (Garlick et al., 1983) which may prove to be a problem in the chromatographic assay for glucitollysine as well (Schmid et al., 1984). Previous attempts at raising polyclonal antibodies have only rarely been successful (Javid et al., 1978), but have been useful in studies of diabetes and pregnancy and cord blood studies (Peterson et al., 1979; Jovanovic et al., 1980; Jovanovic et al., 1982). The use of phytic acid to bind to hemoglobin A thus producing a change in absorbance which does not occur when hemoglobin is glycosylated appears attractive since it could be automated if problems with standards and calibration are solved (Walinder et al., 1982).

Most appealing is the estimation of serum fructosamine utilizing the principle that the Amadori rearrangement product formed by the condensation of glucose and proteins acts as a reducing agent in alkaline solution (Jovanovic et al., 1982). An assay has been described which uses nitroblue tetrazolium and 1-deoxy, 1-morpholinofructose standards (Johnson et al., 1982). As described the method is rapid, relatively inexpensive, amenable to automation and appears useful as a screening test for diabetes with 88% sensitivity and 9% false positives (Baker et al., 1983). The assay can only be used on sera and other reducing substances such as ascorbate or glutathione may interfere. Preliminary studies in pregnancy are encouraging (Roberts et al., 1983).

A number of investigators have now evaluated measurement of serum glycosylated proteins as a means of monitoring hyperglycemia (Johnson et al., 1982; Baker et al., 1983; Kennedy et al., 1982; Nakayama et al., 1982, 1982a; Gragnoli et al., 1982; Manda et al., 1982; Murtiashaw et al., 1983; Mehl et al., 1983; Jones et al., 1983). To be clinically useful, the measurements should have the same constraints in terms of accuracy and precision as for quantitation of glycosylated hemoglobins. Most authors agree that serum or plasma protein glycosylation measurement provides an "intermediate" index of hyperglycemia. Whereas normalization of blood glucose is followed by normalization of hemoglobin A_{1c} in approximately 8 weeks (Jovanovic et al., 1980), glycosylated serum values reach a stable plateau in 3–5 weeks (Baker et al., 1983; Manda et al., 1982; Jones et al., 1983).

A number of studies have been performed comparing one methodology with another (Blouquet et al., 1983; Klenk et al., 1982; Little et al., 1983; Abraham et al., 1983; Middle et al., 1983; Johnson et al., 1982; Walinder et al., 1982; Peterson et al., 1984; Murtiashaw et al., 1983; Mehl et al., 1983; Jones et al., 1983; Yatscoff and Braidwood, 1982; Lee et al., 1982; Dahl-Jorgensen and Larsen, 1982; Hammons et al., 1982; Castagnola et al., 1983; Mortensen et al., 1983). Almost all methodologies appear to be able to perform with the appropriate precision if conditions of assay and handling are optimized. Storage conditions markedly effect the values obtained especially in methods which rely on physical property changes induced by glycosylation (Peterson et al., 1984; Cachon et al., 1982; Little et al., 1983). Proper handling of specimens with appropriate controls is therefore mandatory. Without references or standards, accuracy cannot be evaluated. Nevertheless each method correlates well with others and therefore can be used clinically. An automated method with references and standards preferably expressed as moles of glucose per mole hemoglobin or protein is sorely needed.

A number of investigations have been performed regarding the significance and elimination of the labile fraction (NATHAN et al., 1982; BISSE et al., 1982; MAQUART et al., 1982; SHENOUDA et al., 1982; JURY et al., 1983; MORTENSEN and MARSHALL, 1983). The labile fraction affects the results in methods which separate hemoglobin by charge except for isoelectric focussing (MORTENSEN and MARSHALL, 1983) where the labile glucopyranose ring is more anodal than hemoglobin A_{1c}.

Dialysis for 4 hours at 37°C, 18 hours at 4°C, lysis at pH 5 with or without semicarbazide and aniline all appear to be successful means of eliminating the labile adduct. Dialysis at 22°C for 18 hours was also recommended (SHENOUDA et al., 1982) but found to lead to small differences and therefore was not felt to be worth the effort in clinical situations. However, under these latter conditions artefactual increases in HbA_{1a+b} will occur compensating for the decrease in HbA_{1c} (PETERSON et al., 1984) and therefore measurements of HbA_1 will not change as dramatically as anticipated. Elimination of the labile fraction becomes most important where the values are elevated since the amount of labile fraction is proportional to the total fast fraction and the mean amplitude of glycemic excursion (MORTENSEN and MARSHALL, 1983; DANEMAN et al., 1982; TIBI et al., 1982; UKENA et al., 1982). This has led some authors to recommend elimination of the labile adduct only if values are above 12% (UKENA et al., 1982) or not at all (SHENOUDA et al., 1982; JURY et al., 1983; TIBI et al., 1982). Since the

labile adduct provides information which is different from the stable fraction and failure to eliminate the adduct provides a "yea or nea" test of glucose "control", the clinical situation will determine the best way of handling this adduct.

Certainly for studies which are using the measurement of glycosylated hemoglobin to document long term glycemia, the elimination of the labile adduct would appear judicious. The removal of the labile adduct may be especially important in pregnancy since HUISMAN et al. (1982) found that in non-diabetic pregnant women, the unstable fraction was significantly higher than in the nonpregnant individual. Removal of the labile fraction resulted in a lower reference range for pregnancy.

Our own studies (unpublished observations) have shown that the removal of the labile adduct from hemoglobin A_{1c} is parallel in pregnant and nonpregnant female controls. However, the formation of a labile adduct following incubation with 20 mM glucose is less when blood from pregnant individuals is used than when nonpregnant female controls are used. It was also found that the hemoglobin A_{1a+b} fraction in the blood of pregnant individuals showed an artefactual increase after 2 hours incubation in saline and after 5 hours storage as whole blood. These latter changes were not seen in the nonpregnant female controls.

These studies and others emphasize the need to establish norms and procedures specific for the study of the pregnant diabetic population. WORTH and coworkers (1985) recently performed a longitudinal study of glycosylated hemoglobin in normal pregnancy using two methodologies (ion exchange chromatography and colorimetric). Both methods were in agreement in showing a nadir at 17 weeks, a peak at delivery, and a fall postpartum.

Studies on the role of glucosylation in the problems of pregnancy complicated by diabetes mellitus

A number of studies have shown that glycosylation of various proteins or tissues occurs *in vitro*. The documentation that these reactions occur concomitant with or are causative of pathological consequences has been more difficult.

A number of studies have documented changes in the erythrocyte coincident with glucosylation of hemoglobin. The hypothesis that glucosylation of hemoglobin might contribute to pathology through an effect on oxygen affinity (Ditzel, 1976) does not appear to be applicable since the whole blood oxygen affinity in persons with diabetes was not increased when compared with normals (Madsen and Ditzel, 1984). A number of studies have confirmed glucosylation of erythrocyte membrane proteins (McMillan and Brooks, 1982; Compagnucci et al., 1983) and altered physical properties of the erythrocyte and blood viscosity in diabetes (McMillan and Brooks, 1982; Compagnucci et al., 1983; Kanada and Otusuji, 1983; Poon et al., 1982; Juhan-Vagua and Vague, 1982). Whether these phenomena are related remains conjectural since insulin infusion seems to improve erythrocyte deformability within hours (Juhan-Vagua and Vague, 1982).

Glucosylation reactions may play a role in the hyperaggregation of platelets seen in diabetes which may in turn contribute to the placental pathology seen in pregnancy complicated by diabetes. Erythrocytes from diabetic individuals led to increased aggregation of platelets from controls (Juhan-Vagua and Vague, 1982) and in rat studies platelets stimulated with glucosylated collagen showed hyperaggregation when compared with controls (Le-Pape et al., 1983). Le Pape et al. (1983 a) also studied pepsin extracted collagen and an acid soluble glycoprotein purified from placentae of normal and diabetic human samples. They found that diabetic samples had a significant increase in ketoamine-linked glucose whereas both amino acid and carbohydrate composition were unaffected. The excess nonenzymatic condensation of glucose on free amino groups was found to increase platelet aggregating potency of these proteins independent of any modification in fiber morphology. In our own studies of placental morphology, a significant correlation was found between glucosylated hemoglobin values in diabetic pregnant women and villous capillary basement membrane thickening of placentae (Firpo et al., 1981).

To date, nonenzymatic glycosylation reactions have not been linked to pathology of the fetus of the diabetic mother. In fact it is controversial whether there is increased glucosylation of fetal hemoglobin in the infant of the diabetic mother. Roberts et al. (1983) found maternal fructosamine at 29 weeks gestation correlated significantly with both fasting blood glucose levels at the time and birthweight ratio. The levels of fructosamine in cord blood were significantly higher in gestational diabetic than in normal pregnancies. Zeller et al. (1983) could not detect a difference in glucosylation between normal and diabetic cord blood samples using a thiobarbituric acid assay of glucosylation. Sosenko et al. (1982) found a good correlation between maternal and cord blood glucosylated hemoglobin values using a thiobarbituric acid assay. Glucosylated hemoglobin in cord blood also correlated with C-peptide levels and hypoglycemia in the neonate. Cord blood glucosylation at term did not correlate with infant macrosomia. Using affinity chromatography Weare et al. (1984) have also found a good correlation between cord blood glucosylated hemoglobin and maternal postpartum glucosylated hemoglobin in 20 subjects. Levels of glucosylation in cord blood appear to be about 60% of those seen in maternal blood (Sosenko et al., 1982; Weare et al., 1984) by both the thiobarbituric acid and affinity chromatography methods. It would therefore appear that excess nonenzymatic glucosylation does occur in the fetus during gestation in the hyperglycemic maternal milieu. Whether this process results in pathology in the fetus remains to be determined.

Studies on the clinical utility of glucosylated protein

Measurements during pregnancy

The major utility of measuring glucosylated proteins during gestation appears to be in their ability to identify pregnancies at high risk. The use of glucosylated hemoglobin or serum proteins as a diagnostic test for gestational diabetes awaits the development of assays with the appropriate accuracy and precision and rigid standards of diagnosis. Nevertheless, determination of serum fructosamine levels may be useful in the diagnosis of gestational diabetes pending confirmatory studies (ROBERTS et al., 1983).

The question of "normal" glucosylated hemoglobin or protein levels during gestation also requires resolution. Assays such as affinity chromatography (WEARE et al., 1984) and radioimmunoassay (PETERSON et al., 1984) for glucosylated hemoglobin which are highly specific would appear to confirm (LIND and CHEYNE, 1979) that norms for pregnancy are slightly but significantly lower than for nonpregnant individuals. Nevertheless, the point remains controversial. The controversy may in part reside in the relative precision of the assays themselves and in part in the fact that the labile fraction is not as well defined and storage artefacts may be more prevalent in specimens from pregnant persons (vide supra).

WRIGHT et al. (1983) found that in 58 consecutive pregnancies in insulin-dependent diabetic women, glycosylated hemoglobin levels were abnormally high in 78% at the time of booking for antenatal care. Spontaneous abortion was the outcome in 15 prenancies, 10 occurring before the 15th week of gestation. Glycosylated hemoglobin levels were significantly higher in those women who aborted spontaneously than in women who delivered successfully (12.8% vs 11.2%). In addition to confirming the use of glucosylated hemoglobin as a means of identifying pregnancies at risk, these studies confirm the inadequacy of diabetic control in the first trimester in most centers.

Two studies appear to confirm that elevated glucosylated hemoglobin values during the first trimester are indicative of a higher prevalence of malformations in the infants of diabetic mothers (MILLER et al., 1981; YLINEN et al., 1981). The levels of elevation which leads to an increased risk for malformations appear to be quite high in that the mean glucosylated hemoglobin value in early gestation for mothers of the malformed infants was approximately four or more standard deviations above the normal mean. As noted above, the relationship of elevated glycosylated hemoglobin or proteins during pregnancy to macrosomia in the infant remains controversial (CAMERON, 1979; COEN et al., 1980; WIDNESS et al., 1978). Part of the problem may lie in the timing of the drawing of the sample and the acute attempts at term to normalize blood glucose. Our own results would confirm that glucosylated hemoglobins correlate well with infant birth weight even in White Class R-F pregnancies if the mean of weekly values from 30 weeks gestation are considered.

Conclusions

The monitoring of glucosylated hemoglobin or serum proteins as an index of glycemia during gestation appears warranted as a means of identifying the pregnancy at risk for complications induced by hyperglycemia including: 1. spontaneous abortion and malformations (in the first trimester), 2. perinatal loss (in the second trimester), and 3. macrosomia and infant morbidity (in the third trimester). The use of glycosylated proteins as a diagnostic test for gestational diabetes appears premature and awaits the definition of diagnostic criteria and references and standards for the assay systems. The hypothesis that glucosylation reactions may in and of themselves contribute to maternal or fetal pathology in pregnancy complicated by diabetes remains provocative but untested.

9

Glycosylated Haemoglobins and Neonatal Macrosomia

P. *Czekelius* and J. *Rollmann*

Centre of Gynaecology and Obstetrics of the Philipps University of Marburg, Federal Republic of Germany

Glycosylated haemoglobins (GHb)

The coupling of glucose to haemoglobin is dependent on time, concentration and pH. The optimum pH is 7.4 (NIEDERAU et al., 1980). The process is not catalyzed enzymatically and is relatively slow. Although aldimine is formed by attachment of the sugar to the N-terminal β-chain of haemoglobin, glucose can be split off again by an increase in blood acidity or, for example, by processes occurring during dialysis (NIEDERAU et al., 1980). A stable ketoamine is only formed by a prolonged process of molecular transformation according to AMADORI (Fig. 26).

Accordingly, short-term blood sugar increases do not lead to the formation of irreversible glycohaemoglobins.

Glucose is not the only glycosylating sugar. By means of column chromatography, GHb can be separated — as was first shown by TRIVELLI et al. (1971) — into the fractions shown in Table 1.

The diagnostically most important and quantitatively largest fraction is formed by HbA_{1c}.

The heterogeneity of GHb is of limited importance for clinical problems, since there exists a very close correlation between HbA_{1c} and GHb as total fraction (DUNN et al., 1979). There is strong correlation with glucosuria (GABBAY et al., 1977), fasting blood sugar levels (O'SHAUGHNESSY et al., 1979) and the mean blood glucose concentration of the preceding 2–3 months (DUNN

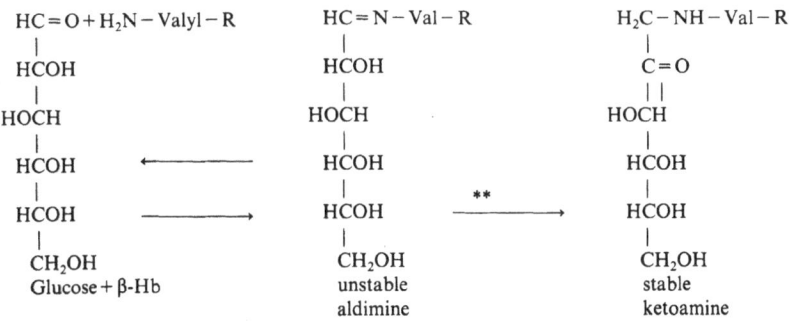

** Transformation according to Amadori.

Fig. 26. Reaction equation of the formation of HbA_{1c} according to BUNN et al. (1975)

Table 56. GHb values and Hb fractions of normal subjects (modified according to McDONALD et al., 1978)

Haemoglobin component	Percentage of total haemolysate ± s	Supposed glycosyl residue (BUNN et al., 1978)
HbA_{1a1}	0.19 ± 0.02	Fructose-di-phosphate
HbA_{1a2}	0.19 ± 0.04	Glucose-6-phosphate
HbA_{1b}	0.48 ± 0.15	Deamidation product[a] of HbA_{1c}
HbA_{1c}	3.3 ± 0.30	Glucose

[a] (KRISHNAMOORTHY et al., 1977).

et al., 1979; MADSEN et al., 1982). The best information can be given for the glycaemic condition of the preceding 8 weeks. For this reason, measurements of total GHb have become generally accepted for monitoring the metabolic condition of diabetic patients over prolonged periods of time.

GHb and pregnancy

The above mentioned characteristics of glycosylated haemoglobins also apply to pregnancy. An influence on the level of maternal GHb values by gravidity itself has been suggested but not yet unequivocally established. The findings compared with non-pregnant women range from reduced (FEIGE and WOESSNER, 1985; (LIND and CHEYNE, 1979; NIEDERAU et al., 1981; WORTH et al., 1985) to uninfluenced (BACIGALUPO et al., 1984; PHELPS et al., 1983; POLLAK et al., 1979; STEEL et al., 1981) to elevated (GRÄFENSTEIN and DUCHNA, 1981). These differences may be due not only to technical problems brought about by co-measurement of non-glycosylated, fetal and acetylated haemoglobins (macro, micro column and ion-exchange chromatography), but may also be due to the heterogeneity of the studied patient populations.

As shown in incubation experiments, the reaction kinetics of fetal haemoglobin are comparable to those of adult Hb (FELDMAN et al., 1984; POON and TURNER, 1981). Chromatographic separation of fetal haemoglobins with DEAE cellulose (SCHWARTZ et al., 1980) yields the following fractions:

- HbF   72.7 ± 10.0%
- acetylated HbF   14.6 ± 4.8%
- HbF_{1a} (glycosylated)   2.7 ± 0.5%

The rest is formed by adult Hb.

Problem formulation

Maternal hyperglycaemia is considered to be the essential cause of fetal macrosomia, neonatal hypoglycaemia and frequent malformations (PEDERSEN, 1977). The early treatment of disturbed carbohydrate metabolism reduces the incidence of these complications (ADASHI et al., 1979; ARTAL et al., 1983; COUSTAN and IMARAH, 1984; DE MUYLDER, 1984).

If one retrospectively examines the obstetric history of female diabetics, the incidence of delivery of macrosomic infants is often increased even many years before diabetes becomes manifest (FITZGERALD et al., 1961; MICKAL et al., 1966; THEILE et al., 1985). Accordingly, delivery of a macrosomic child by a mother with non-manifest diabetes is considered to indicate a potential prediabetic condition of the mother (WHO Expert Committee on DM, 1980). It is suspected that even in this latent phase of the disease subclinically raised blood sugar levels cause fetal macrosomia according to the hyperglycaemia-hyperinsulinaemia theory (PEDERSEN, 1977).

In the absence of more informative anamnestic data, postpartum glucose tolerance tests for the assessment of neonatal macrosomia yield useful results only within very narrow and often not realizable time limits (HEISIG, 1975; LOVE et al., 1964). Bed rest, postoperative fasting, infusion therapy, inadequate calorie supply sub partu additionally influence the measured values.

By measuring glycosylated haemoglobins as percentage of total haemoglobin, the mean blood sugar level of the last few antepartal weeks can be assessed (MADSEN et al., 1982; O'SHAUGHNESSY et al., 1979; WIDNESS et al., 1978). Postpartal blood tests in mothers of macrosomic newborns, however, have led to contradictory findings (BACIGALUPO et al., 1984; BRANS et al., 1982; MAYET et al., 1983; POLLAK et al., 1981; STEEL et al., 1981; WIDNESS et al., 1981; YATSCOFF et al., 1985). The measurement of maternal GHb does not permit a sufficient differentiation between parturients with latent and manifest diabetes (COUSINS et al., 1984; KARPELLUS and BICHLER, 1984). On the basis of the index sign of neonatal macrosomia, it was therefore the purpose of our studies to examine the interdependence between infant and maternal GHb values, to compare them with data from a control population without macrosomia, and to establish correlations with different maternal parameters which might influence neonatal macrosomia.

Methods

170 blood samples of mothers and their newborns taken immediately postpartum were examined by affinity chromatography (Glyc-affin-GHb, Isolab Inc., Akron, Ohio). Glycosylated haemoglobins were determined as percentage of total haemoglobin (ROLLMANN, 1987). The separation of GHb from total haemoglobin occurs in microcolumns and is based on the specific affinity of the boric acid groups in the ion-exchange resin (m-phenylboronate bound to agarose) for the 1,2-cis-diol bonds which are common to all glycosylated haemoglobins. The light absorption of the eluates of glycosylated or non-glycosylated haemoglobins is measured at a wavelength of 405 nm (Hg) in an Eppendorf photometer against distilled water as blank. Calculation of the GHb percentage of total haemoglobin is performed according to the following formula:

$$\% \, GHb = \frac{100 \, x}{x + 10 \, y}$$

x = extinction of the second fraction (GHb)
y = extinction of the first fraction (non-glycosylated haemoglobins)

Three study groups were differentiated as follows:
I. Mothers of normal-weight newborns (10th–90th percentile according to HOHENAUER, 1980) without risk factors for diabetes mellitus (n = 30).
II. Mothers of normal-weight newborns with risk factors for diabetes mellitus (n = 29).
III. Mothers of macrosomic newborns (above the 90th weight percentile according to HOHENAUER) (n = 26).
Mothers with manifest diabetes and mothers with newborns below the 10th weight percentile were excluded from the study.
The following anamnestic data were regarded as risk criteria for the assignment to group II:
a) Hereditary diabetes in a first and/or second degree relative,
b) previous spontaneous abortion or stillbirth,
c) delivery of a child with severe malformations,
d) delivery of a macrosomic child (above 3999 g),
e) glucosuria,

Table 57. Comparison of the mean values of important maternal characteristics

Characteristic	Group I + II	Group III
Age	27.3 years	26.7 years
Height	165.0 cm	167.9 cm
Body weight before pregnancy	64.6 kg	68.4 kg
Weight deviation from Broca normal weight	+ 1.06%	+ 1.07%
Weight gain during pregnancy	12.6 kg	13.2 kg
Parity	1.8	1.9

i) obesity (body weight before gravidity 20% in excess of the Broca normal weight).

The characteristics of the mothers included in the study and the obstetric parameters are given in Tables 57 and 58.

As concerns age, size, body weight before pregnancy, weight deviation from Broca normal weight, weight gain during gravidity and parity, there were no discrepancies between the study groups. With respect to obstetric parameters, the somewhat higher caesarian section rate in group II and the higher rate of male newborns with macrosomia in group III were conspicuous.

For the discrimination of the distribution of values in the significance calculations of mean values, the Kolmogoroff-Smirnoff test and the Kruskal-Wallis test were used; for the comparison of maternal and infant GHb, the single analysis of variance was employed. Correlations between two parameters were examined, with the t-test for matched samples, the linear and Spearman rank correlation coefficients (SACHS, 1973).

f) transitorily raised blood sugar levels before pregnancy,
g) transitorily disturbed glucose tolerance before pregnancy,
h) hydramnion, and

Results

From our studies, the following values were established as normal ranges of maternal and infant GHb:

Postpartal GHb	Percentage of total haemoglobin
GHbM (maternal)	3.4–5.2%
GHbC (infant)	2.0–2.3%

Table 58. Comparison of obstetric parameters of the three study groups

Parameter	Group I	Group II	Group III
Mode of delivery			
Spontaneous delivery	23	21	19
Cesarean section	4	8	5
Vacuum extraction	1	1	1
Forceps delivery	1	0	1
Sex			
Male	13	16	17
Female	16	14	9
Gestational age Arithmetic mean of weeks of pregnancy	40.2	40.3	40.7
Birth weight Arithmetic mean ($\bar{x}$)	3396 g	3346 g	4283 g

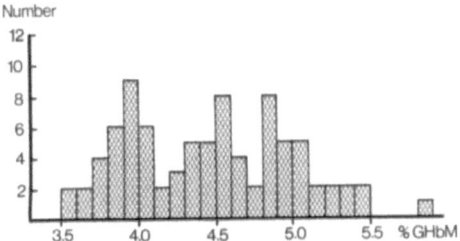

Fig. 27. Frequency distribution of GHbM values

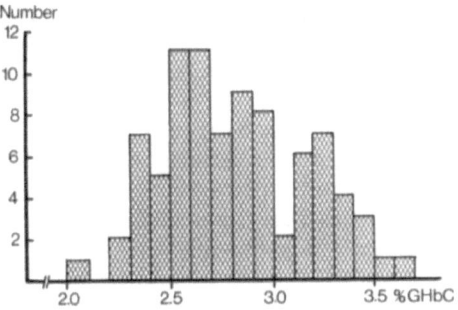

Fig. 28. Frequency distribution of GHbC values

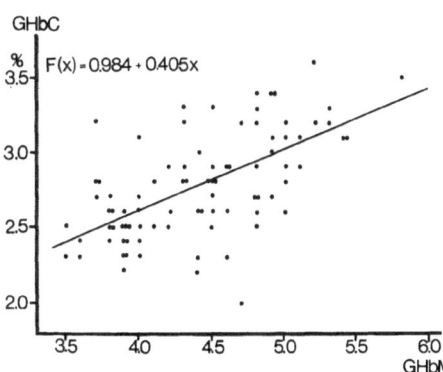

Fig. 29. Interdependence between fetal and maternal GHb values (n = 85; r = 0.61; r = 0.60; p < 0.001)

They are considered to be "definitely normal". Values beyond the $\bar{x} \pm 3$ standard deviation of 2.9–5.8 for GHbM and 1.7–3.6 for GHbC were classified as "definitely pathological". The range of the distribution of all maternal and infant GHb values is shown in Figs. 28 and 29.

The infant glycohaemoglobins were an av-

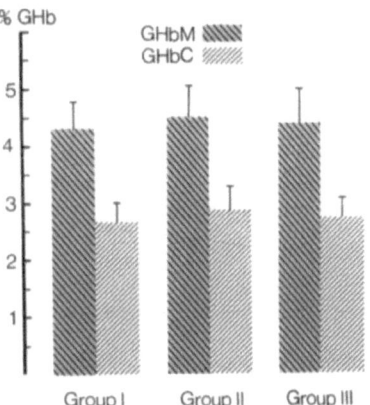

Fig. 30. Mean values and standard deviations of maternal and fetal GHb of the three study groups

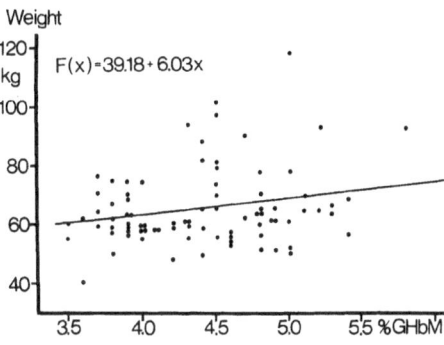

Fig. 31. Correlation between GHbM values and maternal body weight before pregnancy (n = 84; r = 0.24; r_s = 0.19; p < 0.005)

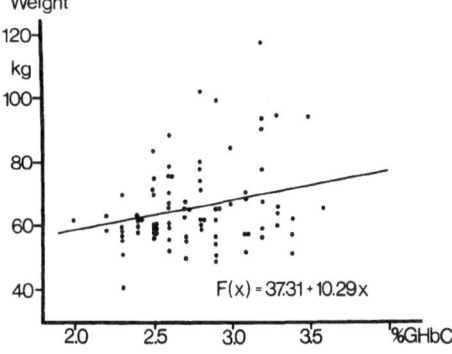

Fig. 32. Correlation between GHbC values and maternal body weight before pregnancy (n = 84; r = 0.27; r_s = 0.24; p < 0.005)

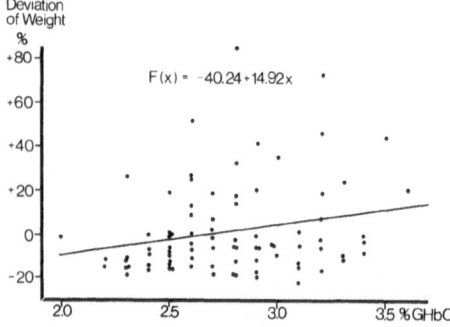

Deviation of Weight %

$F(x) = -40.24 + 14.92 x$

Fig. 33. Correlation between GHbC values and maternal weight deviation from Broca normal weight in % (n = 83; r = 0.25; r_s = 0.24; p < 0.005)

erage of 37% lower than the corresponding maternal haemoglobin levels (p < 0.001). The interdependence of both values in linear regression is shown in Fig. 29.

A statistically significant difference of glycosylated haemoglobins could not be demonstrated for mothers with an without macrosomic newborns, nor between macrosomic and normal-weight infants. The mean values and standard deviations of these measurements are shown in Fig. 30.

There was no correlation with the mother's age, size and weight gain during pregnancy and a modest significant correlation (p < 0.05) of both GHbM and GHbC values with the mothers body weight before gravidity (Figs. 31 and 32).

The interdependence between GHbC values and maternal weight deviation from Broca normal weight (p < 0.05) is shown in Fig. 33. The Spearman rank correlation revealed no significant correlation of both GHbM and GHbC values with the infant's birth weight correlated with gestational age and sex.

Discussion

GHb have been increasingly studied in recent years as a possible method of retrospective assessment of disturbances of carbohydrate metabolism in mothers of macrosomic newborns. A review of the literature (ROLLMANN, 1987) shows that postpartal determination of maternal GHb alone is inadequate as screening method for latent diabetics (BRANS et al., 1982; COUSINS et al., 1984; KARPELLUS and BICHLER, 1984; POLLAK et al., 1981 WIDNESS et al., 1978). Owing to different selection criteria of the subjects, the results can be compared to only a limited extent. WIDNESS et al. (1981) report slightly raised HbA_{1c} values in mothers of macrosomic newborns (above the 95th weight percentile), but do not present detailed information on the method of exclusion of diabetic women. POLLAK et al. (1981) and STEEL et al. (1981) report similar findings. BRANS et al. (1982), MAYET et al. (1983), MANSANI et al. (1983) and BACIGALUPO et al. (1984) obtain a result to the contrary, probably due to different measuring methods. The investigations by BACIGALUPO et al., in particular,

suggest the differing GHb values are mainly related to varying proportions of women with diagnosed and undiagnosed gestational diabetes. Unlike most other authors (BRANS et al., 1982; MAYET et al., 1983; POLLAK et al., 1981; STEEL et al., 1981; WIDNESS et al., 1981), but in accordance with MORRIS et al. (1985), they exclude this possibility by means of a glucose tolerance test during pregnancy. In the present study which was based on the assessment of neonatal macrosomia postpartum as the only indicator, we tried to minimize the assessment risk by specific anamnestic exploration.

Prepartal determination of glycosylated haemoglobins as well do not allow a differentiation between mothers with latent and manifest diabetes (ARTAL et al., 1984; MCFARLAND et al., 1984). BAXI et al. (1984) describe a false negative rate of 36.4% for gestational diabetes. COUSINS et al. (1984) report a higher specificity, sensitivity and a greater prognositc value of oral 1-hour glucose tolerance tests as compared to GHb.

The GHb values of macrosomic newborns

of mothers without manifest diabetes have so far been investigated only in a few studies (FADEL et al., 1981; HALL et al., 1983; THEILE et al., 1985). FADEL et al. (1981) categorized 70 mothers into healthy women, mothers with anamnestic diabetic risks, diabetics of White class A and patients with manifest diabetes, without being able to find a correlation with the infant's birth weight. YATS-COFF et al. (1985) summarized the findings for diabetic and non-diabetic mothers of macrosomic newborns using, as we did, affinity chromatography as the laboratory method. With this method, determination of both maternal and fetal GHb and hence a direct comparison of both values is possible. The assay is independent of the presence of nonglycosylated or acetylated fetal haemoglobin, it has a low temperature sensitivity, and the time required for analysis is short. YATS-COFF et al. as well did not find significantly raised postpartal glycohaemoglobin levels after delivery of a macrosomic infant.

As can be seen from our studies, the GHb values of newborn infants were about 1/3 lower than those of their mothers. This corresponds to the data in the literature (Table 4). The reasons for this are the elevated proportion of young fetal erythrocytes with reduced degree of glycosylation (FITZ-

GIBBONS et al., 1976), their shorter survival compared with those of adults (OSKI and NAIMAN, 1972), the higher percentage of acetylated haemoglobin (SCHWARTZ et al., 1980) and thus the protection of the terminal amino group from the glycosylation process, as well as the fetal blood sugar level which is 20–30% lower than that of the mother (HEISIG, 1975). Assuming a facilitated diffusion of glucose through the placenta (HEISIG, 1975), a close interdependence between fetal blood glucose levels and maternal blood glucose levels is to be expected.

The good correlation ($r_s = 0.60$; n = 85; p < 0.001) between maternal and fetal GHb values confirms this interdependence.

Similar findings have also been reported by other authors:

POON and TURNER (1981)
$r_s = 0.40$; n = 24; p < 0.05
SOSENKO et al. (1982)
$r = 0.61$; n = 63; p < 0.001
ZELLER et al. (1983)
$r = 0.64$; n = 29; p < 0.001
WORTH et al. (1983)
$r = 0.60$; n = 25; p < 0.01

In contrast, YATSCOFF et al. (1985) surprisingly do not find this correlation without giving an explanation for it.

Probably owing to the narrow age range in our populations (93% of the mothers were younger than 35 years), our results did not reveal an effect of the mother's age on the glycosylation rate of her haemoglobin, as is described by GRÄFENSTEIN and DUCHNA (1981) or COUSINS et al. (1985). Likewise, we were not able to find a correlation with maternal body size. In accordance with the findings of NIEDERAU et al. (1980), we found, however, a slight correlation with the mother's body weight before gravidity.

In particular, the paired GHb values of very heavy-weight mothers (above 80 kg) and their infants were found to be in the upper normal range. The shift of the mean glycaemic condition towards raised blood sugar levels in obesity might be relevant in this respect.

Table 59. Mean percentage of fetal GHb values as a function of maternal GHb values in different studies

Author	Fetal GHb values in % of maternal GHb values
HALL et al., 1983[a]	59%
WORTH et al., 1983[a]	70%
Own findings[a]	63%
SOSENKO et al., 1982[b]	69%
ZELLER et al., 1983[b]	60%
GOUÉDARD et al., 1984[b]	63–78%
YATSCOFF et al., 1985[b]	60%

[a] Mothers with non-manifest diabetes.
[b] Mothers with non-manifest diabetes *and* diabetic mothers.

Mansani et al. (1982) find a correlation between GHb values and maternal weight gain during gravidity. We cannot confirm this finding. It is possible, however, that in our patient population an intentionally induced reduction in food intake by intense prenatal care caused this negative result.

Recently, Golz et al. (1986) compared the HbA$_{1c}$ determination in mothers of macrosomic newborns with the result of the oral glucose tolerance test within 48 hours postpartum. There was no interdependence of the two test results. Weiss (1986) doubted whether glycosylation of fetal haemoglobin would occur at all with transitorily and moderately raised blood sugar levels. The onset of HbF and insulin formation occurs during about the same gestational phase (10th week of pregnancy). The fetus would thus be able to offset blood sugar peaks immediately by its own insulin production. Insulin transfer from the mother is so small as to be without functional importance (Milner, 1971). In this way, the absent rise in glycosylated haemoglobin in puerperae with non-manifest diabetes despite neonatal macrosomia could be explained.

Since our findings show no correlation of both maternal and infant glycosylated haemoglobins with the symptom of neonatal macrosomia, the determination of glycosylated haemoglobin after delivery of excessweight infants is in our opinion not suitable to confirm or exclude gestational diabetes.

Summary

Maternal hyperglycaemia is an essential cause of fetal macrosomia, neonatal hypoglycaemia and frequent malformations. Early treatment reduces the incidence of fetal complications. By means of the determination of glycosylated haemoglobins, the mean blood sugar level of the last few antepartal weeks can be assessed. Previous studies of postpartal blood tests in mothers of macrosomic newborns have resulted in contradictory findings; the determination of maternal GHb alone does not permit a sufficient differentiation between puerperae with latent and manifest diabetes. Therefore, we investigated the values of glycosylated haemoglobins of mothers and their newborns, assessing their correlation with each other and with neonatal macrosomia.

In 85 pairs of maternal and infant blood samples obtained immediately postpartum from patients with and without diabetic risk (n = 30 and n = 29, respectively) as well as mothers of macrosomic newborns (n = 26) the percentage of glycosylated haemoglobins was determined by affinity chromatography. The infant GHb values were by an average of 37% lower than those of their mothers. They showed a linear correlation with maternal GHb values as well as a correlation with maternal weight deviation from Broca normal weight, whereas no correlation was found with maternal weight gain during gravidity, body size and age. For both maternal and infant GHb determinations, a significant correlation with maternal body weight before pregnancy could be demonstrated. There was no correlation of glycosylated haemoglobins with the infants' birth weight corrected for gestational age and sex. The determination of GHb for the diagnosis of gestational diabetes in cases of neonatal macrosomia is in our opinion not suitable as a screening method.

10

Recurrence of Gestational Diabetes in Subsequent Pregnancies

Virginia R. Lupo and *S. J. Stys*

Department of Obstetrics and Gynecology, University of Cincinnati Hospitals, Cincinnati, Ohio, USA

Introduction

Gestational diabetes mellitus (GDM) was originally defined by O'SULLIVAN and MAHAN (1964) as glucose metabolism exceeding two standard deviations beyond normal values during pregnancy. Although a number of clinical risk factors have since been associated with the development of these abnormalities (O'SULLIVAN et al., 1973), the significance of a previous history of abnormal carbohydrate metabolism in pregnancy is unclear. A clinical observation of an occasional patient with gestational diabetes in pregnancy, followed by normal diabetes screening in a subsequent pregnancy, prompted a formal assessment of the recurrence rate of gestational diabetes when another pregnancy occurred.

Materials and methods

Medical records from the University of Cincinnati Hospital from 1980–1984 were examined, and 28 women were identified with a diagnosis of gestational diabetes who went on to have a subsequent pregnancy and delivery. Gestational diabetes is strictly defined, as the onset of abnormal carbohydrate metabolism in pregnancy, with no known antecedent abnormality before pregnancy (Second International Workshop-Conference on Gestational Diabetes Mellitus 1985: Summary and Recommendations). All pregnant women in this population are screened for diabetes at 28 weeks gestation, with women having clinical risk factors also screend at their initial prenatal visit (LAVIN et al., 1981). Screening consists of a 50 gram oral glucose load with a one hour venous plasma result obtained. A level of ⩾ 150 mg/dl prompted a formal glucose tolerance test until 1983, when the cut-off was dropped to 135 mg/dl.

Criteria for a diagnosis of GDM were at least two abnormal values on a 3 hour oral glucose tolerance test (GTT) following an overnight fast and a 100 gram oral glucose load, with normal defined at < 105 mg/dl fasting, < 190 mg/dl at one hour, < 165 md/dl at two hours, and < 145 mg/dl at three hours. Values were obtained on venous plasma, utilizing the glucose oxidase method. During subsequent pregnancies, oral GTTs were performed if a glucose challenge test (GCT) following a 50 gram glucose load exceeded 135 mg/dl in the third trimester. Infant charts were also examined.

Recurrence rates were analyzed with respect to maternal age at delivery, parity, race, maternal weight at delivery and weight gain during pregnancy, the need for insulin, the sum of deviation of the two most abnormal values on the GTT from normal, and the mean body weight index (weight in Kg ÷ height in meter^2), using weight at the beginning of pregnancy. Neonatal factors assessed were sex, birth weight, length of gestation and birth weight percentile > 90%. Means and standard deviations were calculated and significance was assessed by means of Student's t test.

Results

Twenty-eight females with a diagnosis of GDM who then went on to have a subsequent pregnancy and delivery were identified. GDM recurred in 10 of 28 women (36%). Maternal characteristics at the time of the first pregnancy are shown in Table 60. Women in whom diabetes recurred in a subsequent pregnancy were younger at the time of the index pregnancy (24.9 vs. 29.5 years, p < 0.01) and had a lower mean body weight index (22.5 vs. 31.3, p < 0.01) than women who did not manifest diabetes in their next pregnancies. The degree of carbohydrate abnormality, as manifest by the sum of the two most abnormal GTT values and the requirement for insulin therapy to maintain weekly

Table 60. Maternal characteristics in initial diabetic pregnancy[a]

	Recurring[b]	Nonrecurring	p
Number	10 (36%)	18 (64%)	
Age	24.9 ± 5.3	27.5 ± 6.4	<0.01
Parity	1.7 ± 1.8	1.6 ± 2	NS
Race (% black)	40%	40%	NS
MBWI[c]	22.5 ± 5.2	31.3 ± 7.7	<0.01
Maternal wt gain in pregnancy (lbs)	21.2 ± 12.7	20.8 ± 20.2	NS
Maternal wt at delivery (lbs)	170.5 ± 49.4	188 ± 52.3	NS
Sum of 2 most abnormal GTT values	72 ± 75.5	61 ± 43	NS
Insulin required	$^3/_{10}$ (30%)	$^2/_{18}$ (11%)	NS

[a] Mean ± standard deviation
[b] Patients in whom gestational diabetes recurred in next pregnancy
[c] Mean Body Weight Index = wt in kg ÷ height in M^2. Weight is at beginning of pregnancy

Table 61. Maternal characteristics in subsequent pregnancy[a]

	Recurring[b]	Nonrecurring	p
Number	10 (36%)	18 (64%)	
Age	27 ± 5.6	29 ± 6.3	NS
Parity	2.8 ± 1.9	2.6 ± 2	NS
Interval since last delivery (mos)	26.8 ± 11.6	20 ± 8.7	< 0.02
MBWI[c]	22.4 ± 6	31.5 ± 8	< 0.02
Maternal wt gain in pregnancy (lbs)	25 ± 13	27.5 ± 20.4	< 0.05
Maternal wt at delivery (lbs)	175.1 ± 55	192 ± 61	< 0.05

[a] Mean ± standard deviation
[b] Patients in whom gestational diabetes was recurring
[c] Mean Body Weight Index = wt in kg ÷ height in M^2. Weight is at beginning of pregnancy

Table 62. Infant characteristics in initial diabetic pregnancy and subsequent pregnancy

	Initial Pregnancy			Subsequent Pregnancy		
	Recurring[a]	Nonrecurring	p	Recurring[c]	Nonrecurring	p
Number	10	18		10	18	
% male	60%	50%	NS	40%	55%	NS
Birthweight in grams	3453 ± 668[b]	3312 ± 609	NS	3480 ± 724	3456 ± 564	NS
% LGA	50%	28%	NS	50%	28%	NS
Length of gestation	39 ± 1.5[b]	38.8 ± 1.9	NS	38.9 ± 1.6	38.8 ± 2.1	NS

[a] Infants of mothers whose subsequent pregnancy was complicated by a recurrence of gestational diabetes
[b] Mean ± standard deviation
[c] Infants of mothers in whom diabetes was recurring

fasting blood glucose levels < 105 mg/dl and 2-hour postprandial glucose levels < 120 mg/dl, was comparable in the two groups.

In the subsequent pregnancies, described in Table 61, women in whom diabetes recurred were thinner (175.1 vs. 192 lb, $p < 0.05$), gained slightly less weight (25 vs. 27.5 lb, $p < 0.05$), again had a lower mean body weight index (22.4 vs. 31.5, $p < 0.02$) and had longer intervals since their previous deliveries (26.8 vs. 20 months, $p < 0.02$), overcoming the differences in age by the time of delivery (27 vs. 29 years, $p > 0.05$).

None of the infant characteristics examined correlated with the recurrence of diabetes (Table 62).

Discussion

The recurrence of gestational diabetes in a subsequent pregnancy has not previously been described beyond O'SULLIVAN's observation that 23 of 63 patients (40%) with moderate or severe gestational diabetes proved to have diabetes at least as severe in their next pregnancy (O'SULLIVAN and MAHAN, 1964). Recurrence of diabetes did not correlate with the degree of abnormality of carbohydrate metabolism in the initial pregnancy, although the risk of development of overt diabetes in the remote future did correlate with the degree of abnormal carbohydrate metabolism during pregnancy.

The strong correlation of recurring diabetes with the lower mean body weight index in the present study is intriguing in light of the heterogeneity in gestational diabetics demonstrated by CHENEY et al. (1985). These workers found a significant difference in the insulin response to meal stimulation between lean and obese gestational diabetics. It was felt that lean gestational diabetics tended to

develop a relative glucose intolerance based on an insulin deficiency, and not on the basis of increased insulin resistance seen in obese gestational diabetics. Over time, when another pregnancy occurs, the insulin deficiency of lean patients appears to persist more reliably than the insulin resistance of obese patients, as shown by the higher recurrence of the disease in leaner patients. Both lean and obese diabetics had a delayed insulin response to a 400 kcal meal, but the magnitude of the insulin response showed obese women to be hyperinsulinemic in general, with lean women excreting much less insulin.

It is tempting to speculate on the role of self-modification of dietary intake in women who become pregnant after a previous diabetic pregnancy. Patients were not advised to return to a diabetic diet unless diabetes was again diagnosed, but it is possible that some did restrict their calories and lowered their fat and sucrose intake voluntarily, enough to prevent a recurrence of chemical diabetes. Such a voluntary restriction in obese patients but not lean ones is possible but does not seem likely.

Finally, the lack of predictability of the traditional clinical risk factors of maternal weight and age in assessing recurrent diabetes strongly supports routine universal screening of all pregnant women for abnormal carbohydrate metabolism in pregnancy.

Summary

Medical records of pregnancies of women with gestational diabetes mellitus (GDM) in a previous pregnancy were reviewed and the recurrence of GDM was assessed. GDM recurred in 10 of 29 (36%) of subsequent pregnancies when a previous pregnancy had been complicated by GDM. Recurrence was significantly correlated with younger age at initial pregnancy (24.9 vs. 27.5 years, $p < 0.01$) and lower mean body weight index (22.5 vs. 31.3, $p < 0.01$). The recurrence of GDM did not correlate with maternal weight, weight gain, degree of abnormality of the GTT or requirement for insulin. We conclude that females who develop GDM when younger and leaner are not "protected" in subsequent pregnancies by the lack of risk factors of weight and age.

D. On Prophylaxis and Therapy in Gestational Diabetes

11

Preconceptional Diabetes Counseling in Gestational Diabetes

J. A. Goldman, D. Dicker, D. Feldberg, A. Yeshaya, and M. Karp

Department of Obstetrics and Gynecology, Golda Meir Medical Center and Tel Aviv University School of Medicine, Petah-Tikva, Israel

Introduction

Gestational diabetes has been known to be associated with an increased perinatal morbidity and mortality (O'SULLIVAN et al., 1973; MESTMAN, 1980; COUSTAN and IMARAH, 1984). A significant number of these pregnancies are complicated by fetal macrosomia with its attendant morbidities of traumatic delivery such as shoulder dystocia, brachial plexus injury, or bony trauma and Caesarean section. Furthermore, the infant of the mother with gestational diabetes is at increased risk for other neonatal complications such as hypoglycemia, hypocalcemia, hyperbilirubinemia and erythemia (MESTMAN, 1980; COUSTAN and IMARAH, 1984; COUSTAN and LEWIS, 1978; Summary and Recomendations of the 2nd International Workshop-Conference 1985). With early detection, appropriate prenatal care and strict metabolic control the perinatal mortality in such pregnancies does not differ from that occurring in the general population (COUSTAN and IMARAH, 1984; COUSTAN and LEWIS, 1978) and the risk of perinatal morbidity might be reduced as well.

We suggest that women with previous gestational diabetes should be followed postpartum at periodic intervals, in order to detect any disturbance of glucose homeostasis early in its course, particularly in preparation for any future pregnancy. This is a time when most of these patients are not under medical supervision.

The purpose of this study was to correlate the glucose homeostasis and the outcome of pregnancies in gestational diabetes with and without preconceptional diabetic counseling.

Material and methods

Between the years 1981–1986, 156 women, with known previous gestational diabetes and normal glucose tolerance between pregnancies, contemplating pregnancy, attended a preconceptional clinic at least two months before conception and were regularly consulted by a diabetological team: (obstetrician, diabetologist, dietician, psychologist and nurse). Patients in this group who did not prove to have gestational diabetes in the present pregnancy were eliminated from this study. Thus all patients in this group had recurrence of gestational diabetes. They were compared to a group of 154 patients with gestational diabetes who attended our clinic at different stages of pregnancy and to a group of 150 normal, pregnant, non-diabetic patients. The groups matched well, the patients being comparable in all aspects demographically and socially. The ethnic composition of the patients was the usual proportion of Ashkenazi and non-Ashkenazi Jews (Caucasian and Oriental origin, respectively). Two patients were advised against pregnancy because of recurrent urinary tract infections, and were advised to conceive only after renal function evaluation and antibiotic therapy.

Of the 154 women with gestational diabetes who attended the clinic at different stages of pregnancy, 32 had unplanned pregnancies. All patients delivered at the time of submission of our report.

Close surveillence of patients was directed towards monitoring for elevations of fasting and postprandial glucose in capillary blood and designed to detect any deterioration in glucose homeostasis as gestation progressed. All women attending prior to conception underwent a 3-hour oral glucose tolerance test with 100 g glucose, repeated in each trimester. Patients with abnormal OGTT prior to conception were eliminated from this study.

Glucose homeostasis was maintained by dietary management consistent with the recommendations for caloric distribution, as proposed by the American Diabetes Association (1985). Metabolic control was monitored by SBGM 4–6 times daily with the use of glucose-meters. Hemoglobin A_1 was measured monthly by cation exchange column chromatography using microcolumn fast hemoglobin test system (Isolab, Akron, Ohio, U.S.A.). The normal range in pregnancy in our laboratory is presented in Table 63. In order to facilitate the very early diagnosis of pregnancy, patients were instructed to use basal body temperature charting through each menstrual cycle. The beta-subunits of HCG pregnancy test was performed whenever four weeks had elapsed since the last menstrual period; if the first test was negative it was repeated at weekly intervals while amenorrhea persisted. Antepartum surveillence included: monthly ul-

Table 63. Metabolic control in 290 gestational diabetics

Variable	Period	Preconception counseling (n = 156)	Others (n = 154)	Normal pregnant (n = 150)	p[b]
Mean blood glucose (mg%)	Initial visit	89 ± 6	—	86 ± 6.0[a]	—
	At conception	87 ± 6	—	89 ± 5.9	—
	1st trim.	111 ± 8	126 ± 9	87 ± 5.6	NS
	2nd trim.	122 ± 8	132 ± 9	91 ± 5.9	< 0.05
	3rd trim.	115 ± 8	140 ± 10	89 ± 6.1	< 0.05
Mean Hb A_{1c} (%)	Initial visit	6.76 ± 0.3	—	6.76 ± 0.3[a]	—
	At conception	6.79 ± 0.3	—	6.73 ± 0.3	—
	1st trim.	7.39 ± 0.3	7.46 ± 0.3	6.78 ± 0.3	NS
	2nd trim.	7.41 ± 0.3	7.82 ± 0.3	6.74 ± 0.3	< 0.05
	3rd trim.	7.39 ± 0.3	8.30 ± 0.3	6.77 ± 0.3	< 0.05
	> 24 weeks				
Mean OGTT blood glucose (mg/100 ml)	Fasting	87 ± 6	92 ± 6	86 ± 6	NS
	1 hr	174 ± 12	181 ± 12	158 ± 11	NS
	2 hr	161 ± 11	168 ± 11	139 ± 10	NS
	3 hr	130 ± 9	134 ± 9	111 ± 8	NS

[a] Normal non pregnant.
[b] Statistical analysis by Students' t test, preconseption counseling vs others.
NS Not significant.

Table 64. Maternal information on 290 gestational diabetics

Variable	Preconception counseling (n = 156)	Others (n = 154)	Normal pregnant (n = 150)	p
Mean maternal age (yr)	26.5 ± 6.2	22.4 ± 2.5	24.6 ± 4.8	< 0.07
Parity	2.6	1.9	2.3	NS
Mean weight gain (kg)	13.4 ± 3.7	18.7 ± 5.2	13.7 ± 3.8	< 0.05
Hydramnion (%)	4.5	5.8	2	NS
Preeclampsia (%)	7.7	23.4	4	< 0.05
Caesarean section (%)	17.3	31.2	8.6	< 0.05

NS Not significant.

Table 65. Neonatal data on 290 infants of gestational diabetes mothers

Variable	Preconception counseling (n = 156)	Others (n = 154)	Normal pregnant (n = 150)	p[a]
Mean gestational age (wk)	39.7 ± 0.5	39.4 ± 0.5	40.2 ± 0.5	NS
Mean birth weight (gr)	3418 ± 635	3870 ± 895	3380 ± 625	NS
Macrosomia (> 4000 gr) (%)	17.3	40.3	2	< 0.01
Mean 1 minute Apgar	8.6 ± 0.3	8.1 ± 0.3	8.8 ± 0.3	NS
Mean 5 minute Apgar	9.6 ± 0.3	9.0 ± 0.3	9.7 ± 0.3	NS
Hypoglycemia (< 30 mg per 100 ml) (%)	10.9	24	1.5	< 0.01
Hypocalcemia (< 8 mg per 100 ml) (%)	4.5	8.4	0	NS
Hyperbilirubinemia (> 15 mg per 100 ml) (%)	8.3	12.3	4	NS
Malformations (%)	0	0.65[b]	0	NS

[a] Statistical analysis by Students' t test.
[b] Ventricular Septal Defect.
NS Not significant.

trasound imaging starting at 8 weeks gestation (PEDERSEN and MOLSTED-PEDERSEN, 1981) for early detection of fetal anomalies and estimation of fetal growth rate for identification of macrosomia (OGATA et al., 1980), daily fetal movements count after 20–22 weeks gestation (SADOVSKI et al., 1974) and weekly electronic fetal heart rate monitoring, as well as biophysical profile, after 32 weeks gestation (MANNING et al., 1980). Timing and route of delivery were individualized according to maternal and fetal conditions.

Results

Table 63 presents metabolic control in the 310 women with gestational diabetes. Physiologic blood glucose levels and normal HbA$_1$ values were found in patients attending our clinic prior to pregnancy and these values were maintained at conception and throughout the pregnancy. In the non-attending group disturbed glucose homeostasis was detected already in the first trimester and deteriorated significantly during gestation. Mean OGTT did not differ significantly in both groups.

Maternal information concerning the 310 gravidae is presented in Table 64. Non-attenders were younger and of lower parity. The weight gain, frequency of preeclampsia and Caesarean delivery in these patients was higher than in the preconception attending group.

Neonatal data are presented in Table 65. Infants of non-attending mothers prior to conception had possibly higher mean birth weight and a significantly higher rate of macrosomia (40.3%; $p < 0.01$). Hypoglycemia was significantly more frequent in infants of non-attenders (24%; $p < 0.01$), hypocalcemia and hyperbilirubinemia were seen in both groups. Congenital anomalies appeared in the offspring of a non-attending mother; none were found among the others. There were no spontaneous abortions and there was no perinatal mortality.

Discussion

Increased perinatal mortality and morbidity has been reported in pregnancies complicated by gestational diabetes (O'SULLIVAN et al., 1973; MESTMAN, 1980; COUSTAN and IMARAH, 1984; COUSTAN and LEWIS, 1978; Summary and Recommendations of the 2nd International Workshop-Conference of Gestational Diabetes Mellitus, 1985).

Early detection, strict prenatal care and adequate metabolic control may reduce the risks of perinatal mortality and morbidity in these patients to levels comparable to those found in the general neonatal population (COUSTAN and IMARAH, 1984; COUSTAN and LEWIS, 1978; TYSON, 1971). In fact gestational diabetes is impossible to diagnose in nonpregnant women; the metabolism not yet being influenced by the stress of pregnancy, and OGTTs during the preconceptional phase are unreliable. Thus, the fact remains that gestational diabetics are not identified for high-risk management, but are initially followed as normal healthy obstetric patients.

The growing recognition that good health for pregnancy begins prior to gestation provides a compelling case for the institution of preconceptional counseling in these patients. Moreover, any disturbance of glucose homeostasis might be detected early in its course, particularly in preparation for any future pregnancy, a time when most of these women are not currently under medical supervision.

The aims of our preconception clinic were as follows:

- To assess individual patients' fitness for pregnancy, special attention being paid to hypertension and infectious disorders.
- To offer advice on contraception counseling in order to ensure that pregnancies were planned.
- To identify and treat gynecological problems.
- To evaluate glucoregulation prior to conception and detect diabetes early in it's course, thus to consider appropriate intervention policies prior and throughout gestation.
- To determine the time of conception.
- To obtain maximum cooperation from patients and their partners. In fact SBGM contributed to achieve their active participation in their own management.

Our study reveals a good correlation between preconceptional counseling and improved glucose homeostasis during pregnancy, as well as significantly favorable perinatal outcome. Attending patients maintained the preconceptional physiologic blood glucose levels and normal HbA_1 values throughout the pregnancy, whereas disturbed glucose homeostasis was detected in the first trimester and deteriorated significantly during gestation in non-attending patients. In fact improved motivation in the former group led to close adherence to their strict prescribed diet, thus excessive weight gain was avoided and glucoregulation improved. It seems that this factor might have played an important role in the causation of more frequent maternal complications in non-attending patients, mainly preeclampsia and a significantly higher rate of Caesarean deliveries. Furthermore, the rate of macrosomia among infants of non-attending mothers was significantly higher, a contributing factor to the elevated Caesarean section rate; also, hypoglycemia, hypocalcemia and hyperbilirubinemia were more frequent.

Our experience with preconceptional counseling in gestational diabetes patients is that it contributed to improved management of pregnancy by achieving better glucoregulation throughout the pregnancy. Improved motivation was in fact noted on the part of our patients, towards better control of blood glucose levels within the normal range of pregnancy, and close cooperation between those most concerned in the management of the pregnancy. It seems that our approach led to the successful management and excellent results in this series. We therefore recommend that all women in the reproductive age group have regular health check-ups and that appropriate evaluation for glucose intolerance should be included as part of a prospective family planning.

12

The Use of Prophylactic Insulin in Women with Gestational Diabetes

D. R. *Coustan*

Brown University Program in Medicine, Providence, Rhode Island, U.S.A.

Introduction

Between 1 and 5% of pregnant women develop a disturbance of carbohydrate metabolism known as gestational diabetes (CARPENTER and COUSTAN, 1986). This abnormality is pathophysiologically similar to type II diabetes in that its main feature is insulin resistance, presumably induced by the contrainsulin hormones of pregnancy, and possibly also by placental insulin degradation. In addition, many investigators have also reported that women with gestational diabetes manifest a reduced insulin response to a glucose load when compared with normal pregnant subjects (BLOHME et al., 1980; FREINKEL et al., 1985; KÜHL et al., 1985).

Diagnosis of gestational diabetes

Whatever the mechanism, this disorder is diagnosed by a glucose tolerance test in which at least 2 of 4 values exceed the mean by two standard deviations or more. Agreement is lacking as to the most appropriate glucose challenge and criteria for abnormality on the glucose tolerance test in pregnancy. In our center, we utilize the 100 gram oral challenge as proposed by O'SULLIVAN and MAHAN (1964). The threshold values have been modified from the original venous whole blood, Somogyi-Nelson technique to the current plasma, glucose oxidase methodology (CARPENTER and COUSTAN, 1982). If two or more of the following values are met or exceeded, the patient is considered to have gestational diabetes:

- Fasting   95 mg/dl
- 1 hour   180 mg/dl
- 2 hours   155 mg/dl
- 3 hours   140 mg/dl

Indications for therapeutic insulin

An abnormal glucose tolerance test is not necessarily accompanied by clinically significant hyperglycemia in the course of a normal day, when no challenge of concentrated glucose is encountered. A portion of women with gestational diabetes, despite following a therapeutic diet regimen, will manifest fasting and/or postprandial hyperglycemia, exceeding the goals for glycemic control in overt diabetics. We presume that such women are at the same increased perinatal mortality risk as similarly uncontrolled overt diabetics (KARLSSON and KJELMER, 1972), and we treat them with therapeutic insulin to reduce their hyperglycemia, and thus their perinatal mortality risk. However, the majority of women with gestational diabetes do not manifest such clinically evident hyperglycemia, and thus do not require insulin to prevent perinatal loss. It is with these patients that the remainder of this discussion will be concerned.

Theoretical causes of fetal macrosomia despite relative maternal euglycemia

Many recently published series' of gestational diabetic pregnancies report that, despite the maintenance of relative euglycemia in the mothers, macrosomic babies are born more frequently than in the general population (Fadel and Hammond, 1982; Miller, 1983; Widness et al., 1985). The Pedersen hypothesis (1977) would suggest that fetal macrosomia is directly related to increased fetal insulin levels, and that this fetal hyperinsulinemia occurs secondary to stimulation of the fetal pancreas by insulin secretogogues crossing the placenta from the maternal to the fetal circulation. If this is true, then it is likely that either

● our definition of "good" metabolic control does not demand low enough blood glucose values,
● we are not measuring glucose often enough in our patients, and thus we do not know how "good" their control really is, or
● insulin secretogogues other than glucose are important causes of fetal hyperinsulinemia.

Any or all of the above possibilities may be important, and evidence exists to support all three. For example, in an uncontrolled series published in 1980, Roversi treated 235 gestational diabetic women, whose fasting blood glucose was ≤ 130 mg/dl, with maximal tolerated doses of insulin. The incidence of macrosomia (> 4000 grams) was 2.6%, and the incidence of operative delivery (primary cesarean section, vacuum extraction, or forceps) was 14.5%. There were also 43 subjects with gestational diabetes but fasting blood glucose levels > 130 mg/dl. These individuals were lumped together with Class B and C individuals, and the incidence of macrosomia in this group was 6%, while operative deliveries were necessary in 28%. Although it is impossible to tell which of these macrosomic babies and operative deliveries were among Class A, as opposed to B and C, individuals, if we assume that all were from the Class A subjects, the overall macrosomia rate among gestational diabetics would only increase to 5%. Of interest in this study is the fact that mean blood glucose levels were approximately 90 mg/dl prior to insulin treatment, and approximately 70 mg/dl with the subjects on the "maximal tolerated dose" of insulin. Thus, a lower definition of "good" metabolic control was associated with a low incidence of macrosomia.

In a recent report, Tallarigo et al. (1986) demonstrated a correlation between the 2 hour plasma glucose determination of the

Table 66. Incidence of large-for-gestational age babies among 249 pregnant women with normal 100 gram oral glucose tolerance tests (Tallarigo et al., 1986)

	1-hour plasma glucose on OGTT:		
	< 100 mg/dl	100–119 mg/dl	120–164 mg/dl
N	151	58	40
Macrosomia (≥ 4000 GMS)	15 (9.9%)	9 (15.5%)	11 (27.5%)[a]

[a] p < 0.01.

Table 67. Effect of increased frequency of glucose monitoring on birth weight (Goldberg et al., 1986)

| | Protocol | |
	Self-monitoring	Control
N	58	58
Macrosomia ($\geq$ 4000 GMS)	5 (9%)[a]	14 (24%)[a]
Large-for-gestational-age ($\geq$ 90%ILE)	7 (12%)[b]	24 (41%)[b]
Required insulin	29 (50%)[c]	12 (21%)[c]

[a] $p < 0.05$.
[b] $p < 0.005$.
[c] $p < 0.01$.

Table 68. Randomized trial of prophylactic insulin (O'Sullivan et al., 1966)

| | Treatment group | | |
| | Gestational diabetic | | Normal control |
	Insulin[+]	No insulin	
N	305	306	324
Babies > 9 pounds	13 (4.3%)[a]	40 (13.1%)	12 (3.7%)

[+] 10 units NPH insulin qAM.
[a] $p < 0.01$.

100 gram oral glucose tolerance test and fetal macrosomia, even among pregnant women with "normal" glucose tolerance tests (Table 66). Thus, even our definitions of gestational diabetes may not be strict enough to detect all pregnancies at risk for fetal macrosomia induced by maternal hyperglycemia.

The second possibility, that more frequent measurement of circulating glucose levels may improve outcome, was addressed in the study of Goldberg et al. (1986). These investigators compared a group of 58 women with gestational diabetes who did daily self glucose monitoring with a historical control group who had circulating glucose levels measured weekly. Insulin was prescribed if fasting blood glucose exceeded 95 mg/dl or one-hour postprandial values exceeded 120 mg/dl. As shown in Table 67, the incidence of macrosomic ($\geq$ 4000 grams) and large-for-gestational-age ($\geq$ 90th percentile) infants was significantly lower in those monitored more frequently. It should be noted that 50% of the home monitored patients, but only 21% of those in the control group, required insulin.

The possibility that glucose is not the only important fetal pancreatic secretogogue is supported by the positive relationship between maternal branched chain amino acid levels and amniotic fluid C-peptide levels in diabetic pregnancies (Persson et al., 1986) and by a correlation between birthweight and maternal levels of alanine, serine and leucine in pregnancies complicated by gestational diabetes (Freinkel and Metzger, 1979).

Prophylactic insulin results of randomized clinical trials

In 1966, O'SULLIVAN and his co-workers tested the hypothesis that administering insulin to women with gestational diabetes who had normal blood glucose levels would lower the incidence of fetal macrosomia (Table 68). In a randomized trial, 305 gestational diabetic women were treated with 10 units of NPH insulin each morning from the time of diagnosis until delivery, while 306 were randomized to an untreated group. There was also a "normal control" group, consisting of 324 nondiabetic pregnant women. Using a birthweight of 9 pounds as the criterion for macrosomia, O'SULLIVAN found that 13% of the untreated gestational diabetic women gave birth to big babies, as opposed to 4% of those treated with prophylactic insulin and 4% of normal controls. Perinatal mortality rates were increased in both treatment groups as compared to normal control subjects, but subsequent reanalysis of the data (O'SULLIVAN et al., 1974) showed a significant reduction in perinatal mortality when gestational diabetic pregnancies in women aged 25 years or more were treated with prophylactic insulin. Unfortunately, data on mode of delivery and birth trauma were not included in this report.

In 1978, we reported another randomized clinical trial in which 72 women with gestational diabetes, diagnosed before the 36th week of gestation, were randomized to treatment with a starting dose of 20 units NPH and 10 units regular insulin plus a diabetic pregnancy diet, to diet treatment only, or to neither of the two modalities (COUSTAN and LEWIS, 1978). Using a birthweight of 8.5 pounds as the criterion for macrosomia, we found that 50% of the untreated women gave birth to large babies, as compared to 36% of the diet-treated and only 7% of the insulin-treated subjects (Table 69). Similar results were apparent when birthweight was expressed as a percentile, correcting for gestational age, gender, birth order, maternal height, and maternal midpregnancy weight. The series was too small to show any difference in mode of delivery or birth trauma. Recently, PERSSON's group (1985) has published another randomized clinical trial, in which 105 women with gestational diabetes were treated with diet and 97 with prophylactic short or intermediate acting insulin in a starting dose of 8–12 units/day. Macrosomia was expressed as birthweight > 90th centile for Swedish standards, and occurred in 13.3% of diet treated subjects and 11.3% of those treated with insulin, a difference which was not statistically significant. Mode of delivery and birth trauma were not listed, but the incidence of other types of neonatal morbidity were similar in the two groups. Although this study did not demonstrate any advantage for prophylactic insulin, it should be noted that patients in both groups did

Table 69. Randomized trial of prophylactic insulin (COUSTAN and LEWIS, 1978)

	Treatment group		
	Prophylactic insulin[a]	Diet	Untreated
N	34	11	27
Birth weight > 8.5 pounds	2 (7%)[b]	4 (36%)	17 (50%)
Birth weight > 75th %ile	3 (11%)[c]	3 (27%)	16 (47%)

[a] 20 Units NPH plus 10 units regular qam.
[b] p < 0.025 vs each of the other groups.
[c] p < 0.01 vs untreated group.

home glucose monitoring six times daily for three days each week, and insulin was begun if fasting glucose exceeded 7 mmol/1 or one hour postprandial values exceeded 9 mmol/1. Thus, these patients were intensively monitored, similar to those in the previously described series of GOLDBERG (1986). Furthermore, approximately 20% of the subjects in each group were entered into the study after the 36th week of gestation, at a time when it seems unlikely that macrosomia could be prevented. Nevertheless, of the three randomized trials in the literature, two support and one refutes the benefit of prophylactic insulin for women with gestational diabetes.

Retrospective data

Because the above data had demonstrated, to our satisfaction, that prophylactic insulin could reduce the incidence of macrosomia among gestational diabetic pregnancies, but had not addressed the probability of a reduction in macrosomia being associated with a parallel reduction in the clinically important variables of operative delivery and birth trauma, we recently looked at 445 consecutive women with gestational diabetes delivering at the Yale-New Haven Medical Center over a 5 year period (COUSTAN and IMARAH, 1984). This was not a randomized trial; the mode of therapy was chosen by the clinician managing the case. All charts were reviewed, and it was found that 115 women were treated with *prophylactic* insulin, initiated with the intent to reduce the likelihood of macrosomia, without regard to maternal glucose levels. An additional 184 subjects were initially treated with diet only, the intent being to add insulin if hyperglycemia occurred despite the dietary therapy. The final 146 subjects were diagnosed as having gestational diabetes, but no specific treatment was prescribed by their clinicians. If a subject from any of the three groups required insulin or an increase in the pre-existing insulin dose because of hyperglycemia, she was still considered in her original treatment group, according to the intent of her clinician. Only patients diagnosed between 20 and 36 weeks' gestation were included in this study.

Birthweight outcomes for the three groups were similar to those reported in the randomized trials described previously, and are depicted in Table 70. Using 4000 grams as a definition for macrosomia, 18% of the untreated, 18% of the diet-treated, and only

Table 70. Retrospective study of prophylactic insulin (COUSTAN and IMARAH, 1984)

	Treatment group		
	Prophylactic insulin	Diet	Untrated
N	115	184	146
Birth weight > 4000 GMS	8 (7%)[c]	34 (18.5%)	26 (17.8%)
Birth weight > 90th %ILE	9 (7.8%)[d]	33 (17.9%)	32 (21.9%)
Operative delivery[a]	17/104 (16%)[c]	52/171 (30%)	39/137 (28%)
Birth trauma[b]	5 (4.8%)[c]	23 (13.4%)	20 (14.9%)

[a] Excluding repeat cesarean sections.
[b] Shoulder dystocia, fracture, soft tissue injury.
[c] $p < 0.05$.
[d] $p < 0.01$.

Table 71. Distribution of potential bias among three treatment groups (COUSTAN and IMARAH, 1984)

	Treatment group		
	Prophylactic insulin	Diet	Untreated
Degree of GTT abnormality (mg/dl ± SD)	75.4 ± 65.8[a]	48.0 ± 33.3	54.4 ± 41.6
Fasting plasma glucose on GTT (mg/dl ± SD)	100.9 ± 19.4[a]	92.3 ± 12.8	92.4 ± 17.5
Gestational age at birth (WKS ± SD)	38.9 ± 1.6	39.2 ± 1.5	39.3 ± 1.5
Midpregnancy weight (LB ± SD)	165 ± 43	160 ± 38	171 ± 44

[a] $p < 0.001$.

7% of the prophylactic insulin treated patients had macrosomic babies. If a corrected birthweight above the 90th centile was used, similar differences were seen.

We next looked at mode of delivery. Primary cesarean sections, midforceps, and vacuum extractions were considered as operative deliveries, while spontaneous and low forceps deliveries were considered nonoperative. Operative deliveries were performed in 28% of the untreated group and 30% of the diet treated group, but only 16% of the prophylactic insulin treated group.

Birth trauma included shoulder dystocia, fracture, cephalhematoma, or other soft tissue injuries. These occcured in 20% of untreated pregnancies, 13% of those treated with diet, and only 5% of those treated with insulin.

Because this was not a randomized trial, we wondered whether other confounding variables might account for the apparent benefit of prophylactic insulin in preventing macrosomia, operative delivery, and birth trauma. For example, perhaps the women treated with prophylactic insulin had gestational diabetes which was less severe, and thus would have had fewer macrosomic babies whatever the treatment regimen. We therefore analyzed the glucose tolerance test results for degree of abnormality, calculated by adding the amount by which the two most abnormal values exceeded the upper limits of normality (Table 71). In fact, the insulin-treated patients had *more severe* gestational diabetes than those in the other two groups, a potential bias which might to expected to work *against* the apparent benefit of prophylactic insulin. Similarly, the fasting plasma glucose level in the glucose tolerance test was significantly higher among the insulin-treated subjects. We therefore concluded that selection of less severe gestational diabetics for insulin treatment did not occur, and thus could not account for the observed outcome differences.

We looked at other confounding variables such as weight gain during pregnancy, midpregnancy weight, gestational age at diagnosis and at delivery, and duration of treatment. None could account for the observed differences in outcome. Finally, we wondered whether it would be possible to identify those pregnancies destined to result in macrosomic babies by looking at mean maternal glucose levels during the third trimester. Among the 169 gestational diabetic subjects with adequate data to evaluate, the mean plasma glucose of the 27 who delivered macrosomic babies was 99 ± 13 mg/dl, while that of the 142 who had babies under 4000 grams was 95 ± 15 mg/dl, not significantly different (Table 72). Thus, we found it impossible to identify candidates for prophy-

Table 72. Effect of potential sources of bias on macrosomia (Coustan and Imarah, 1984)

	Macrosomic		Large-for-dates	
	⩾ 4000 GMS	< 4000 GMS	⩾ 90 TH %ILE	< 90 TH %ILE
Plasma glucose (mean ± SD)	98.9 ± 13.4	94.6 ± 15.4	100.2 ± 14.4	94.3 ± 15.1
Fasting value on GTT (mg/dl)	96.8 ± 15.9	94.1 ± 16.8	96.7 ± 17.8	94.1 ± 16.1

No significant differences.

lactic insulin treatment by looking at maternal glycemia. Whether more frequent measurement of maternal glucose levels, or more stringent goals for metabolic control, would allow more rational selection of patients remains to be proven.

In work described elsewhere in this book, Weiss and his colleagues have suggested that amniocentesis with measurement of amniotic fluid insulin levels could be the basis for therapeutic decision-making in gestational diabetic pregnancy. We are not currently performing routine amniocentesis on pregnancies complicated by gestational diabetes, and so have no data to contribute on this issue. However, as more data become available from centers using this procedure, particularly if patients are randomized, it may become possible to more accurately identify those individuals with gestational diabetes who require insulin to prevent macrosomia. This may reduce the need for general prophylactic use of insulin in gestational diabetes.

Clinical strategy

Our present strategy is to inform newly diagnosed gestational diabetic women about the available data on prophylactic insulin, and then to give them the option of using it or not. Since we have little data indicating a beneficial effect on perinatal mortality rates, we do not pressure our patients to use prophylactic insulin. Some are frightened at the prospect of a daily injection, while others wish to do everything possible to decrease the likelihood of a cesarean section. If they opt for prophylactic insulin, we start at a dose of 20 units NPH and 10 units regular human insulin, mixed in the same syringe and given prior to breakfast each day. Circulating glucose levels are measured 3–4 times daily, at least once each week, and the dose is increased if fasting values exceed 100 mg/dl and/or 2 hour postprandial values exceed 120 mg/dl.

It is also reasonable to offer patients the option of frequent self blood glucose monitoring, with insulin therapy instituted if stringent criteria for normoglycemia are not met. Which option is more acceptable to a given patient may be determined by individual experiences and prejudices.

Risks

It is important to discuss the risks of any therapeutic modality. The obvious risk of insulin is hypoglycemia. In our experience, gestational diabetic women started on insulin therapy at a dose of 20 units NPH and 10 units regular after the 28th week are extremely unlikely to develop documented hypoglycemia. We have seen only two such cases in over 300 women. Nevertheless, we always warn our patients of this possibility, and give them instructions as to how to handle such symptoms. There are no apparent longterm risks to prophylactic insulin therapy. When O'SULLIVAN und MAHAN (1980) looked at 16 year follow-up of their patients from the earlier mentioned randomized trial, in which animal species insulin was used in half the subjects, they found that the incidence of diabetes was the same in both groups. In fact, those women treated with insulin during their pregnancies appeared to have less severe diabetes than did those in the control group. In addition, human insulin is now available, so that theoretical problems with the development of antibodies to animal insulins should be avoidable.

Conclusions

Perinatal mortality is no longer the criterion by which successful treatment of gestational diabetic pregnancy is measured. Current efforts should be aimed at reducing morbidity, particularly that related to fetal macrosomia. A number of options are available, and it is not yet clear which is most advantageous. It is evident, however, that it is no longer acceptable to diagnose gestational diabetes and then ignore the potential problems which may occur if no therapeutic strategy is adopted.

13

Insulin Treatment of Gestational Diabetes. The Basal Bolus Concept

H. M. H. Hofmann, P. A. M. Weiss and *F. Kainer*

Department of Obstetrics and Gynecology, University of Graz, Graz, Austria

Introduction

Gestational diabetes mellitus is caused primarily by peripheral insulin resistance and by a relative lack of the hormone. The gestational diabetic's insulin production can cover her basal insulin requirement – as evident in the often normal fasting blood glucose level – but insulin secretion after carbohydrate intake (bolus production) is delayed and decreased. The resulting pulsatile hyperglycemia causes an oversupply of glucose to the fetus, which eventually reacts to even slight glucose elevations by overproducing insulin. As a result of the materno-fetal concentration gradient, even more glucose is metabolized by the fetus. Because of this syphoning (the extent of which is not known), the fetal glucose turnover can no longer be estimated by the maternal blood glucose level. Hyperstimulation of the fetal pancreas, fetal hyperinsulinism, and biochemical fetopathy (see p. 18 f.) can occur even at seemingly normal maternal blood glucose values.

Consequently, insulin treatment of gestational diabetes is directed towards the needs of the fetus. The fetus often requires tighter metabolic control than the mother, and reacts to an elevated fuel supply increasingly by hyperinsulinemia and subsequently elevated amniotic-fluid insulin levels. Glucose oversupply to the fetus occurs mainly after a meal and should thus be prevented by preprandial administration of regular insulin.

The six-hour duration of regular insulin's action makes necessary the application of at least four separate doses, one of them at night. Such schedules are not readily accepted (and may be ignored) by the patient, and the lack of nocturnal insulin substitution can elicit exaggerated fetal insulin production through enhanced glucose syphoning.

Maternal insulin secretion is markedly inhibited by the administration of exogenous insulin (HOFMANN et al., 1986; see Fig. 18). This makes even the substitution of the basal requirement seem indicated. Administrating a long-acting insulin (Ultratard HM, Novo) at 10 pm covers the basal insulin requirement over almost 24 hours and obviates the need for a further nocturnal dose. The substitution for the basal insulin requirement of a single dose of long-acting insulin and the covering of postprandial requirements by regular insulin simulates physiologic insulin secretion (Fig. 21 A, p. 41).

Patients and methods

We treated 17 gestational diabetics with basal-bolus insulin therapy. Diagnosis and assignment were made according to the modified White classification are described on p. 6.

Eleven patients had impaired glucose tolerance and an elevated amniotic-fluid insulin level (Class AB). In these the fetus presented the indication for insulin treatment. Six patients were Class Bo gestational diabetics who required insulin for a maternal indication additionally (previous oral antidiabetic medication or a mean blood glucose over 130 mg/dl). Table 73 summarizes the demographic information.

Eight blood glucose values within 24 hours (6, 9 am, noon, 3, 6, 9 pm, midnight, and 3 am) at weekly to daily intervals were performed before the initiation of insulin treatment, during its course, and after its discontinuation in the childbed. The measurements were done by an enzymatic method (Gluc DH® Merck) using capillary blood samples.

If the results of the oral glucose-tolerance test or of a 24-hour blood glucose profile were abnormal (p. 24), we began dietary treatment and performed an amniotic-fluid insulin determination between the 28th and 30th gestational weeks. The diet was planned

and the patient instructed by a dietician according to the guidilines discussed on p. 30 f. The patient was also advised to continue the diet after completion of the pregnancy for weight reduction.

Insulin treatment was indicated if the amniotic-fluid insulin exceeded the normal range (Class AB; Fig. 1, p. 6) or if the criteria for Class Bo were met (see p. 6). It consisted of three doses of regular insulin (Actrapid HM, Novo) at the main meals (usually at 6 am, noon, and 6 pm) and one dose of depot insulin (Ultratard HM, Novo) at 10 pm. Actrapid was administered by an insulin applicator (Novo Pen®). The insulin-meal interval averaged 30 min.

Glycosylated hemoglobin (HbA1c) was measured every two weeks by column chromatography (BioRad Laboratories; normal range for pregnancy: $4.32 \pm 0.59\%$). At the same time the fructosamine content was determined by a colorimetric test (Roche; normal range for pregnancy: 2.22 ± 0.22 mmol/l DMF equivalents).

Initial planning of the insulin-schedule

Patients for whom insulin treatment was indicated were admitted to the hospital for five days. Insulin administration was begun after a 24-hour blood glucose profile (ten values) was drawn. Because of the anti-insulin hormones of pregnancy, as well as the usually present obesity and associated peripheral insulin resistance, the insulin requirement of AB and Bo diabetics is high. In addition,

AB diabetics and some Bo cases are begun on insulin during the period when the requirement is highest (30th to 36th week) so that a further increase of insulin is generally not required once an adequate initial dose has been found. The further insulin requirement of Bo diabetics, who are generally recognized earlier in pregnancy, is similar to that of higher White classes (B–D) (Weiss

Table 73. Demographic data of gestational diabetic patients

Class	AB	Bo	Total
Number	11	6	17
Age (years)	30.2 (6.8)	35.5 (7.4)	32.1 (7.3)
Height (cm)	164.2 (7.8)	165.2 (11.4)	164.5 (8.9)
Pregnancy weight (kg)	82.9 (14.4)	77.7 (26.7)	81 (19)
Weight gain (kg)	10.2 (5.3)	11.3 (6.9)	10.6 (5.7)
Multiparae	6	4	10
Previous infant over 4,000 g	2	1	3

Mean (± standard deviation).

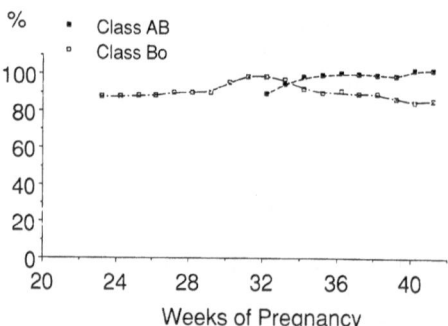

Fig. 34. Dynamics of mean insulin requirement during pregnancy. 100% = insulin requirement at maximum

and Hofmann, 1984). In AB diabetics the insulin requirement does not decrease, even at the end of gestation, while a small such decrease can be observed in Bo patients (Fig. 34). Fig. 34 also shows the insulin requirement of Bo patients during pregnancy.

Insulin dosage is increased gradually in AB and Bo patients to avoid maternal hypoglycemia and consequently exaggerated hypoglycemia in the fetus due to its hyperinsulinemia. The initial dose depends on the patient's weight, the preceding 24-hour blood glucose profile, and the amniotic-fluid insulin value. The average total requirement is about 1 unit per kg actual body weight. The indication for insulin therapy should be

reassessed if the requirement is less than 30 U. On the other hand, extremely obese women can have a requirement greatly exceeding 1 U/kg. Daily-doses of over 160 U are not uncommon. These high doses are needed to overcome peripheral insulin resistance.

The administration of exogenous insulin reduces the patient's own insulin secretion by about half. This is reflected in the decline of plasma C-peptide levels during insulin therapy to half the initial values, only to reassume the initial levels once insulin is discontinued in the childbed (Fig. 18). As expected, the serum insulin concentration rises during exogenous insulin administration and declines to its initial level after insulin is discontinued (Fig. 18).

Excess fetal insulin secretion is inhibited only when the maximum therapeutic insulin dose has been reached, thus alleviating the glucose oversupply. Depending on the initial amniotic-fluid insulin level, the normal range of amniotic fluid insulin is attained after two to four weeks. Table 74 shows the daily insulin schedule. The basal requirement (Ultratard) comprises about 30% of the total. This is considered when calculating the distribution at the beginning of treatment. When switching from an oral antidiabetic, the oral medication must be discontinued and only small insulin doses given in

Table 74. Basal-bolus therapy: the daily distribution of insulin

	Regular insulin[a]		Depot insulin + Total		
	6 a.m.	noon	6 p.m.	10 p.m.	in 24 hours
Class AB	30.8%	19.2%	20.2%	29.8%	
Units	19.8 (5.7)	12.4 (4.7)	13.0 (4.9)	19.2 (5.3)	64.3
Range (units)	8–36	4–24	6–24	10–32	28–106
Class Bo	32.8%	20.0%	19.1%	28.1%	
Units	24.8 (12.4)	15.1 (7.7)	14.4 (7.3)	21.2 (7.3)	75.3
Range (units)	8–48	4–28	4–32	10–48	28–156

Mean (± standard deviation).
[a] Actrapid HM, + Ultratard HM.

the first few days to avoid a potentiating or overlapping effect and consequent hypoglycemia. Hypoglycemia is rare, even at very high daily insulin doses. In our series of 17 gestational diabetics, only six cases of mild hypoglycemia (sweating) were observed in 205 therapeutic weeks. Most occurred in the first few days of the treatment; the subjective symptoms were alleviated by a glass of milk. There was no case of hypoglycemic coma.

Despite high insulin doses, our patients showed no excess weight gain during pregnancy (Table 73). In fact, their weight gain was less than the mean weight gain of metabolically healthy gravidae. However, the prepregnancy weight of most gestational diabetics is clearly above normal (Table 73).

During her initial hospital stay, the patient is instructed intensively in the use of the insulin applicator (Novo Pen®), the insulin syringe, self-checking her blood glucose, and the diet. Planning the insulin schedule on an outpatient basis can be done for patients who have undergone instruction in a previous pregnancy. When an adequate insulin schedule has been worked out, and when the patient is thoroughly familiar with the treatment, she is discharged from the hospital. Her blood glucose is monitored by self-checks at home (ten values once a week), and at ambulatory hospital or office visits. She is readmitted to the hospital only if pregnancy-related complications arise.

Depending on the estimated risk of the pregnancy, the patient is readmitted to the hospital at term or up to one week earlier for intensive prepartal monitoring. After the onset of periodic labor contractions, insulin is discontinued and the blood sugar checked hourly. Low values occur often during delivery and are managed by the infusion of a 5%-glucose solution. If a cesarean section is planned, the evening dose of long-acting insulin (Ultratard HM) is omitted and the morning dose adjusted to the current blood glucose value.

No further insulin treatment is given in the childbed. The 24-hour blood glucose profile is checked and the mother is encouraged to continue her diet and to lose weight after she has finished nursing. Even after previous oral antidiabetic medication, insulin is no longer necessary in the childbed. Nursing usually keeps the maternal blood glucose within acceptable limits, so that it is only after nursing, and only if the blood glucose rises again, that renewed oral medication is required.

To recognize the advent of IDD early, the patient is advised to undergo further 24-hour blood glucose profiles once a month. Preconceptional insulin therapy is urgently recommended for a Class Bo diabetic who is diagnosed before becoming pregnant (see also the chapter by GOLDMAN and colleagues).

Results

Table 75 summarizes the data concerning glucose tolerance and insulin therapy. As expected, groups AB and Bo differ in the initiation and duration of insulin therapy. Figs. 35 and 36 show the course of the blood glucose during gestation. Fig. 35 shows the mean blood glucose and SD of Class AB patients. Once insulin is started, the MBG falls and the SD decreases. After insulin has been discontinued in the puerperium, the MBG rises again during dietetic therapy (Fig. 37).

Fig. 36 shows the course of the MBG values

in the Class Bo patients. Here too insulin keeps the MBG in the normoglycemic range. In contrast to the AB patients, the SDs are larger. The lowest single blood glucose values were about 50 mg/dl in both groups. Hypoglycemic periods with danger of fetal retardation were rare (ABELL et al., 1976).

Fig. 37 summarizes the MBG values before and during insulin therapy as well as in the childbed. The differences are significant in both groups (AB $p < 0.05$, Bo $p < 0.01$, t-Test).

Fig. 38 shows the HbA1 c values before and

Table 75. Data on gestational diabetic patients

Class	AB	Bo	Total
Number	11	6	17
First examination (weeks)	23.5 (9.5)	18.5 (7.8)	21.8 (8.9)
oGTT (1 gm/kg) fasting[+]	99 (18)	93[a] (12)	97 (17)
at 1 hour[+]	184 (39)	182[a] (11)	184 (35)
at 2 hours[+]	122 (27)	173[a] (56)	130 (36)
Onset of insulin therapy (weeks)	32 (2.1)	20.7 (8)	28 (7.4)
Duration of insulin therapy (weeks)	7.7 (3.2)	20 (8.8)	12.1 (8.2)

Mean (± standard deviation).
[a] oGTTs performed only in two cases.
[+] mg/dl.

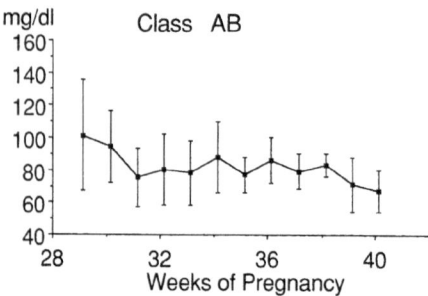

Fig. 35. Mean blood glucose values and SD in 11 Class AB diabetics in the course of pregnancy

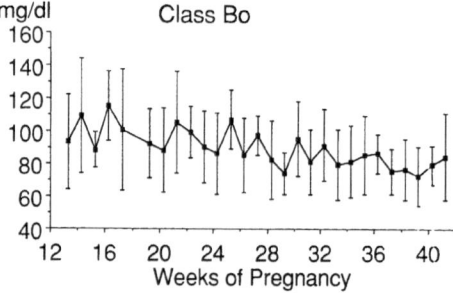

Fig. 36. Mean blood glucose values and SD in 6 Class Bo diabetics in the course of pregnancy

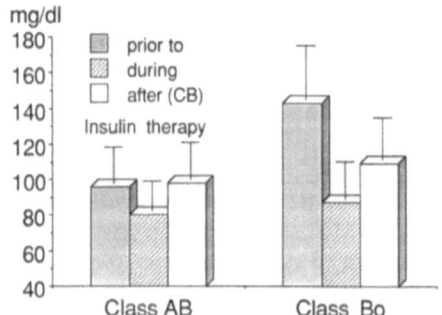

Fig. 37. Mean blood glucose prior to and during basal-bolus-insulin therapy as well as in childbed

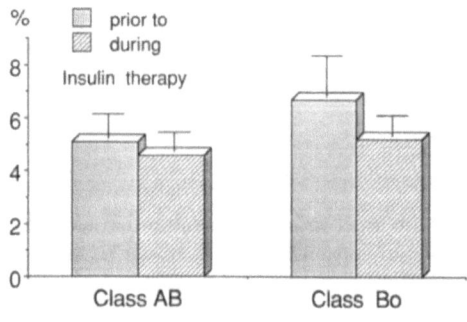

Fig. 38. Mean HbA1 c level and SD prior to and during basal-bolus insulin therapy

during insulin therapy. In Class AB, the HbA1 c values before insulin therapy differed significantly from the normal range, but there was no significance between the values before and during insulin therapy. This is because a relevant metabolic disorder in gestational diabetes begins in the second half of gestation. Since the "memory-like

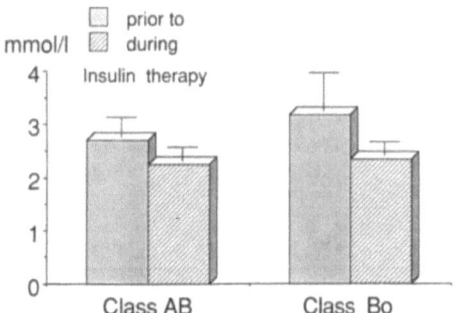

Fig. 39. Mean fructosamine level and SD prior to and during basal-bolus-insulin therapy

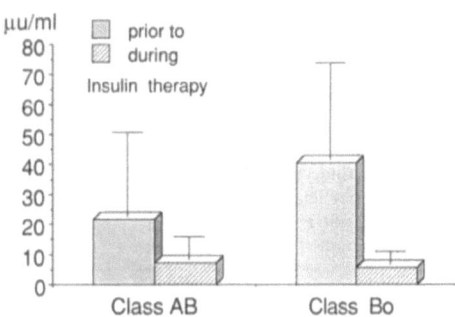

Fig. 40. Mean amniotic-fluid insulin level and SD prior to and during basal-bolus-insulin therapy

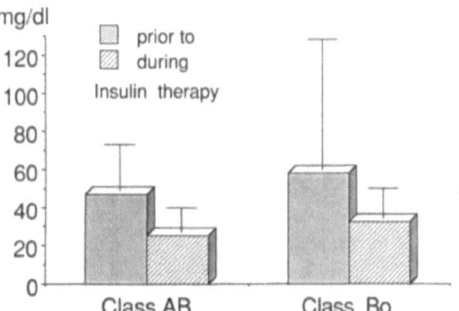

Fig. 41. Mean amniotic-glucose level and SD prior to and during basal-bolus-insulin therapy

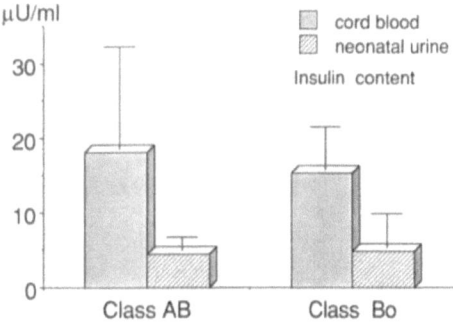

Fig. 42. Mean insulin level and SD in cord blood and neonatal urine

system" of glycohemoglobin surveys the last four to six weeks, the glycosylation of hemoglobin is not yet completed at the time of diagnosis.

Fructosamine, the glycosylation product of the amines, is a "memory-like system" reviewing about 2 weeks, so that this value is more useful for evaluating metabolic changes than is the HbA1c. This is confirmed by the significantly elevated fructosamine level and its decline after the initiation of insulin treatment (Fig. 38 (AB $p < 0.01$, Bo $p < 0.05$, t-Test). The glycohemoglobin level of Class Bo diabetics is markedly higher than of AB patients since the metabolic disorder becomes relevant earlier in gestation or even precedes it. In addition, the average duration of insulin treatment of these patients is longer so that a

marked decline of HbA1c becomes apparent. Also, in Bo diabetics, the fructosamine content is significantly reduced during insulin therapy ($p < 0.05$, t-Test). Comparing the course of HbA1c and fructosamine during pregnancy, one recognizes that the fructosamine level responds to insulin therapy sooner and that it falls to the normal range earlier.

Neither of the glycosylated proteins is a reliable indicator of fetopathy since, as with the oGTT, the values of diabetics with and without hyperinsulinemic children overlap. As discussed on p. 25, only the amniotic-fluid insulin level is proof of fetopathy and thus specific for the decision of whether diet or insulin treatment should be begun.

Fig. 40 and 41 show the insulin and glucose content of the amniotic fluid before and dur-

Table 76. Complications in pregnancy and mode of delivery (%)

Class	AB	Bo	Total
Number	11	6	17
Preeclampsia	5 (45)	2 (33)	7 (41)
Urinary tract infection	3 (27)	2 (33)	5 (29)
Preterm labor (Beta-mimetic agents)	0	1 (17)	1 (6)
Cesarean section	3 (27)	3 (50)	6 (35)

Table 77. Indication for cesarean sections

Class	Indication	Section
AB	Multiple sclerosis, preeclampsia	primary s.
AB	Advanced maternal age, breech presentation, preeclampsia	primary s.
AB	St. p. cesarean section, breech presentation preeclmpsia	primary s.
Bo	Preeclampsia, prim. labor dystocia, intrauterine asphyxia	secondary s.
Bo	Breech presentation, premature rupture of the membrans, advanced maternal age	primary s.
Bo	Two previous neonatal deaths	primary s.

Table 78. Gestational diabetes. Neonatal data

Class	AB	Bo	Total
Number	11	6	17
Males	4	4	8
Weight (g)	3,399 (490)	3,412 (402)	3,403 (448)
Height (cm)	49.4 (2.1)	49.2 (2.1)	49.3 (2.0)
Apgar score			
at 1 minute	9 (0)	8.7 (0.5)	8.9 (0.3)
at 5 minutes	10 (0)	10 (0)	10 (0)
pH in the cord blood			
arterial	7.23 (0.08)	7.24 (0.05)	7.24 (0.07)
venous	7.34 (0.07)	7.32 (0.06)	7.33 (0.07)
Duration of pregnancy	38.8 (2)	39.7 (1)	39.1 (1.8)
Preterm delivery	1	0	1
> 4,000 g	1	0	1
> 90%-ile	2	1	3

Mean (± standard deviation).

ing insulin treatment. Both parameters show a significant decline under insulin (p < 0.01, Wilkoxon-test).

The insulin levels in the cord blood (CBl) and in the newborn's urine were within normal limits, reflecting a satisfactory prepartal fetal metabolic condition (Fig. 42). Only one high CBl value was found, presumably a result of tocolytic therapy with beta-mimetic agents. The corresponding urinary insulin value on the first day was normal, indicating that this was not true hyperinsulinism.

Tables 76 and 77 list the route of delivery, the complications of pregnancy, and the indications for cesarean section. Five of the six sections were performed because of high obstetric risks and thus are not attributable to diabetes alone (Table 77).

Table 78 presents the fetal outcome. Neither in the AB nor Bo diabetics was a case of diabetogenic fetopathy with its corresponding morbidity found (see p. 26, 52 f.). There was no difference from the fetal outcome of metabolically healthy women.

In addition to the obstetric results, the following advantages of the basal-bolus insulin schedule, administered by the Novo Pen, are apparent:

● Appropriate scheduling of the daily insulin requirement imitates physiologic insulin secretion. Food intake is no longer necessary to counter unphysiologic insulin peaks and can thus be managed more liberally.

● The Novo Pen helps the patient accept the frequent insulin injections since it is easy to use and virtually painfree.

● Erroneous doses are avoided since the patient no longer has to calculate and draw up the needed insulin dose herself.

E. On the Fetus and Newborn
of Gestational Diabetic Women

14

Fetal Consequences of Maternal Diabetes

F. A. Van Assche, L. Aerts, and J. Verhaeghe

Department of Obstetrics and Gynecology, University Leuven, Leuven, Belgium

Introduction

Maternal diabetes compromises the fetal development and behaviour. Congenital malformations, unexpected intrauterine death, macrosomia and sometimes intrauterine growth retardation are known features. Functional and morphologic changes are present at both cellular and tissue levels. Although lethal consequences still occur, fetal and neonatal outcome have dramatically improved since tight diabetic control has been instituted before and during pregnancy.

During recent years progress in understanding the fetal consequences of diabetes and pregnancy was also made possible by the use of experimental models. Unfortunately, interspecies comparison is sometimes unfeasible. We have accumulated experience with streptozotocin induced diabetes and with the diabetic BB rat.

Human diabetes as well as experimental diabetes in animals involves an abnormal intrauterine milieu of the developing fetus. The function and structure of the fetal endocrine pancreas is mostly affected (VAN ASSCHE, 1970; AERTS and VAN ASSCHE, 1977).

The human fetal endocrine pancreas

The morphology of the endocrine pancreas

The fetal endocrine pancreas comprises about 5 percent of the total pancreatic volume. The pancreas derives from two buds of the primitive gut, a dorsal and a ventral primordium; these two buds fuse into one pancreas. The dorsal primordium develops into the tail and the superior part of the head of the pancreas; the ventral primordium develops into the lower part of the pancreatic head (PICTET and RUTTER, 1972).

The endocrine pancreatic system is formed out of the ductuli. In the first step there is formation of knots; afterwards these knots become independent from the ductuli, resulting in formation of endocrine islets. In these primitive islets the different cell types are intermingled. From about 16 weeks of fetal life we have the typical mantle islets with a central core of B cells (insulin producing) and a mantle of non B cells.

Identification of the different cell types is possible with electronoptical and immunocytochemical methods or with a combination of both methods (VAN ASSCHE et al., 1982). The B cells have large granules. The core can be electron dense (dark granules), sometimes even crystalized, or more electron lucent (light granules). Beta granules contain insulin. The A cells have smaller granules than the B cells. The core is electron dense with a surrounding closely fitting membrane. Some A cells have larger granules than other A cells. A granules contain glucagon. D cells have granules heterogenous in size and electrondensity; they contain somatostatin. P.P. (pancreatic polypeptide) cells have small and electrondense granules, containing pancreatic polypeptide. The lower part of the pancreatic head, originated from the ventral primordium, contains a high proportion of PP cells and is called the PP rich zone.

A fifth cell type (D_1 cell) has been described by us (VAN ASSCHE et al., 1982) with very small granules. The topography of this cell type is in close relationship with nerve fibers. The hormonal content of these granules is still unknown.

The fetal endocrine pancreas in diabetes

The islets of fetuses of diabetic mothers are more numerous and larger than islets of normal fetuses. Innumerable small islets composed of only a few cells can be demonstrated. The islet capillaries are congested. The B cells have an enlarged and hyperchromatic nucleus. In about 30 percent of the cases an infiltration by eosinophilic polymorphs is present in and around the islets. An increased amount of endocrine tissue (volume density of endocrine tissue) is pres-

ent in fetuses born to diabetic mothers compared with control fetuses. An increased volume density of endocrine tissue is also present in erythroblastosis fetalis and in anencephalics born to diabetic mothers, but only when a functional hypothalamo-hypophysis has been active.

An increased percentage of B cells is present in the islets of fetuses of diabetic mothers, but not in fetuses affected by erythroblastosis. B cell hyperplasia is also present in anencephalics from diabetic mothers, provided that in these fetuses a functional hypothalomo-hypophyseal system was present.

Not only is the volume density of fetal B cells increased in maternal diabetes, but also that of the PP cells. An increased amount of PP cells can be explained by the higher glucose content of the amniotic fluid, stimulating via the gastro intestinal tract the development of the PP cells (MILNER et al., 1984).

Data are summarized in Table 79.

The maternal diabetic environment results in an increased transfer of glucose and amino acids to the fetus. This results in an increased amount of hyperactive B cells, leading to an increased synthesis and secretion of insulin. Fetal hyperinsulism is responsible for fetal overgrowth.

Table 79. Morphometric data of the human fetal endocrine pancreas

	Volume density of endocrine tissue	Percentage B cells
Normal controls (n = 40)	5.1 ± 1.6	40 ± 7.5
Maternal diabetes (n = 10)	12.9 ± 4.2	63.8 ± 8.9
Erythroblastosis (n = 10)	9.0 ± 4.5	40 ± 8
Anencephalics without functional H. H. system of a diabetic mother (n = 8)	5.0 ± 1.6	38 ± 9.7
Anencephalics with a functional H. H. system of a diabetic mother (n = 7)	11.6 ± 4.9	59.2 ± 6.9

The cause of an increased amount of endocrine tissue, with a normal proportion of B cells, in infants affected by erythroblastosis is still unknown. The responsible factor(s) produce an equilibrated islet growth.

The study of the anencephalic provides interesting data. The basal development of the endocrine pancreas is normal. However, B cell hyperplasia and increased insulin secretion are only present in anencephalics with a functional hypothalamic-hypophyseal system born to diabetic mothers. Furthermore only those anencephalics are macrosomic.

Fetal consequences in experimental diabetes induced by streptozotocin

The antibiotic streptozotocin has a B cell cytotoxic effect and a slight carcinogenic effect (RAKIETEN et al., 1963; EVANS et al., 1965). Streptozotocin induces diabetes in a variety of animals. The cytotoxic action is not fully understood, however the non-B cells are not affected. Streptozotocin injected before conception affects reproduction even when insulin is substituted. Diabetes induced by subtotal pancreatectomy has also been shown to reduce fertility, attributed to changes in the oestrus (FOGLIA et al., 1970). In a preliminary study we have shown that the changes in the fetal endocrine pancreas were not different when streptozotocin was injected before pregnancy or on the day of mating.

Fetuses of *mildly diabetic* mothers show normoglycemia, hyperinsulinemia and hypoaminoacidemia. In their pancreas an increased amount of endocrine tissue is present; the B cells look hyperactive. These fetuses have an increased body weight. Fetuses of *severely diabetic* mothers on the contrary are characterized by hyperglycemia, hypoinsulinemia and low amino acid levels. The endocrine tissue in the pancreas is hyperplastic; the B cells are severely degranulated and show signs of exhaustion (AERTS and VAN ASSCHE, 1977).

An important question can be asked concerning the long lasting effect of the endocrine pancreas developed in an abnormal intrauterine environment. In 1979 AERTS and VAN ASSCHE (1979) showed that there are longterm consequences in the offspring of diabetic rats. A few years later the Phoenix group confirmed this effect in offspring born to diabetic Indians (PETTITT et al., 1982).

In our previous experimental study with streptozotocin (STZ) injected pregnant rats (first generation) we found that adult offspring (second generation) of these mildly diabetic rats developed gestational diabetes;

Table 80. Fetuses of the third generation at day 20 of gestation: Group 1. Control mothers; father born to a diabetic mother; Group 2. Mother to a diabetic mother; control father; Group 3. Mother and father born to a diabetic mother; Group 4. Mother and father are both controls

	Group 1	Group 2	Group 3	Group 4
Glycemia, mg/dl	37 ± 8 (14)	34 ± 14 (16)	44 ± 11 (14)	32 ± 9 (14)
Body weight, g	1.8 ± 0.3 (18)	1.9 ± 0.2 (31)	1.8 ± 0.2 (19)	2.0 ± 0.1 (18)
Volume density endocrine tissue	2.8 ± 1.5 (14)	5.3 ± 3.4[a] (16)	4.0 ± 0.6[a] (19)	2.0 ± 0.7 (18)
Volume density granulated B cells	0.6 ± 0.4 (23)	0.4 ± 0.3[a] (22)	0.4 ± 0.1[a] (14)	0.6 ± 0.3 (14)

Results are expressed in mean ± standard deviation; number of observations are in parentheses.
[a] Indicates a significant difference with group 4.

their fetuses (third generation) showed islet hyperplasia and B cell degranulation to the same degree as second generation fetuses of diabetic mothers (AERTS and VAN ASSCHE, 1979).

At least two questions remained unanswered in our experimental work; *First:* It could be possible that STZ (injected on the day of conception) had produced an effect on the genetic material of the offspring. In order to answer that question we mated from the second generation: 1. a control mother with a father born to a diabetic mother; 2. a mother born to a diabetic mother with a control father; 3. a mother and a father both born to a diabetic mother and 4. a mother and father who are both controls. In this experimental design 30 mg of streptozotocin was injected the day of mating in the first generation which had induced a mild diabetes (non fasting glycemia 256 ± 96 mg/dl versus 134 ± 37 mg/dl in the controls).

From the results shown in Table 80 it is clear that changes in the endocrine pancreas (islet hyperplasia and B cell degranulation) in fetuses of the third generation occur only when the mother of the second generation is an offspring of a diabetic mother.

From this new set of experiments we can conclude that even mild diabetes during pregnancy has an effect until the third generation offspring: this transmission is not of genetic origin, but must be due to diabetic influence during intrauterine life; indeed the changes in the fetal pancreas and the disturbed glucose tolerance are found only when the mother of the second generation was born to a diabetic mother, and not when the father was born to a diabetic mother.

A *second* important question is which mechanism can be responsible for the changes in the impaired response to an infusion with glucose? In order to answer this point plasma glucose and plasma insulin levels were measured in second generation female offspring during glucose infusion.

The first generation animals were injected on the day of mating with 22.5 mg streptozotocin inducing a mild diabetes (glyc. 179 ± 17 mg/dl) or with 30 mg streptozotocin inducing a severe diabetes (glyc. 685 ± 40 mg/dl); the glycemia of the controls was 141 ± 9 mg/dl.

The second generation offspring of these mildly and severely diabetic mothers, and the controls, were infused during three hours

Table 81. Plasma glucose and insulin concentrations and I/G ratios during the glucose infusion experiment in control rats (C), in offsprings of mildly diabetic rats (MDF) and in offspring of severely diabetic rats (SDF)

	0'	30'	60'	180'	240'	n
Glycemia						
C	110 ± 14	334 ± 45	334 ± 47	330 ± 44	132 ± 16	11
MDF[a]	115 ± 5	380 ± 40[a]	406 ± 89[a]	335 ± 46	147 ± 18	9
SDF[b]	115 ± 10	333 ± 51	332 ± 37[b]	360 ± 52	150 ± 19[a]	10
Insulinemia						
C	16 ± 7	96 ± 42	118 ± 74	214 ± 115	31 ± 17	11
MDF	28 ± 6[a]	94 ± 26	84 ± 23	125 ± 50[a]	34 ± 11	9
SDF[a, b]	25 ± 9[a]	119 ± 71[a, b]	118 ± 52[a, b]	239 ± 60[b]	53 ± 19[a, b]	10
I/G ratio						
C	0.15 ± 0.08	0.30 ± 0.18	0.37 ± 0.25	0.69 ± 0.45	0.23 ± 0.12	11
MDF	0.25 ± 0.06	0.25 ± 0.07	0.22 ± 0.08	0.35 ± 0.14[a]	0.23 ± 0.68	9
SDF[a]	0.22 ± 0.10	0.59 ± 0.17[a, b]	0.57 ± 0.18[a, b]	0.67 ± 0.16[b]	0.35 ± 0.12[a, b]	10

Results are expressed as mean ± standard deviation.
n The number of animals.
[a] Indicates a significant differences (p < 0.05) with the corresponding C group.
[b] Indicates a significant difference between SDF and MDF.

with a thirty percent glucose solution at a rate of 1.5 ml per hour, and blood samples were taken at 0, 30, 60, and 180 minutes, and at 60 minutes after stopping the infusion.

The data before, during and after glucose infusion for the three groups are given in Table 81.

After 24 hours of starvation with similar plasma glucose concentrations in the three groups, plasma insulin concentrations and I/G rations are equally elevated in the offspring of mildly and severely diabetic rats. By definition this situation must be considered as an insulin resistant state, which can be due to different factors at the prereceptor, receptor or postreceptor stage.

Offspring of mildly diabetic mothers respond to glucose infusion with almost similar insulin levels as the controls. Due to their basal insulin resistant state insulin secretion is insufficient to maintain glycaemia within the normal range.

A defect in glucose reception or in the transmission of the glucose stimulus to the secretory response (the stimulus secretion coupling) might be the origin of this deficient adaptation to the increased insulin demands after glucose stimulation. However other secretagogues (amino acids) could be involved. The offspring of severely diabetic mothers, in contrast do manage to maintain their glycemia levels within a normal range during the whole infusion time. In order to do so they need much higher insulin levels especially during the first hour of infusion. Insulin resistance, as seen in the fasting state, continues in the infusion stress situation.

Since all animals are reared and treated under the same standard conditions, and the only difference between them is the degree of maternal diabetes during pregnancy, the reason for the inpairment of glucose tolerance in adult life must be found in the abnormal intrauterine milieu during fetal life.

At the end of gestation, fetuses of mildly diabetic mothers display normoglycemia, hyperinsulinemia and increased sensitivity to stimuli, while fetuses of severely diabetic mothers in contrast display severe hyperglycemia, hypo-insulinemia and insensitivity to insulin stimuli (KERVRAN et al., 1978; AERTS and VAN ASSCHE, 1977).

In the same period, insulin receptors appear to be abundant and fully functional in the peripheral tissues of normal rat fetuses, while maturation of post receptor insulin "handling" (internalization and degradation) occurs only in these last days of intrauterine life (SODOYEZ-GOFFAUX et al., 1982). Insulin deficiency, as seen in fetuses of severely diabetic mothers, has been shown to induce insulin resistance in alloxan diabetic dogs (REAVEN et al., 1977). Hyperinsulinemia, as seen in fetuses of mildly diabetic mothers, can produce insulin resistance probably by a post receptor effect (MARSHALL and OLEFSKY, 1980). On the other hand, longstanding hyperglycemia, even mild, can impair islet function and B cell sensitivity to glucose and induce insulin resistance in man (DIMITRIADIS et al., 1985).

Alterations in number and sensitivity of insulin receptors, as well as some post receptor effects, are known to be reversed after normalisation of the inducing factors in adults. However it might be that in this critical period at the end of gestation, during extensive development of the endocrine pancreas, during extension of the insulin reception system and during maturation of the biochemical processes in the peripheral tissues, lasting damage can be caused to any of these structures by an abnormal intrauterine environment.

Therefore tight diabetic control during pregnancy not only reduces fetal and neonatal complications, but can also be of benefit for the further development of the offspring.

Table 82. Reproductive data at day 21 of pregnancy in the BB rat

	Diabetic (n = 17)	Control (n = 13)	p
Non-fasting glucose (mg/percent)	646 ± 33	189 ± 13	< 0.0001
Maternal weight at conception (g)	246 ± 8	250 ± 9	NS
Maternal weight gain during pregnancy (g)	75 ± 10	103 ± 5	< 0.05
Number of living fetuses per mother	10 ± 1	9 ± 1	NS
Number of resorptions per mother	2.3 ± 0.4	1.3 ± 0.5	< 0.01
Percent resorptions of total offspring	17.5 (n = 173)	12.5 (n = 111)	
Fetal weight (g)	3.40 ± 0.04	4.10 ± 0.05	< 0.0001
Placental weight (g)	0.59 ± 0.01	0.43 ± 0.08	< 0.0001
Fetal : placental weight ratio	6.07 ± 0.37	9.36 ± 0.49	< 0.0001

Data are presented as mean ± SEM.

Table 83. Fetal plasma and amniotic fluid, insulin and glucose

	Diabetic	Control	p
Plasma	(n = 13)	(n = 11)	
Insulin (μU/ml)	57.3 ± 5.1	95.3 ± 8.6	< 0.001
Amniotic fluid	(n = 17)	(n = 13)	
Glucose (mg/dl)	365 ± 42	36 ± 2	< 0.0001
Insulin (μU/ml)	12.0 ± 0.9	23.7 ± 1.9	< 0.0001

Data presented as mean ± SEM. (Statistical analysis performed by Mann-Whitney's μ-test.)

The fetal endocrine pancreas in the diabetic BB rat

The diabetic BB rat has been recently used by us in our experimental work. Diabetes in the BB rat occurs on an auto-immune basis; both cellular and hormonal immunologic disturbances have been recognized.

In our colony diabetes occurs at the mean age of 110 days. At the beginning of the disease lymphocytic infiltration of the islets (insulitis) is present. After a few days an end stage is produced with almost total destruction of the B cells, unmeasurable insulin levels and high glucose, lipid and amino acid levels in the peripheral circulation. Most of the animals are ketosis prone and need insulin for survival. Reproduction is severely affected, certainly in untreated animals. Table 82 shows the reproductive data.

Maternal weight gain is reduced in the diabetic animals. On the fetal side the lower fetal weight is the most striking phenomenon. The same feature has been found in severe streptozotocin induced diabetes (Aerts and Van Assche, 1977). Hydramnios is a constant finding and is proportional to the degree of diabetes.

Table 83 shows hyperglycemia in fetuses of diabetic rats, while plasma insulin and amniotic fluid insulin are decreased.

The fetal endocrine pancreata show degranulated and exhausted B cells. In the pancreas of fetuses of diabetic BB rats, clusters of lymphoid material are present.

An increased incidence of congenital anomalies associated with maternal diabetes is also an important finding. Recently (VERHAEGE et al., 1986) we have demonstrated important changes in the fetal and maternal Ca and vit D metabolism.

Conclusion

The fetal endocrine pancreas is severely affected in human diabetes as well as in experimental diabetes in the rat.

There seems to be a contradiction in the findings in the human situation compared to those in the experimental situation. Both in streptozotocin induced severely diabetic rats and in diabetic BB rats a degranulation of the fetal B cells was observed; moreover the fetal B cells had morphologic signs of exhaustion. Although this finding might be explained by species differences, it is also true that severe experimental diabetes was associated with glycemia levels up to 600 mg/dl. We were able to demonstrate fetal B cell degranulation in the human in one exceptional case of maternal diabetes with glycemia levels over 400 mg/dl (VAN ASSCHE et al., 1983). In mildly diabetic animals, fetal B cells were hyperactive associated with an increased plasma insulin level and increased fetal body weight.

Since a marked effect on the fetal B cells is demonstrated, the question of the long term consequences is important. We have clearly shown that a longterm effect exists which is only transmitted through the maternal line. We also suggest that insulin resistance originates during intrauterine life.

Our experimental work is well supported by human data. PETTITT et al. (1981) have shown a longterm diabetic effect in Pima Indians associated with obesity (insulin resistance). MARTIN et al. (1985) have found that the incidence of gestational diabetes is transmitted mostly through the female line. The diabetic pathogenesis of the BB rat is comparable to the human situation. Furthermore fetal consequences are similar. This experimental model opens exciting perspectives for a better knowledge of maternal and fetal complications in diabetes. However, a BB rat colony is a very expensive experience. Other disadvantages are the difficulty in obtaining a mild diabetic state and the heterogenity in the study groups.

Summary

Maternal diabetes affects the development and function of the fetal B cells. Fetal B cell hyperplasia explains hyperinsulinism and overgrowth in the human.

Experimental diabetes in the rat (streptozotocin, diabetic BB rat) has also a dramatic effect on the fetal B cell. In streptozotocin induced diabetes the changes in the fetal B cells have a long lasting effect operating through the female line. Pregnancy in the diabetic BB rat opens exciting perspectives for a better knowledge of maternal and fetal complications.

15

Macrosomia and Birth Trauma in Infants of Diet Treated Gestational Diabetic Women

J. L. Kitzmiller, L. A. Hoedt, E. P. Gunderson, T. S. Theiss, C. L. Ceresa, and A. M. Kitzmiller

Department of Obstetrics, Gynecology, and Reproductive Sciences, Children's Hospital of San Francisco, University of California, San Francisco, U.S.A.

Introduction

It is well recognized that untreated gestational diabetes is associated with an excess rate of stillbirths and a high frequency of large for dates (macrosomic) infants. The latter can lead to birth trauma for infants delivered vaginally, or high rates of cesarean sections. Treatment of gestational diabetes should be designed to prevent these complications. Standard treatment has consisted of a well balanced diabetic diet. What is not clear is whether all gestational diabetic women should be automatically treated with insulin. Management by diet alone with good control of blood glucose values has been shown to result in normal perinatal mortality (GABBE et al., 1977; ADASHI et al., 1979). However, perinatal morbidity may remain excessive with just dietary therapy. For example, in Gabbe's series the incidence of fetal macrosomia in the well controlled diet treated women was 20%.

KALKHOFF (1985) reviewed investigations of the effect of insulin therapy compared to diet treatment alone on the outcome of gestational diabetic pregnancies (Table 84). ROVERSI et al. (1980) and METZGER et al. (1981) observed low frequencies of large babies in women treated with diet plus insulin, and both investigators found that a fasting blood glucose (FBG) > 105 mg/dl was associated with a higher macrosomia rate among diet

treated women. O'SULLIVAN and co-workers reported a randomized controlled trial in 1975. The macrosomia rate was 14.8% in 108 infants of diet treated women, compared to 5.9% in 101 infants of women placed on insulin, without significant change of maternal blood glucose. OPPERMAN et al. (1975), GYVES et al. (1980), and COUSTAN et al. (1984) observed rates of macrosomia in diet treated gestational diabetics of 39%, 32%, and 18.5% respectively. The frequency of macrosomia was lower in the insulin treated group in all of these studies. The limitation of these studies was that little information was given as to the effectiveness of the dietary therapy in producing normoglycemia in the diet treated groups, and it is doubtful that women in those groups were monitored as intensively as those on insulin.

One reason that insulin therapy of all gestational diabetic women has not been universal was the reluctance of many physicians to use beef or pork insulin for temporary treatment of diabetic women, since almost all the women would develop insulin antibodies, and some of them might have severe insulin resistance if insulin therapy was needed for overt diabetes later in life. This problem has been obviated by the availability of human insulin for the temporary treatment of gestational diabetic women.

Table 84. Therapeutic results of insulin therapy in gestational diabetes mellitus

		Pregnancies (% large babies)	
		Diet only	Diet plus insulin
O'SULLIVAN	1975	108 (15%)	101 (6%)[a]
OPPERMANN	1975	128 (39%)	23 (4%)[a]
GYVES	1980	156 (32%)	27 (?)
COUSTAN	1984	184 (18.5%)	115 (7%)[a]

[a] $p < 0.05$
Modified from KALKHOFF, 1985.

With this background we sought to determine the perinatal outcome in a program based on intensive diet therapy for gestational diabetic women, in which insulin treatment was reserved for women who did not achieve normoglycemia on the diet.

Material and methods

Pregnant women were screened for carbohydrate intolerance by a 50 gram oral glucose loading test at 24–28 weeks gestation. If the one hour plasma glucose concentration exceeded 140 mg/dl, a full three hour 100 gram oral glucose tolerance test (GTT) was performed after three days of extra carbohydrate (approximately 70 grams) in the diet. If any two of the following values were exceeded (fasting 105 mg/dl, one hour 185 mg/dl, two hour 160 mg/dl, or three hour 140 mg/dl) a diagnosis of gestational diabetes was made and diet therapy was begun. The diet was based on 25–35 kcal/kg ideal body weight. Calories were distributed into three main meals and three snacks per day. The eating plan was designed to give 38–45% complex carbohydrate, 25–35% fat, and 20–28% protein. Types of food suggested for the diet were individualized for patient needs.
Patients underwent followup interviews with the nutrtionists to determine dietary adherence. Effectiveness of the diet on producing postprandial normoglycemia was tested by at least weekly postprandial blood glucose tests. Postprandial normoglycemia was defined as plasma blood glucose < 130 mg/dl one hour after eating, or < 105 mg/dl if tested two hours after eating (O'SULLIVAN

et al., 1974). If weekly postprandial values were borderline, or elevated, the women were trained in self blood glucose monitoring with reflectance colorimeters, and were instructed to test after each main meal, at least three times each day. If postprandial values were then above the norm insulin therapy was begun with human insulin, using a mixture of regular and intermediate insulin injected in the morning. Some patients added a second injection of regular and intermediate insulin in the evening. All insulin treated women performed self blood glucose monitoring, testing at least four times each day, FBG and postprandial (PPBG). In this group insulin doses were adjusted at least weekly at achieve FBG < 100 mg/dl and PPBG < 140 mg/dl.
For data analysis, gestational diabetic women were classified as A_1 if FBG in the GTT was < 105 mg/dl, as A_2 if FBG was 105–129 mg/dl, and as class B if FBG was ⩾ than 130 mg/dl. Macrosomia was defined as birth weight above the 90th centile for gestational age and sex, using norms established for babies delivered in California (WILLIAMS et al., 1982). These babies are indicated as large for gestational age (LGA). More severe macrosomia was defined as birth weight ⩾ 4200 grams. Birth weight ra-

Table 85. Maternal age and weight distribution at onset of pregnancy in 2 treatment groups of gestational diabetic women

Pregnancies	Diet only (87)	Insulin added (38)	Total (125)
Maternal age			
< 25 years	7 (8.0%)	1 (2.6%)	8 (6.4%)
25–29 years	16 (18.4%)	7 (18.4%)	23 (18.4%)
30–34 years	39 (44.8%)	19 (50.0%)	58 (46.4%)
≥ 35 years	25 (28.7%)	11 (28.9%)	36 (28.8%)
Maternal weight			
< 150 pounds	58 (66.7%)	16 (42.1%)	74 (59.2%)
150–179 pounds	15 (17.2%)	10 (26.3%)	25 (20.0%)
≥ 180 pounds	14 (16.1%)	12 (31.6%)[a]	26 (20.8%)

[a] $p < 0.05$.

tio (BWR) was defined as the actual birth weight divided by the 50th centile value for gestational age and sex. Neonatal hypoglycemia was defined as serum glucose < 30 mg/dl during the first 24 hours of life. Differences of means of normally distributed variables between groups were tested by unpaired t-tests. Differences in frequencies of characteristics or outcomes among groups were tested by contingency table analysis, either chi square or Fisher's exact test, depending upon the lowest expected number in each cell.

Results

Gestational diabetes was diagnosed in 125 pregnant women. Normoglycemia was attained with diet treatment alone in 87 women, and insulin treatment was added in 38 subjects. The distribution of maternal age and weight at onset of pregnancy is given in Table 85. Only 6.4% of subjects were less than age 25, and 28.8% had reached age 35 or greater. Maternal weight at the beginning of pregnancy exceeded 150 lbs in 40.8% of the gestational diabetic women, and a significantly higher frequency of overweight women required insulin therapy.

The relationship of FBG level on the GTT to the frequency of insulin treatment is illustrated in Table 86. Fasting blood glucose was less than 105 mg/dl in 65% of the subjects and only 10% of them required insulin. FBG was between 105–129 mg/dl in 25%, and 56% of them required insulin ($p < 0.01$). All of the women with FBG greater than 130 mg/dl required insulin therapy. Also in Table 86, mean maternal weight

at the beginning of pregnancy was significantly higher in the group requiring insulin therapy, but there was no significant difference in maternal age or maternal weight gain during pregnancy between these two groups. Of the women weighing < 150 lbs at the beginning of pregnancy, 21.6% required insulin therapy, compared to 43.1% of the overweight group ($p < 0.01$). On the other hand, insulin therapy was not more frequent in those women who gained more than 24 lbs during gestation. (37% in < 24 lb weight gain group, 23% in the ≥ 24 lb weight gain group.)

The characteristics of women with FBG 105–129 mg/dl on the GTT are listed in Table 87 and compared to women with normal FBG. There is no difference in maternal age between the two groups, but the group with elevated FBG had higher weight at onset, but significantly lower weight gain during gestation. The much higher frequency of insulin therapy has been noted for this

Table 86. Characteristics of 2 treatment groups in a population of gestational diabetic women

	Diet only (87)	Insulin added (38)	Total (125)
FBG			
< 105 mg/dl	73	8	81 (65%)
105–129	14	18	32 (25%)
≥ 130	0	12	12 (10%)
Maternal age, years	31.8 ± 4.7	32.1 ± 4.2	
Maternal weight, onset of pregnancy, pounds	142 ± 34	168 ± 52[a]	
Maternal weight gain, pounds	24.8 ± 11	22.0 ± 14	

[a] $p < 0.01$.
FBG fasting blood glucose on oral glucose tolerance test.

Table 87. Characteristics of gestational diabetic women with normal compared to elevated fasting blood glucose on the glucose tolerance test

	Fasting blood glucose on GTT	
	< 105 mg/dl	105–129 mg/dl
N	79	31
Maternal age	31.6 ± 4.6	32.8 ± 4.0
Maternal weight, onset	137 ± 36	177 ± 43[b]
Maternal weight gain	24.8 ± 9.5	19.4 ± 14.7[a]
% on insulin	10%	58%[b]
Birth weight ratio	1.01 ± 0.15	1.1 ± 0.15[a]

[a] $p < 0.05$.
[b] $p < 0.01$.

Table 88. Frequency of large for dates infants according to treatment required for gestational diabetes and to fasting blood glucose on the glucose tolerance test

FBG	LGA/total infants (%)	
	Diet only (87)	Insulin added (38)
< 105 mg/dl	8/73 (11.0%)	2/8 (25)
105–129	2/14 (14.0%)	5/18 (27.8%)
≥ 130	0/0	2/12 (16.7%)
	10/87 (11.5%)	9/38 (23.7%)

group, and the mean BWR for the infants of mothers in the insulin treated group was higher.

The lack of influence of FBG at the GTT on the frequency of macrosomia in both treatment groups is indicated in Table 88. LGA infants were observed in 11% of the diet treated women with normal FBG, compared to 14.0% in infants of the diet treated women with elevated FBG on the GTT. The reader should recall that the frequency of LGA in the control population by definition should be 10%. In the group which did not maintain normoglycemia with the diet therapy and required insulin treatment, frequency of macrosomia was 25% in those who had a normal FBG on the GTT, compared to 23.3% in those who had had an elevated FBG. The overall rate of macrosomia was 23.7% in the group in which insulin treatment was required, compared to only 11.5% in the group in which diet treatment sufficed to produce normoglycemia.

Tables 89 and 90 show the influence of maternal weight at the beginning of pregnancy on the frequency of macrosomia in the diet and insulin treated groups, respectively. Of the 30 women overweight at onset in whom diet therapy produced normoglycemia, 20.1% of the infants were LGA, compared to only 6.9% in the group weighing < 150 lbs (p 0.05). The same association was seen in the insulin treated group. On the other hand, maternal weight gain during gestation of > 24 lbs was not associated with higher frequencies of macrosomia in either treatment group.

By stepwise multiple linear regression analysis of maternal age, weight, and weight gain, the major influence on BWR in both the diet and insulin treatment groups was maternal weight at delivery giving correlation coefficients of 0.35 (p < 0.01) and 0.33 (p < 0.05), respectively. Surprisingly, in this

Table 89. Influence of maternal weight on fetal macrosomia in normoglycemic diet treated gestational diabetes woman

	AGA	LGA
Maternal weight at onset		
< 150 lbs	51	4 (6.9%)
≥ 150 lbs	23	6 (20.1%)[a]
Maternal weight gain		
< 24 lbs	37	4 (9.8%)
≥ 24 lbs	40	6 (13.0%)

[a] p = 0.05.

Table 90. Influence of maternal weight on fetal macrosomia in insulin treated gestational diabetes woman

	AGA	LGA
Maternal weight at onset		
< 150 lbs	15	1 (6.3%)
≥ 150 lbs	14	8 (36.4%)[a]
Maternal weight gain		
< 24 lbs	17	7 (28%)
≥ 24 lbs	12	2 (14.3%)

[a] p = 0.04.

sample of gestational diabetic women, there was no influence of maternal age on the frequency of macrosomia.

Gestational age and route of delivery are listed on Table 91 for women in each treatment group. Preterm deliveries occurred in 9.6% of this gestational diabetic population,

Table 91. Treatment for gestational diabetes, gestational age, and route of delivery

Gestational age	Diet only (87)	Insulin added (38)	Total (125)
< 37 weeks	9	3	12 (9.6%)
37–39	35	28	63 (50.4%)
40–41	39	7	46 (36.8%)
42	4	0	4 (3.2%)
Cesarean section	29 (33%)	16 (42%)	45 (36%)
Vaginal delivery	58	22	70
Shoulder dystocia	2 (3.4%)[a]	3 (13.6%)[a]	

[a] % of vaginal deliveries.

Table 92. Treatment for gestational diabetes and perinatal outcome

Infants	Diet only (87)	Insulin added (38)	Total (125)
LGA (> 90th centile)	10 (11.5%)	9 (23.7%)	19 (15.2%)
> 4200 gm	4 (4.6%)	4 (10.5%)	8 (6.4%)
Birth trauma	2 (2.3%)	3 (7.9%)	5 (4.0%)
Neonatal hypoglycemia (< 30 mg/dl)	3 (3.4%)	9 (23.7%)	12 (9.6%)

36.8% delivered at 40–41 weeks, and 3.2% were post dates, delivering at 42 weeks gestation. As expected, a lower percentage of insulin treated women were allowed to continue pregnancies to 40 weeks gestation. The frequency of cesarean section delivery was 33% in the diet treated group, compared to 42% in those who required insulin therapy. For comparison, the cesarean section rate during the years of this study in non-diabetic women at Children's Hospital of San Francisco was 23%. Of the 58 women in the diet treated group who delivered vaginally, there were two instances of shoulder dystocia (3.4%). This compared to three occurrences (13.6%) in the 22 vaginal deliveries of women who were treated with insulin.

Numbers of babies with macrosomia, birth trauma and neonatal hypoglycemia are listed on Table 92. Of the 87 infants of the diet treated mothers, only 4.6% had birth weights exceeding 4200 grams, compared to 10.5% of the 38 infants of women who required insulin therapy. During the time of this study, 4.9% of 1224 infants of non-diabetic women had birth weights > 4200 gm. In the diet treated group there were two instances of birth trauma (2.3%). There was one brachial palsy in one of the cases of shoulder dystocia and a cephalhematoma in another infant. In the insulin treated group there were three cases of birth trauma (7.9%). One infant who suffered shoulder dystocia at delivery developed transient brachial palsy, another infant had a cephalhematoma, and the third infant had a swollen ear from the forceps delivery which preceded shoulder dystocia. Neonatal hypoglycemia was seen in 3.4% of the infants in the diet treated group compared to 23.7% in the infants of the women who required insulin therapy.

Discussion

In this study diet therapy was successful in achieving postprandial normoglycemia in 70% of gestational diabetic women. Diet therapy in this group with normoglycemia produced low rates of macrosomia, birth trauma, and neonatal hypoglycemia. However, the frequency of large for dates babies was above normal in the overweight women who were normoglycemic on the diet. A similar result was reported by HOLLINGSWORTH and co-workers (ALGERT et al., 1985). This suggests the need for further studies to determine the factors contributing to macrosomia in this group, and the need for a large

randomized trial of diet plus insulin versus diet alone, with comparable monitoring of each group, to find out if insulin therapy will reduce the macrosomia rate in the overweight normoglycemic gestational diabetic women.

As expected, women requiring insulin therapy because the diet did not achieve normoglycemia had higher rates of cesarean section, macrosomia, and neonatal morbidity. Excessive maternal weight was also associated with almost all of the cases of macrosomia in the insulin treated group.

The results of this study are supported by

the randomized controlled trial of PERSSON and co-workers (1985). In their study 105 women were randomized to diet therapy and 97 to diet plus insulin. Of the diet treated group 14% needed insulin to achieve acceptable blood glucose control (FBG < 126 mg/dl and PPBG < 162 mg/dl). The two treatment regimens disclosed no differences regarding achieved degree of maternal blood glucose control or infant size at birth. Frequency of LGA infants was 13% in the diet treated group in the PERSSON study, which was very similar to our result.

The importance of monitoring of PPBG to assess the adequacy of the diet in gestational diabetic women was illustrated in the recent study of GOLDBERG and co-workers (1986). In a group of 46 diet treated gestational diabetic women who received only weekly monitoring of blood glucose, 26% of infants weighed more than 4000 grams at birth, compared to only 10.3% in diet treated women who maintained normoglycemia using self blood glucose monitoring, testing themselves several times each day.

We conclude that despite FBG > 105 mg/dl. intensive dietary therapy with rigorous follow-up to assess adherence to diet and the achievement of postprandial normoglycemia was successful in reducing perinatal morbidity for these gestational diabetic women to a minimum, unless women were overweight. These data should be confirmed by a large randomized controlled trial in which careful attention should be paid to all the demographic and metabolic variables which might influence birth weight and other neonatal complications.

16

Fetal Hemoglobin in the Infant of Gestational Diabetic Mothers (IGDM)

P. Doménech[1], X. Pastor[1], O. Cruz[1], F. Botet[1], R. Jiménez[1], and J. L. Aguilar[2]

[1] Department of Pediatrics and [2] Hematology Laboratory, Hospital Clínic, Faculty of Medicine, University of Barcelona, Spain

Introduction

The normal changeover in the hemoglobin pattern during the perinatal period has been reported to be delayed in the infants of diabetic mothers (IDM) (PERRINE et al., 1985; BARD and PROSMANNE, 1985). The same finding has been observed in the infants of mothers with chronic lung disease (BROMBERG et al., 1956) and in intrauterine growth delay (BARD, 1974).

Several factors such as increased erythropoietin (EPO), chronic hypoxia, and metabolic immaturity have been proposed to explain this fact. In the IDM some authors consider hyperinsulinism to be responsible for alteration of the globin-gene switch, based on experimental models (PERRINE et al., 1985).

An elevation in hemoglobin F (HbF) has been noted during hematologic stress associated with high levels of EPO. This could be explained by a direct effect of EPO, producing a burst-promotion activity on erythroid progenitor populations (DE SIMONE et al., 1982).

High EPO levels have been reported in the umbilical cord plasma of newborns delivered from mothers with several abnormal conditions: post dates pregnancy, preeclampsia, Rh isoimmunization and diabetes mellitus, intrauterine growth retardation and after the acute stress of normal labor (FINNE, 1964,

1968; WIDNESS et al., 1981, 1984). There is now evidence that EPO does not cross the placental barrier in any sense (ZANJANI and GORDON, 1971); thus chronic intrauterine hypoxia is probably responsible for the increased EPO levels in these cases.

This would also suggest that chronic intrauterine hypoxemia is an important prenatal factor in fetuses whose mothers have diabetes mellitus, but the ultimate mechanism is not well understood. It may be difficult to explain chronic hypoxia with a large weight for gestational age.

The presence of fetal hypoxia in human diabetic pregnancy was suggested early, in 1954 by BERGLUND and ZETTERSTROM, on the basis of increased erythropoiesis and decreased oxygen saturation in cord blood of IDM. However the presence of vascular disease was not excluded in these patients.

Experimental models on fetal sheep show a positive correlation between hyperglycemia, hyperinsulinemia, hypoxia and EPO (WIDNESS et al., 1981; CARSON et al., 1980; PHILLIPS et al., 1982). There is evidence that acute erythroid expansion and increased production of HbF are associated.

Some authors affirm that HbF γ chains are potentially β chains that do not achieve their normal structure because of a blockade in the metabolic pathway of β chain synthesis, due to immaturity of the enzymatic machinery (Davidsen, 1974).

Increased erythropoiesis, low protein concentration for gestational age, and elevated transferrin levels are features of the IDM pointing to relative immaturity.

Supraphysiological insulin concentration stimulates erytropoiesis independently from EPO.

Some authors have suggested that chronic intrauterine fetal hypoxemia induced by maternal diabetes offers an unifiying explanation linking venous polycythemia, hyperbilirrubinemia, fetal acidosis and late fetal demise in this disorder (Cornblath and Schwartz, 1976; Phillips et al., 1985; Shannon et al., 1986).

A prospective study has been conducted since 1985 in our department to assess differences in the HbF content in IGDMs and compare them with normal newborns and another group of hyperinsulinemic macrosomic non IDM newborns.

Patients and methods

A total of 59 term newborns (gestational age between 37 and 42 weeks) were included in the study, none of them had neonatal asphyxia or fetal distress. This exclusion was done to avoid confounding by other factors. The newborns were divided in three groups: 11 normal newborns, 32 IGDM and 16 macrosomic non IDM. All the mothers were under control during pregnancy and screened for glucose intolerance. The diagnosis of gestational diabetes (GD) was done according to the criteria of O'Sullivan and Mahan revalidated by the National Diabetes Data Group (1979). Maternal control was assessed by glycosylated hemoglobin (HbA$_1$) determined by ion-exchange chromatography (Glyco Hb-kit Helena Lab.). HbF was obtained from the femoral vein on the fourth day of life and measured by Singer technique (Singer et al., 1965) (alkali resistance). Hormonal studies were performed in all the newborns at cord and 60 min of life. Glucose was determined by polarography in an autoanalyzer (Astra-4®). Radioimmunoassay was used to determine the levels of insulin (IRI) (INSIK-I ®) and C-peptide (CPR) (Daiichi ®), and molar ratios were computed to avoid bias due to differences in the glycemic levels. The insulin receptors, number and affinity, were assayed according to the Gambhir technique (Gambhir et al., 1978) and the estimations obtained from the SCATNOL program for non-linear regression fittings. Statistical analysis was performed in an IBM PC/XT microcomputer using the SSPS/PC + ® package (1986).

Results

There was a highly significant difference in the HbF content among the three groups (Fig. 43). The highest levels were found in the IGDM (64.7 ± 2.46) while lower values existed in the macrosomic non IDM (54.13 ± 3.1) and normal newborns (55.1 ± 3.1) (ANOVA: F = 5.44; p < 0.01). There were not significant difference between the two last groups.

The molar ratio IRI/glucose in cord blood also showed significant differences (Fig. 44), being greatest in the macrosomic non IDM (5.8 ± 0.94) * 10^{-8} followed by IGDM (3.02 ± 0.36) * 10^{-8} and normal newborns (1.93 ± 0.33) * 10^{-8} (ANOVA: KW = 13.74: p < 0.001). The differences were due to a true hyperinsulinism because there were no differences in the glucose content. The same relationship held at 60 min of life, the IRI/glucose levels were: macrosomic non IDM (7.67 ± 1.62) * 10^{-8}, IGDM (3.47 ± 0.37) * 10^{-8} and normal newborns (1.88 ± 0.63) * 10^{-8} (ANOVA: KW = 13.22; p < 0.0013). Similar results were ob-

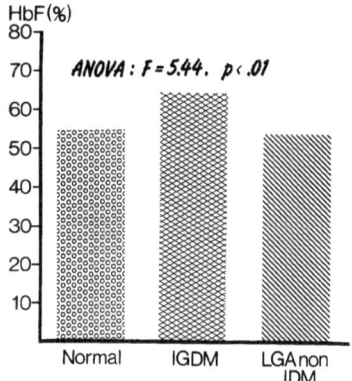

Fig. 43. Fetal hemoglobin content in the different neonatal groups

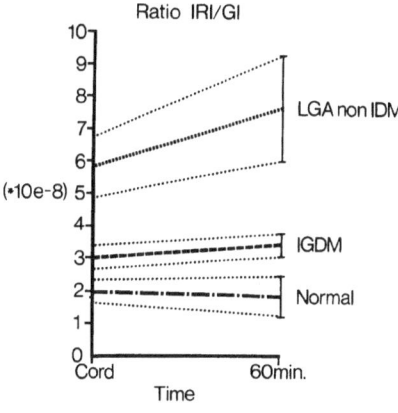

Fig. 44. Molar ratio IRI/glucose at cord and 60 minutes of life in the different neonatal groups

tained from statistical analysis of the molar ratio of CPR/glucose.

With respect to insulin receptor study, the macrosomic non IDM had a greater number (102 ± 27) than IGDMs (67 ± 9) ($F = 2.7$; $t = 2.64$; $p < 0.02$). No statistical difference was found in the affinity but there was a trend toward a higher affinity in the IGDMs (0.23 ± 0.06) than macrosomic non IDMs (0.13 ± 0.05).

No correlation at all was found between the HbF content and IRI or IRI/glucose ratio. In the group of IGDMs there was a poor negative correlation between HbF and maternal HbA_1 ($r = -0.374$; $n = 32$; $p < 0.05$).

Discussion and conclusions

High levels of HbF in IGDMs have been recently published (PERRINE et al., 1985; BARD and PROSMANNE, 1985). The differences probably would be greater if the absolute amount of HbF was measured, due to the polycythemic condition of IDM. But it is difficult to maintain the hypothesis that hyperinsulinism *per se* is the only explanation, because the macrosomic non IDMs are more hyperinsulinemic and have a greater number of insulin receptors that IGDMs, while their HbF level is lower.

The receptor findings support the theory of an *up-regulation* during the newborn period, just the opposite to the adult. *Up-regulation* of insulin receptors in circulating blood cells is detected in diabetic adults and IDMs. The influence of this abnormality on erythropoiesis is unknown. The metabolic rate of insulin, the insulin receptor density on the cell surface, and the absence of modulation in the insulin receptors are not similar in the fetus to the adult. It is possible that an insulin effect upon fetal erythropoiesis is limited to the fetus and not aplicable to a postnatal patient with abnormalities in β-chain (PERRINE et al., 1985).

Other indirect data in IGDMs suggests that a lower HbF content is found in the fetuses of poorly controlled mothers with higher HbA_1, where intrauterine hyperinsulinemia is more likely according to the universally accepted PEDERSEN hypothesis. However the correlation coefficient is poor and HbA_1 can not be considered a predictive variable on an individual basis. This could be related to its low sensitivity in gestational diabetes (ALBUTT et al., 1985), or a finding by chance.

The chronic experimental model of lambs, catheterized and glucose infused, offers insights into the metabolic derrangements observed in infants born to mothers with gestational and insulin-dependent diabetes. Studies in fetal sheep, using either glucose or insulin infusions, show a decreased blood oxygen content and elevated EPO values (Widness et al., 1981; Carson et al., 1980; Phillips et al., 1982; Widness et al., 1986, 1986 a). There is evidence that fetal ovine glucose excess stimulates oxidative metabolism with an increase in the metabolic rate of glycolysis with resulting increase of fetal oxygen consumption (Phillips et al., 1985). Severe, fetal lamb hyperglycemia may result in fetal hypoxia with metabolic acidosis and fetal demise. The metabolic goal of the stimulation of fetal oxidation may be related to increases in fetal activity such as respiration, excessive fetal growth and other unidentified factors. Insulin or catecholamines may be mediators of some of these events (Phillips et al., 1985).

Data on human diabetic pregnancy show that the percentage of oxygen saturation of cord blood was inversely correlated with cord blood C-peptide level and with the relative birth weight ratio of the neonate (Teramo et al., 1986). Other results correlate EPO with cord plasma insulin level, but not with the hematocrit of the IDM (Widness et al., 1981). Amniotic fluid EPO is increased in infants of diabetic and hypertensive mothers, and in the absence of labor, reflects

closely plasma EPO. This could be used as antepartum indicator of chronic fetal hypoxemia (Teramo et al., 1987). Umbilical cord blood EPO level is generally normal in IDM delivered by mothers in whom good glycemic control is maintained throughout gestation (Shannon et al., 1986). On the other hand erythropoiesis seems to be qualitatively abnormal in IDMs, as evidenced by the delay in the fetal globin switch. It has also been determined that the response of erythroid progenitors to EPO is intrinsically normal in IDMs, and supports the hypothesis that polycythemia in these infants is a secondary phenomenon (Shannon et al., 1986). Also the erythropoiesis in IDMs may be ineffective as determined by the index of total bilirubin production (VECO): Some authors (Bucalo et al., 1984) found increased pulmonary excretion of carbon monoxide in IDMs and suggested that this might reflect ineffective erythropoiesis. This is an explanation for the observation that cord blood EPO values do not correlate with hematocrit in IDM.

In this way our group has initiated research on EPO levels and hypoxia parameters in the infants of gestational diabetic mothers and other White classes, and in macrosomic infants. These data will be correlated with the delayed modulation pattern of hemoglobins in the perinatal period.

17

Somatometric Study in the Infant of Gestational Diabetic Mother

X. Pastor[1], P. Doménech[1], J. M. Jorba[1], A. Martínez-Gutierrez[1], J. Figueras[1], R. Jiménez[1], and R. Casamitjana[2]

[1] Department of Pediatrics, Faculty of Medicine, University of Barcelona, Spain
[2] Hormonal Laboratory, Hospital Clínic, Barcelona, Spain

Introduction

The infant of a gestational diabetic mother (IGDM), has been considered as the prime example of a macrosomic newborn. In the strict sense any baby with a birthweight greater than two standard deviations above the mean for gestational age can be considered macrosomic or large for gestational age (LGA). However, not all LGA are IGDM and, on the other hand, tight control during diabetic pregnancy decreases the neonatal overweight (OLOFSSON et al., 1984; KITZMILLER et al., 1978).

Many of the clinical problems of the IGDM are related to macrosomia. Among them, shoulder dystocia, fetal distress and obstetric trauma may lead to irreversible sequelae. Several studies have demonstrated that fetal insulin is a factor of primary importance in fetal growth (MILNER, 1979; SUSA and SCHWARTZ, 1985). The "Pedersen hypothesis" (hyperglycemia-hyperinsulinism) as well as a greater fuel supply could explain the overweight in the IGDM (PEDERSEN et al., 1954; HOLLINGSWORTH, 1984). The genesis of the same outcome in the LGA non IDM is not so well understood.

Therefore a prospective survey has been conducted during 1985 in our department to assess the somatometric state of a group of IGDM whose mothers underwent good metabolic control along pregnancy. The results have been compared to normal newborns and macrosomic non IDMs. The study has included hormonal evaluation, the consideration of maternal and neonatal somatometry and an evaluation of metabolic control by means of glycosylated hemoglobin (HbA_1) measurements.

Patients and methods

A total of 69 newborns have been included in the study. All of them were term-newborns (gestational age between 37 and 42 weeks) and none had neonatal asphyxia or fetal distress. These exclusions avoid confounding based on other factors. Twenty-two were normal newborns (NORM), 32 were IGDMs and 15 were macrosomic non IDMs. All the mothers were cared for during pregnancy and screened for glucose intolerance. The diagnosis of gestational diabetes (GD) was done according to the criteria of

O'Sullivan and Mahan revised by the National Diabetes Data Group (1979). Those mothers with GD were managed by a diabetologist on an outpatient basis and caloric intake was restricted. Maternal control was assessed by clinical follow-up, weight gain and repeated measures of glycosylated hemoglobin.

The somatometric parameters considered were the weight (kg), length (cm), head circumference (HC) (cm), thickness of subcutaneous tricipital (TSF) and subscapular (SSF) skinfolds (mm), body surface area (BS) (m^2) computed by the Haycock (1978) formula relating weight and length: $BS = 0.024265 * length$ $(cm)^{0.3964} * weight$ $(kg)^{0.5378}$, body mass index (BMI) as $BMI = weight (kg)/[length]^2$, and deviation from 50th percentile: $DEV 50^{th} p = [weight-50^{th}p)/50^{th}p] * 100$. Deviation from ideal weight for height and age, was computed in the mother.

The HbA$_1$ was measured by ion-exchange chromatography. Glucose was determined by polarography in mg/dl, while insulin (IRI) and glucagon (IRG) were measured by radioimmunoassay (RIA) using INSIK-I[®] kit and the Dr. Unger 30 K[®] antibody respectively. The levels of these three substances were established in maternal blood, cord blood and neonatal blood at 60 min of life. Molar ratios were calculated to avoid bias due to difference in the glucose levels among groups. The insulin receptor number and affinity on erythrocytes were assayed according to the Gambhir technique (1978) and the estimations obtained from SCATNOL, a microcomputer program for nonlinear regression fittings.

Statistical analysis was performed in an IBM[®] PC/XT microcomputer using the SPSS/PC+[®] package. The significance level was established under 0.05. All the values are expressed as mean ± standard error of mean.

Results

No significant differences were found between normal newborns and IGDM in the following parameters: weight (NORM = 3.307 ± 0.208; IGDM = 3.562 ± 0.601), length (NORM = 50.36 ± 1.38; IGDM = 50.97 ± 2.66), HC (NORM = 35.2 ± 1.36; IGDM = 35.1 ± 1.74), BS (NORM = 0.23 ± 0.009; IGDM = 0.24 ± 0.026) and BMI (NORM = 14.5 ± 0.402; IGDM = 14.9 ± 1.04).

Highly significant differences were found in the width of the skinfolds, SSF (NORM = 4.07 ± 0.66; IGDM = 5.60 ± 1.98. F = 8.03; MWz = 2.734; p < 0.006) and TSF (NORM = 4.06 ± 0.75; IGDM = 5.98 ± 1.64. F = 7.31; MWz = 3.278; p < 0.001) and also in the deviation ratio from 50th percentile (NORM = −3.2 ± 1.31; IGDM = +6.87 ± 2.87. F = 4.44; T = 5.21; p < 0.0001) (Fig. 45).

Obviously, LGA non IDM showed significant differences with the two precedent groups in each parameter except for the TSF with the IGDM. In order to relativize the results a linear correlation was performed between SSF, TSF and weight in the three

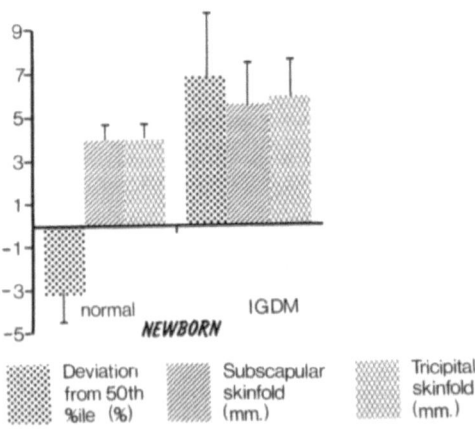

Fig. 45. Significant differences between normal newborns and IGDM in the somatometric parameters

Subscapular Skinfold
Width in mm

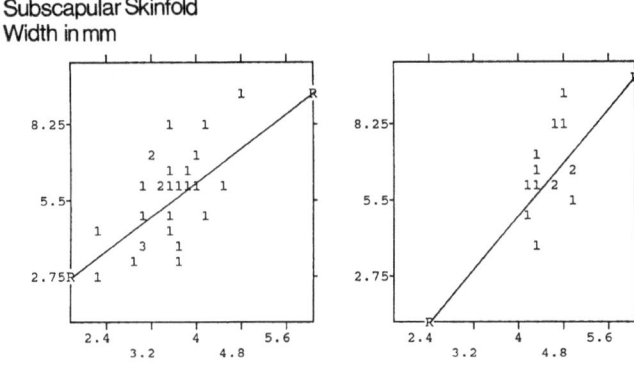

IGDM   Macrosomic non IDM
Neonatal weight in kg

Fig. 46. Correlation between weight (X axis) and the thickness of the subscapular skinfold (Y axis) in IGDM ($r = 0.61$, $p < 0.0006$) and macrosomic non IDM ($r = 0.45$, n.s.)

Tricipital Skinfold
Width in mm

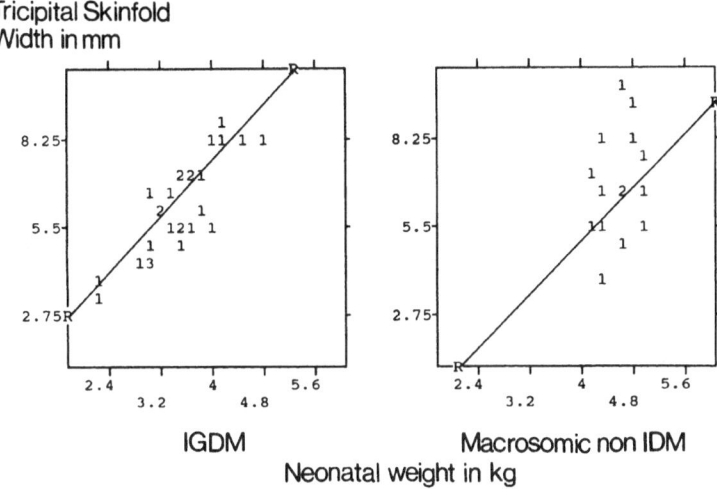

IGDM   Macrosomic non IDM
Neonatal weight in kg

Fig. 47. Correlation between weight (X axis) and the thickness of the tricipital skinfold (Y axis) in IGDM ($r = 0.83$, $p < 0.00001$) and macrosomic non IDM ($r = 0.35$, n.s.)

groups. A highly significant relationship only was found in the IGDM (SSF: $r = 0.61$, $p < 0.0006$; TSF: $r = 0.83$, $p < 0.00001$) (Figs. 46 and 47). No correlation at all was found between the neonatal weight and such other parameters as maternal weight, weight increase during pregnancy, or HbA_1 levels.

Hormonal studies demonstrated a higher IRI/Glucose ratio in cord blood in the IGDM ((3.11 ± 0.3) $* 10^{-8}$] and in LGA non IDM newborns ((5.80 ± 1.1) $* 10^{-8}$] with regard to normals ((1.82 ± 0.4) $* 10^{-8}$), (ANOVA: KW $= 13.74$; $p < 0.001$). The same relationship provided at 60 min of life: IGDM ((3.4 ± 0.35) $* 10^{-8}$), LGA non IDM ((7.7 ± 1.8) $* 10^{-8}$) and normals ((1.6 ± 0.34) $* 10^{-8}$) (ANOVA: KW $=$

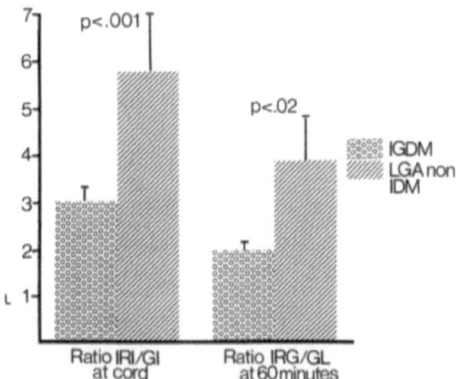

Fig. 48. Main hormonal differences between IGDM and LGA non IDM

13.22; p < 0.0013). These differences are due to a true hyperinsulinism because the glucose levels in both situations are not statistically different. The IRG/Glucose ratio in cord blood was similar in the three groups but at 60 min of life there was a significant difference between IGDM $((1.95 \pm 0.19) * 10^{-8})$ and LGA non IDM $((3.90 \pm 0.91) * 10^{-8})$ $(F = 11.4; MWz = 2.26; p < 0.02)$ (Fig. 48). Insulin receptor studies showed a greater number in LGA non IDM (102 ± 27) than IGDM (67 ± 9) $(F = 2.70; t = 2.64; p < 0.02)$. Affinity was not different.

The mothers of LGA non IDM were similar to GD mothers in the number of pregnancies, weight, BS, BMI, SSF, TSF, glucose and IRG at delivery. The only significant differences were the age (LGA non IDM = 26 ± 5.2; GDM = 32.4 ± 5.1), height (LGA non IDM = $160 \; 8c \, 4.8$; GDM = $156 \; 8c \, 6.2$), weigth gain in pregnancy (LGA non IDM = 14.9 ± 6.4; GDM = 8.3 ± 6.7), and the circulating insulin levels at delivery (LGA non IDM = 71.6 ± 40; GDM = 31.8 ± 24).

Discussion and conclusions

Birthweight alone is not a good parameter to differentiate normal newborns from infants of well controlled GD. However, differences still are found when we relate the weight with its mean for gestational age to the subcutaneous fat measurement looking for the thickness of the skinfolds. The increase in the fat content is a clear final effect of fetal hyperinsulinemia as has been proposed (Farquhar, 1976). Thus, this measurement is still useful in order to evaluate the impact of maternal GD upon the fetus (Kerner et al., 1982).

The most striking result is that concerning LGA non IDM newborns. In these babies the hyperinsulinemia is greater than IGDM. The IRI/Glucose ratio is nearly twice that seen in IGDMs and the number of insulin receptors is slightly higher; however the absence of correlation between the width of the skinfolds and the weight suggests a different kind of growth. The higher glucagon response in the face of a decrease in the glucose levels suggests a better equilibrium between the insulin and glucagon. This delay in the IRG response to glucose lowering in the IDM has already been reported by other authors (Bloom and Johnston, 1972; Kühl et al., 1982). Our results show an impairment of the response as early as at 60 min of life, and it is more pronounced when the mother had high HbA₁ levels in late pregnancy. The studies of multiple regression disclosed the value of IRG at 60 min of life as the most relevant hormonal parameter during the first hour concerning hypoglycemia (Pastor, 1987). A more balanced insulin/glucagon ratio in the fetal period could explain in part the "harmonic" growth pattern in the LGA non IDM (Sperling, 1982). Nevertheless many other factors as somatomedins and their receptors inside a genetic background can be also related to this phenomenon.

The HbA₁ failed to predict the overweight in the IGDMs. According to other reports (Weiss et al., 1984; Bacigalupo et al., 1984). this could be related to its low sensitivity (Albutt et al., 1985; Buchanan et al., 1985). As to the results in the mothers, the similarities between GD and those mothers giving birth to LGA non IDMs let us consider that GD might be underdiagnosed in the last group. But the absolute normality of the screening tests for glucose intolerance, the absence of clinical symptoms of GD, the same weight increase as the mothers of normal newborns, and the differential findings in their newborns are enough to include them in a different group. However the risk of potential diabetes in further pregnancies has to be borne in mind.

18

The Impact of Diet-induced Ketosis During Pregnancy on the Offspring

R. Steldinger[1], B. Weber[2], J. Kneer[2], and H. Hättig[2]

[1] Department of Obstetrics and Gynaecology and [2] Department of Paediatrics, Klinikum Charlottenburg, Freie Universität Berlin

Introduction

During pregnancy there is an increased maternal tendency towards ketogenesis. The ketogenic effect can become even stronger when, in addition to this pregnancy-related influence on energy metabolism, maternal diabetes – Type I, Type II or Gestational Diabetes – is present. Similar metabolic effects are also to be expected with a ketogenic weight reduction diet during pregnancy. The ketosis of the mother also influences fetal metabolism, since maternal ketone bodies reach the fetus transplacentally. Two groups of investigators (CHURCHILL et al., 1969; NAEYE and CHEZ, 1981) have described the association between long-term elevated intrauterine ketone body exposition and retarded intellectual development of such children.

Case Report

We report the effects of a self-prescribed ketogenic diet in a now 41-year-old woman with Type II diabetes mellitus and obesity (height 156 cm, body weight varying between 75 and 90 kg). Following a number of unsuccessful treatment regimens for body weight reduction, including various dietary programmes, and after using oral antidiabetic agents and insulin (maximal dosage: 40 IU daily for 2 years), she started Atkins' diet (60–80 g fat, 15–20 g carbohydrate and 100–150 g protein daily). On this diet, without any further medication, she achieved a remarkable body weight reduction and normal blood sugar values at the price of persistant ketosis. Contrary to medical advice, the patient continued her dietary regime during her three pregnancies. She refused insulin therapy. She first presented to the Department of Obstetrics and Gynaecology of the Freie Universität Berlin during the 21st week of gestation of her first pregnancy. Althouth the physicians in charge had strong misgivings regarding possible complications for mother and child with this ketogenic diet, they had to tolerate the patient's decision. She was willing to monitor blood glucose five times daily by means of a reflectometer or test strips (Haemo-Glucotest 20-800), in order to be able to detect and respond to metabolic disturbances. These blood sugar measurements were checked for correctness and precision by the hospital laboratory at various times. During all three pregnancies, the median blood glucose levels were found to remain below 100 mg/dl. During the second and third pregnancies, glycosylated haemoglobin values (HbA_1, stable form, micro-

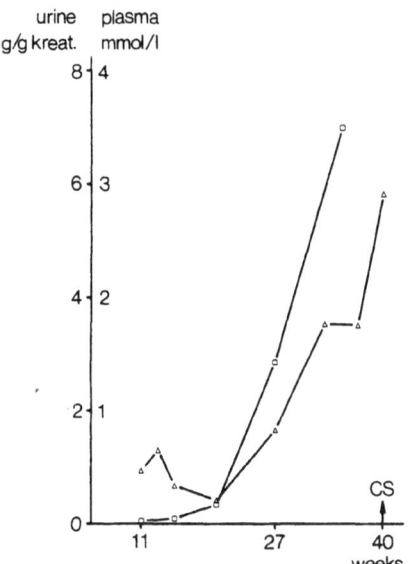

Fig. 49. Concentration of 3-β-hydroxybutyrate in plasma (triangles) and urine (squares) during the second pregnancy. *CS* Caesarean section

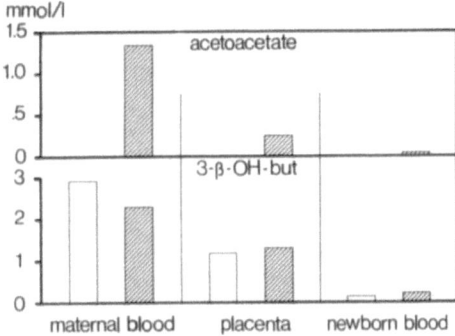

Fig. 50. Concentration of ketone bodies (mmol/l) in mother, placenta, and newborn 30 minutes postpartum during the 2nd (white bars) and 3rd (hatched bars) pregnancy

column method) ranged from 6.2% to 8.4%. Persistant ketonuria was noted (Ketostix + + +). The maternal triglyceride and cholesterol concentrations were continuously found to be above the normal range. During the course of the second and third pregnancies, ketone bodies were repeatedly measured in blood and urine by gas-chromatography. Aceto-acetate (a-a) and 3-beta-hydroxybutyrate (3-β-OH-but) levels stayed within the upper range of normal but rose continuously from the third trimester of gestation until delivery (Fig. 49, Table 93).

During the second and third pregnancies, the ketone bodies were determined in maternal blood, shortly before anaesthesia was administered for Caesarean section, in the placenta and the umbilical cord vein, and in the new borns' blood 30 minutes after birth. In contrast to the 10-fold increases of 3-β-OH-but (2920 and 2300 μmol/l plasma) and a-a (1340 μmol/l plasma) in the mother, the postnatal values in the newborns felt to normal (140 and 220 μmol/l plasma for 3-β-OH-but, 14 μmol/l plasma for a-a). The placental and

umbilical cord vein values ranged somewhat between maternal and fetal values (1180 and 1295 μmol/l plasma for 3-β-OH-but and 250 μmol/l plasma for a-a) (Fig. 50, Table 94).

The physical status of this woman was normal during her pregnancies except for her obesity. Ophthalmological examinations revealed no pathological retinal findings. The patient's body weight increased during every pregnancy from approximately 75 kg to 90 kg. Her first pregnancy was complicated near term by mild gestosis (maximal blood pressure under stress of 150/95 mm Hg, without proteinuria). During the second pregnancy, a urinary tract infection was treated with antibiotics. During the third pregnancy, the patient received tocolytic treatment with Fenoterol between the 32nd and the 36th week of gestation attempting to prevent premature labour. All three pregnancies were terminated by Caesarean section. While the first of these interventions was performed because of impending fetal asphyxia (irregular cardiotocographical findings), the second and third Caesarean sections seemed advisable in this obese patient following the previous one. After the third cesarean section sterilization was performed by ligation of the fallopian tubes.

The intrauterine growth patterns of all three fetuses were found to be normal when monitored by ultrasonic examinations at 4-week intervals. Hydramnios was not detected in any of the pregnancies. Determinations of HPL and estriol performed beyond the 30th week of gestation revealed values within the normal range. All but one cardiotocographies (see above) performed during these three pregnancies gave no indication of fetal distress. All deliveries were performed at term and resulted in normal newborns: 1. girl (8. 11. 1980), body weight 3600 g, length 49 cm, APGAR 8/10/10; 2. boy (20. 1. 1982), body weight 4040 g, length 51 cm, APGAR 9/10/10; 3. boy (11. 5. 1983), body weight 4000 g, length 50 cm, APGAR 9/10/10. The two first-born children showed neither clinical nor laboratory signs of hypoglycaemia. The third-born child exhibited a low blood sugar of 25 to 45 mg/dl six hours postpartum. After feeding the newborn with an oligosaccharide (Dextroneonat®), further blood sugar values were normal (45–90 mg/dl).

All three children underwent repeated check-ups at the Children's Hospital of the Freie Universität Berlin. They are now 6, 4 and 3 years old. They showed normal psychomotor development, body length and weight increasing along Tanner's centile curves.

The intellectual development of these children has so far been normal, too. The two elder ones were tested using a German modification of the Wechsler Pre-School and Primary Scale Intelligence Score, the youngest one was examined by means of an active vocabulary test, of which no normal ranges for the third year of life are yet available, and by the McCarthy's Scale of Children's Abilities. All children showed results within the upper limits of normal for both speech and performance, the two elder ones exhibiting excellent results also in mathematical comprehension. The two elder children were somewhat slow and deliberate, indicating resistance to the pressure of time.

Discussion

The unusual decision of an obese diabetic woman to maintain a strongly ketogenic diet during three pregnancies was the basis for determinations of ketone bodies (3-beta-hydroxybutyrate 3-β-OH-but) and acetoacetate (a-a), in mother and infants and for the

Table 93. Concentration of ketone bodies (overnight fasting) during the 2nd and 3rd pregnancy with 'Atkins diet' in comparison to pregnant diabetic women with calorie-restricted diet and healthy pregnant women with free diet (Coetzee et al. 1980)

		Diabetics 1000 kcal n = 36	Nondiabetics n = 10	2nd pregnancy		3rd pregnancy
				trimester		
Plasma [μmol/l]	3-β-OH-but	220 + 180	170 + 210	1st:	340–650	
				2nd:	210–830	340
				3rd:	1765	
				partus:	2920	2300
	a-a	84 + 60	68 + 50	2nd		490
				partus:		1340
		diabetics 1600 kcal n = 23			0.34–0.45	0,45
Urine [mmol/l]	3-β-OH-but	0.05–7.9		1st:		
				2nd:	1.2–17.1	1.72
				3rd:	86.4	
	a-a	0.02–11.6		2nd:		2.64

3-β-OH-but 3-β-hydroxybutyrate, *a-a* Acetoacetate

Table 94. Concentration of ketone bodies during the 3rd pregnancy at delivery (c.s.) [μmol/l plasma]

	Mother	Placenta	Ratio mother/ placenta	Umbilical cord (u.c.)	Ratio mother/u.c.	Ratio placenta/u.c.	Newborn
3-β-OH-but	2 300	1 295	1.8	1 070	2.1	1.2	220
a-a	1 340	250	5.4	430	3.1	0.6	40
Σ	3 640	1 545	2.4	1 500	2.4	1.0	260
Contr. (Σ)*	960			470	2.0		430
Sectio (Σ)*	1 410			640	2.3		
Ketonuria (Σ)*	1 810			780	2.3		

* Paterson et al. 1967.
3-β-OH-but 3-β-hydroxybutyrate, *a-a* Acetoacetate, Σ sum of 3-β-OH-but + a-a

longitudinal follow-up of her three children. Measurements of ketone body concentration in the mother showed a continuous rise during the third trimester of gestation until delivery. In accordance with reports in the literature, her urinary excretion of 3-β-OH-but was found to be low as long as blood concentrations remained below 500 μmol/l (normal: < 1 mmol/l). With higher values excessive ketone body excretion occurred (Table 93). Comparison of ketone body concentrations in maternal blood, placenta and fetal blood indicate a passive diffusion along the concentration gradient between maternal and fetal circulations.

Similar findings have been reported by others. KIM and FELIG (1972) measured directly proportional rises of 3-β-OH-but and a-a concentrations in maternal blood and amniotic fluid in metabolically healthy pregnant women after a fast of 84–90 hours in midpregnancy (16th–22nd week of gestation). SEEDS et al. (1979) observed a slight correlation between elevated maternal glucose levels and elevated 3-β-OH-but concentrations in amniotic fluid after an overnight fast during the third trimester of pregnancy. Amniotic insulin concentrations, on the other hand, tended to be within the upper limits of normal, indicating that elevated ketone body concentrations in amniotic fluid are of fetal and not of maternal origin. Ketone bodies apparently pass the placenta freely. In ketonuric patients undergoing cesarean section, maternal and cord blood concentrations of ketone bodies were related 2:1 (PATERSON et al., 1967) (see Table 94). From these data they suggested a simple passive diffusion via the placenta along the concentration gradient. Findings of SABATA et al. (1968) demonstrating only slight differences between the ketone body concentrations of venous and arterial cord blood, on the other hand, imply that the fetus usually assimilates only minor amounts of ketone bodies to meet his energy requirements. The findings reported in this paper confirm the transplacental transport of ketone bodies from mother to fetus. However, it cannot be concluded from these findings that an excessive supply has a negative effect on fetal metabolism or that ketone bodies are utilized more intensively. Furthermore, the placental metabolism of ketone bodies should be kept in mind, of which ketogenesis from acetyl-COA is a well known feature. Very likely, the utilization of ketone bodies by placenta is more important for its energy metabolism than that of glucose (COETZEE et al., 1980; SHAMBAUGH et al., 1977). It has not yet been clarified to what extent fetal development is endangered by elevated ketone body concentrations (NAEYE and CHEZ, 1981; CHURCHILL et al., 1969; RUDOLF and SHERWIN, 1983). On the basis of present knowledge, essentially two questions regarding possible consequences of hyperketonuria on the fetus may be asked. Firstly, is there a possibility of increased assimilation and utilization of ketone bodies in the brain and in other organs without changing cerebral function or O_2 utilization? Secondly, do hitherto unknown effects on cerebral development occur with permanent elevations of ketone body concentrations in fetal blood?

The general question whether such disturbances do, in fact, occur in the fetal metabolism or whether the placenta serves as a regulatory barrier to prevent ketone body overload on the fetus was of special interest in this case.

Most nutrients are transported via the placenta by passive diffusion. The transport of glucose is probably facilitated by a carrier, whereas amino acids are transported actively against a concentration gradient. The transport mechanisms for free fatty acids and ketone bodies are still unclear. In women fasting for 84–90 hours prior to planned abortion, no increase of free fatty acid concentrations in the amniotic fluid was observed, while the 3-β-OH-but levels rose to 30 times and the a-a levels to 10 times normal during increases of maternal plasma levels of more than 4 mmol/l and around 1 mmol/l, respectively. From these findings it may be concluded that fatty acids may not, but ketone bodies apparently do, readily pass the placenta. Still, the newborns examined in this study did not develop severe ketosis. In spite of remarkable increases of the plasma ketone bodies in the mother during the third trimester of gestation and high values at delivery, the subsequent plasma concentrations of the newborns remained within normal limits. The ratio of the total ketone body concentrations in the maternal blood versus placenta and umbilical cord blood was found to be 2.4. This is in line with previous findings of Paterson et al. (1967) examining ketonuria during delivery by Caesarean section.

Several groups of authors published data on ketone body utilization in the fetal brain (Adam et al., 1975; Page and Williamson, 1971; Coetzee et al., 1980; Shambaugh et al., 1977; Cornblath, 1975). The molar assimilation of 3-β-OH-but in isolated, perfused, human fetal brains (12th–21st week of gestation) was found to be approximately 50% higher than that of glucose, indicating that ketone bodies may provide about one third of the energy supply to the fetal brain, Adam et al. (1975) and Page and Williamson (1971) demonstrated all enzymatic activities essential for ketone body oxidation to be present in the fetal brain following 32 weeks of gestation. Therefore, 3-β-OH-but may substitute for glucose at least partially as a substrate of oxidative brain metabolism. Usually, this occurs only during hypoglycaemia. Very likely, the substrate level is of greater importance than the enzymatic activity for determining the utilization of ketone bodies, as suggested by in vitro studies in different tissues of fetal rats (Coetzee et al., 1980; Shambaugh et al., 1977). The oxidation rate of 3-β-OH-but in the fetal brain is approximately 70% higher than that of glucose. Furthermore, CO_2 production from glucose or lactate is markedly reduced when 3-β-OH-but is added as an energy substrate. Cornblath (1975) reported a retarded maturation of phosphofructokinase in brains of fetal rats exposed to a fat-enriched maternal diet during pregnancy, and a different developmental pattern of glyolysis, compared to controls. In the fetal brains, lower concentrations of fructose-1, 6-diphosphate and pyruvate and, accordingly, higher concentrations of lactate were found. This indicates a reduction of anaerobic glycolysis and elevated cytoplasmatic redox potentials in newborns of ketonic mothers, which could also explain the high ratio of 3-β-OH-but/ a-a and lactate/pyruvate found in the newborns dealt with in this report. The 3-β-OH-but/a-a ratio in the newborn was measured to be 5, compared to that found in the mother which was 1.7. Accordingly, the lactate/pyruvate ratio was 80 in the newborn and 12 in the mother, indicating normal ratios in the mother and elevated ratios in the newborn. Speculatively, this finding represents a disproportionate NADH oxidation in the infant.

At present, little is known about the extent

to which fetal and infant development may be affected by high maternal ketone body concentrations. CHURCHILL et al. (1969) reported a significant diminution of intelligence in children of diabetic mothers with significant ketonuria during pregnancy. In contrast, STEIN et al. (1975) reported normal physical and intellectual development of Dutch recruits born in 1944 and 1945 although their mothers had been exposed to famine during pregnancy.

Summary

Three children of an obese Type II diabetic mother were exposed in utero to a highly ketogenic diet, chosen by the mother and proven to be effective in controlling both diabetes and overweight, long before the first pregnancy. All developed normally in utero. Despite excessively high levels of ketone bodies in maternal blood at the end of pregnancy and during delivery, low concentrations were measured in one of the newborns shortly after birth. Since ketone bodies pass the placenta freely by passive diffusion, this observation may indicate that the fetal (and placental?) utilization of this energy source may be enhanced during maternal ketosis. Appropriate adaptations of enzymatic activities, as observed in analogous animal experiments, apparently also occur in the human fetus.

The follow-up of these three children for up to 6 years revealed no abnormalities of physical and psycho-mental development, indicating that permanent ketosis in pregnancy (without severe abnormalities of carbohydrate metabolism) does not necessarily affect the fetus.

F. On Oral Contraception in Gestational Diabetes

19

Oral Contraceptives in Women with Previous Gestational Diabetes: Influence on Glucose Metabolism

S. O. Skouby and O. Andersen

Diabetes Center, Department of Obstetrics and Gynecology Y, Rigshospitalet, Copenhagen, Denmark

Introduction

One of the earliest described metabolic effects following intake of oral contraceptives was that upon glucose tolerance (WAINE et al., 1963; WYNN, 1966). Today a considerable amount of evidence substantiates that traditional brands of oral contraceptives with 50 µg estrogen or more and high progestational content decrease glucose tolerance and cause a rise in plasma insulin levels (KALKHOFF, 1975; SPELLACY, 1976). Despite these findings no increased risk of developing clinical diabetes has been found in epidemiological studies (WINGRAVE et al., 1979).

The relative impact of estrogens and progestogens on glucose metabolism has been a matter of discussion from the very beginning of the oral contraception era. The problem was first elucidated in the late 1960s and the early 1970s. Oral glucose tolerance tests (OGTT) were performed before and after 6 months' treatment with mestranol and ethinyl estradiol and no influence of the hormonal treatment was found (SPELLACY et al., 1972). The effect of the progestogens was later found to mimic the diabetogenic effect of the combined compounds (SPELLACY et al., 1976).

In the late 1960s and in the beginning of the 1970s, oral contraceptives containing more than 50 µg estrogen were removed from the market and low dose preparations with an estrogen concentration of 35 µg or less and reduced and varied progestogen content were introduced. This happened after it became evident that both the steroid structure of the hormonal constituents as well as the amount of steroid were of significance for undesired effects on glucose- and lipid metabolism (SKOUBY, 1986).

Influence of low-dose oral contraceptives on glucose metabolism in normal women and women with previous gestational diabetes mellitus

Only a limited number of studies have been published on the effect of low-dose oral contraceptives on glucose metabolism. Generally two types of products have been investigated, both containing 30 or 35 µg estrogen combined with either an estran (e.g. norethisterone) or a gonane (e.g. levonorgestrel). There has been conflicting evidence on the effects on glucose tolerance of these compounds. It has been postulated (SPELLACY et al., 1979; WYNN et al., 1979) that products containing levonorgestrel are associated with more marked alterations of glucose and/or insulin levels than products containing norethisterone, but not all investigations support these findings (SADIK et al., 1985). We have administered a monophasic preparation of 30 µg ethinyl estradiol and 150 µg levonorgestrel (LNG) to healthy non-diabetic women and found a slight decrease in fasting glucose values shortly after treatment was started. No changes in fasting glucose, insulin or glucagon values or in glucose tolerance were observed after 2 and 6 months (SKOUBY et al., 1982).

It has been suggested (SPELLACY, 1982) that the women at greatest risk of developing a deterioration of glucose tolerance during intake of oral contraceptives are those with impaired glucose tolerance in the past, for example during pregnancy, or those with a strong family history of diabetes. In women with previous gestational diabetes mellitus (GDM) an overall incidence of 44% of deterioration of glucose tolerance has been reported during intake of the traditional brands of oral contraceptives (BECK and WELLS, 1969; SZABO et al., 1970). We have compared the effect of the monophasic preparation of 30 µg ethinyl estradiol and 105 µg LNG in lean normal non-diabetic women with the results observed in normal weight women with previous GDM (SKOUBY et al., 1982). In the women with previous GDM significantly elevated glucose values were observed before treatment, as opposed to the values in the non-diabetic controls. After hormonal intake for 6 months the insulin response to oral glucose increased significantly in the women with previous GDM, but no deterioration of glucose tolerance was observed. The glucagon response was similar in the women with previous GDM and in the controls. The findings in our study are in accordance with observations made in a small number of women with borderline abnormal GTT before and during intake of a monophasic low dose ethinyl estradiol/norethisterone preparation (SPELLACY et al., 1979). Hence, the reduced hormonal content in low-dose type oral contraceptives seems to have a clinical significance in women with a previous GDM.

Oral contraceptives in normal women and women with previous GDM: Effect of estrogen/progestogen ratio on glucose tolerance

The strategies used to reduce the metabolic side effects of oral contraceptives have included reduction of the steroid dose, development of new steroids and combinations of reduced steroid doses and different administration regimens. The triphasic ethinyl estradiol/levonorgestrel pills were all responses to this view using the lowest possible dose of both hormones without a concomitant unacceptable uterine bleeding pattern (UPTON, 1983). The progestagen is given incrementally, 50 µg for the first six days, 75 µg for the next five days and 125 µg for the last ten days. The ethinyl estradiol dose is 30 µg daily except during the middle five days when it is 40 µg. The ethinyl estradiol ratio is 1.08 during one 3-week treatment period whereas the levonorgestrel ratio is only 0.61 when compared to the most widely used low dose monophasic ethinyl estradiol/ levonorgestrel combination.

As the progestogen in oral contraceptives seems to be mainly responsible for the effect on glucose metabolism we have compared the influence of a monophasic and a triphasic low-dose ethinyl estradiol/LNG compound on glucose tolerance in normal weight non-diabetic women and in women with previous GDM during a 6-month treatment period. OGTTS were performed in the secretory phase (days 21–28) of the control cycle and during treatment, in the last phase of the tablet intake (days 15–21). Before treatment the women with previous GDM had significant elevated fasting glucose and impaired OGTT when compared to the controls. No differences in areas under plasma glucose and insulin curves (AUC) were

Table 95. Glucose areas (min × mmol/1) (mean ± SEM) calculated from oral glucose tolerance tests performed in women with previous gestational diabetes (Prev. GDM) (n = 10) and in normal women (n = 10) before and during intake of triphasic and monophasic oral contraceptives (EE/LNG)

Months	Normal		Prev. GDM	
	Monophasic	Triphasic	Monophasic	Triphasic
0	1038 ± 41	944 ± 44	1372 ± 100[a]	1218 ± 49[a]
6	1039 ± 60	966 ± 44	1474 ± 117[a]	1302 ± 86[a]

[a] GDMs vs. normals; $p < 0.05$.

Table 96. Insulin areas (min × nmol/1) (mean ± SEM) calculated from glucose tolerance tests performed in women with previous gestational diabetes (Prev. GDM) (n = 10) and in normal women (n = 10) during intake of triphasic and monophasic oral contraceptives (EE/LNG)

Months	Normal		Prev. GDM	
	Monophasic	Triphasic	Monophasic	Triphasic
0	53 ± 5	63 ± 17	49 ± 9	53 ± 5
6	65 ± 6[a]	60 ± 9	69 ± 11[a]	63 ± 5

[a] 0 vs. 6 months; $p < 0.05$.

Table 97. Insulin receptor binding to monocytes in nondiabetic women (n = 6) and previous gestational diabetic women (Prev. GDM) (n = 6) before and during intake of a triphasic oral contraceptive (EE/LNG)

	Pretreatment		6 Months	
	Controls	Prev. GDM	Controls	Prev. GDM.
10^2 × B/F (L/10^{10} cells)	6.3 (4.3–11.2)	5.4 (3.8–8.0)	4.9[a] (2.9–7.6)	8.0 (4.1–9.0)

Binding data are given as the ratio of bound/free insulin (B/F) (median and ranges), at tracer insulin concentration, normalized to 10^{10} monocytes.
[a] 0 vs. 6 months; $p < 0.05$.

found, within the groups of normal women and previous GDMs, between those receiving the monophasic and triphasic compounds respectively (Table 95 and 96). In both the women with previous GDM and in the normal women administration of the triphasic preparation resulted in no significant increase in the insulin response to oral glucose, whereas the low dose monophasic preparation significantly increased the insulin response (Table 96). Our data therefore suggest that in both normal non-diabetic women and in women with previous GDM beneficial effects on glucose metabolism are obtained when the progestational content in low-dose oral contraceptives is further reduced as in the triphasic preparations.

Low dose oral contraceptives and insulin receptor binding in normal women and women with previous GDM

The observed increase in plasma insulin during intake of low-dose oral contraceptives, without a concomitant increase in plasma glucose, points to a condition of insulin resistance. For these reasons, it appears possible that the influence on glucose metabolism associated with the use of oral contraceptives may be mediated by a decrease in the binding of insulin to receptors in peripheral tissues.

Insulin receptors are specific glycoproteins situated on the surface of the cell. The first direct demonstration of insulin receptors was performed 15 years ago (Freychet et al., 1971) and further experiments established the relation between the binding reaction and the subsequent changes of intracellular metabolic pathways (Kahn, 1985). Several studies have been performed on insulin receptor binding during intake of oral contraceptives although no investigators previously have performed longitudinal investigations. Nearly all receptor studies have been *in vitro* measurments on circulating cells, first and foremost erythrocytes and mononuclear leukocytes. Most often Scatchard-analysis has been applied to quantify insulin binding capacity and affinity. De Pirro et al. (1981) found after administration of monophasic compounds (ethinyl estradiol, 50 µg or 30 µg plus 250 µg norgestrel or 150 µg levonorgestrel and 50 µg or 30 µg ethinyl estradiol plus 500, 1000 or 400 µg norethisterone), cessation of the physiological variation of insulin receptor concentration and affinity. We have measured the influence of the triphasic estradiol/levonorgestrel compound on insulin receptor binding to monocytes in non-diabetic women and in women with previous GDM (Skouby et al., 1986) to examine, if the encouraging clinical results previously obtained with the same contraceptive compound (Skouby et al., 1985) could be substantiated by estimations of the insulin antagonistic effects at receptor levels. Insulin binding was measured using well-established methods (Pedersen, 1984). Insulin binding was expressed as the ratio of bound/free insulin and the data were corrected for non-specific binding. Before the hormonal treatment the women with previous GDM had significantly impaired glucose tolerance when compared to the healthy controls, but no differences in insulin receptor binding were observed (Table 97). Glucose tolerance and the insulin response to oral glucose remained unchanged in both groups during the treatment period. In the control subjects a significant decrease in insulin receptor binding was observed after hormonal intake for 6 months whereas the insulin receptor binding remained unchanged in the women with previous GDM (Table 97). The differences in insulin receptor binding between the two groups of women are not easily interpretable, but our data suggest that in women with previous GDM no direct association exists between impairment of glucose tolerance and insulin receptor binding during intake of low-dose oral contraceptives.

G. On the Further Fate of Women Who Had Gestational Diabetes

20

Follow-up Studies in Women with Gestational Diabetes mellitus. The Experience at Los Angeles Country/ University of Southern California Medical Center

J. H. Mestman

Department of Obstetrics and Gynecology and Medicine, Los Angeles Country/University of Southern California Medical Center, Section of Endrocrinology and Metabolism, Good Samaritan Hospital, Los Angeles, California, U.S.A.

Introduction

Gestational Diabetes Mellitus (GDM) has been defined as abnormal carbohydrate metabolism diagnosed for the first time during pregnancy; most women are asymptomatic and the diagnosis is made by the routine use of an oral glucose tolerance test (FREINKEL, 1985). In most patients the fasting serum glucose is normal and remain so throughout pregnancy, but in about 10–15% fasting hyperglycemia is first diagnosed or develops with progression of gestation. In the majority of patients, carbohydrate metabolism returns to normal immediately following delivery. The incidence of GDM has been reported to be between 2–13% of all pregnancies (MESTMAN et al., 1971; HADDEN, 1980; SEPE et al., 1985); it varies according to age, obesity, race, and family history of diabetes. A few studies available on the natural history of gestational diabetes indicate a high prevalence of overt diabetes mellitus years after the initial event (MESTMAN et al., 1972; STOWERS et al., 1985; O'SULLIVAN, 1984; METZGER et al., 1985) (Table 98). Furthermore, early studies in Pima Indians, suggest that the offspring of gestational diabetic mothers have a much higher incidence of diabetes than children of control mothers (PETTITT et al., 1985). In a previous publication (MESTMAN et al., 1972), we reported an incidence of 90% of abnormal oral Glucose Tolerance Test (GTT), in a group of 51

Table 98. Follow-up of patients with gestational diabetes

Author	Year	Follow-up years	Total patients	Overt DM (%)	Total DM (%)
O'SULLIVAN	1984	22–28	615	25	50
STOWERS et al.	1985	22	112	7	35
METZGER et al.	1985	1	113	38	57
MESTMAN et al.	1972	5	232	23	56

pregnant women with pregnancy fasting hyperglycemia up to 5 years after their pregnancies; the incidence of abnormal glucose tolerance test, was 45%, in those women with an abnormal pregnancy GTT but fasting euglycemia. The incidence of overt diabetes was 59% and 13% respectively. If indeed the incidence of diabetes mellitus is as high as these studies suggest, recognition of abnormal carbohydrate metabolism in pregnancy could represent the best and early way to prevent future diabetic complications. In the present study a group of women who has had an abnormal GTT during their pregnancy were contacted between 12 and 18 years later.

Materials and methods

We have interviewed 89 women, 12 to 18 years following a pregnancy in which a GTT was abnormal. Tests were performed during pregnancy; normal values for blood (serum) glucose were as follows: fasting, 100 (110); 1 hour, 170 (200); 2 hours, 130 (150) and 3 hours, 120 (130). These values are close to the ones proposed by O'SULLIVAN and recommended by the Second International Worshop-Conference on Gestational Diabetes (FREINKEL, 1985). For the purpose of this study women were classified into 3 groups according to the results of the pregnancy oral GTT (Table 99): Group I (5 women) with an impaired GTT (IGTT) defined as only one abnormal value in the GTT, Group II (67 women) with an abnormal GTT, but normal Fasting Serum Glucose (FSG), [also known as Class A (MESTMAN, 1985; Class A 1 (METZGER et al., 1985)); and group III (17 pregnant women) with an abnormal fasting serum glucose [Class B (WHITE, 1949), Class A 2-B 1 (METZGER et al., 1985)].

Of the total of 89 GDM women, sixty-five were hispanic, 20 were black, 3 caucasian and one american indian. Of the 89 patients with GDM, 58 (65.2%) of them had developed overt diabetes, and 31 (34.8%) were considered non-diabetics (Table 100). The incidence of overt diabetes in the hispanic women was 60% (39 out of 65); it was 80% (16 out of 20) in the black women. Blacks have a significantly higher incidence of overt diabetes mellitus ($p < 0.01$).

Of the 5 patients with IGTT, 3 (60%) became diabetic; 38 of the 67 women with abnormal GTT but normal FSG were diabetic (56.7%); and all of the 17 women with gestational fasting hyperglycemia had developed overt diabetes (Table 101).

Table 100. Incidence of overt DM according to race

Race	#	Non-diabetics	Diabetics
Hispanic	65	26 (40)	39 (60)
Black	20	4 (20)	16 (80)[a]
Others	4	1 (25)	3 (75)
	89	31 (34.8)	58 (65.2)

[a] $p < 0.01$.

Table 99. Grouping of patients according to the original pregnancy glucose tolerance test

Post-partum	Pregnancy GTT	Nomenclature
Group I	One abnormal plasma glucose value	IGTT[a]
Group II	Abnormal GTT—normal FSG	Class A[b]—A 1[c]
Group III	Abnormal fasting serum glucose	Class B[b] A 2 B 1[a]

[a] Impaired glucose tolerance test.
[b] WHITE P (1949); MESTMAN (1985).
[c] METZGER et al. (1985).

Table 101. Incidence of overt DM in women with GDM according to their original pregnancy GTT

Group		Total number	Non-diabetic	Diabetic
I		5	2 (40)[a]	3 (60)
	H	3	1 (33.3)	2 (66.6)
	B	2	1 (50)	1 (50)
II		67	29 (43.3)	38 (56.7)
	H	52	25 (48)	27 (52)
	B	12	3 (25)	9 (75)
	O	3	1 (33)	2 (66)
III		17	—	17 (100)
	H	10	—	10 (100)
	B	6	—	6 (100)
	O	1	—	1 (100)
		89	31 (34.8)	58 (65.2)

( )[a] = (%), *H* Hispanic, *B* Black, *O* Other.
H & B: I vs. III, $p < 0.05$.
H: II vs. III, $p < 0.01$.
B: II vs. III, p N.S.

Table 102. Subsequent overt DM in women with gestational diabetes mellitus

Pregnancy	Post-partum		Subsequent overt diabetes			
	Group	[b]	[b]	Diet	Oral agents	Insulin
IGT	I	5	3	—	1	2
Abnorm GTT Normal FSG	II	67	38	11 (28.9)[c]	15 (39.4)	12 (31.6)
Elevated FSG	III	17	17	—	3 (17.7)	14 (82.3)[a]
Total		89	58	11 (18.9)	19 (32.7)	28 (48.2)

[a] $p < 0.01$.
[b] Number of patients.
[c] (%).

Diabetic treatment in relation to the original pregnancy GTT is shown in Table 102. Of the 3 women out of 5 with IGTT who developed overt DM, 2 were on insulin therapy and one on oral hypoglycemic agents. Of the 38 women in Group II with overt DM, 11 (28.9%) were on diet alone, 15 (39.4%) on oral hypoglycemic agents and 12 (31.6%) were on insulin therapy. Of the 17 women in Group III only 3 (17.7%) were on hypoglycemic agents and the other 14 (82.3%) were on insulin therapy. A significantly higher number of women in Group III were on insulin therapy ($p < 0.01$). Therefore of the 58 women developing overt DM, 11 (18.9%) were on diet, 19 (32.7%) were on oral agents and 28 (48.2%) were on insulin therapy.

Age at the time of the original pregnancy GTT, and subsequent treatment of overt diabetes was compared (Table 103). Women who subsequently developed overt DM were

Table 103. Gestational diabetes, mean age and subsequent development of overt diabetes

Total patients	Non diabetics	Total diabetics	Treatment		
			Diet	Oral agents	Insulin
89	31	58	11	19	28
Mean age (years)	31.0[a]	33.1	32.8	33.5	33.2

[a] p N.S.

Table 104. Incidence of arterial hypertension in GDM women developing overt diabetes

	Number of women	Hypertension	(%)
Non-diabetic	31	4	(12.9)
Diabetic	58	26	(44.8)
Hispanics	39	17	(43.5)
Blacks	16	7	(43.7)[a]
Others	3	2	(66.6)

[a] p N.S.

Table 105. Incidence of arterial hypertension according to diabetic treatment and race

	Total	Hypertension			
		# (%)	Hispanics	Blacks	Others
Non-diabetic	31	4 (12.9)	3	1	—
Diabetics	58	26 (44.8)	17	7	2
Diet	11	4 (36.3)	1	1	2
Oral agents	19	8 (42.1)	6	2	—
Insulin	28	14 (50.0)[a]	10	4	—

[a] p N.S.

grouped according to their diabetic treatment: diet alone, oral hypoglycemic agents and insulin therapy. The age at the time of the original GTT during pregnancy showed that patients who developed overt DM, were older than non-diabetic women (33.1 v. 31.0 years), although the difference was not statistically significant. The age of the women at the time of the original pregnancy GTT was similar regardless of the type of subsequent treatment.

Arterial hypertension was a common finding in women with overt DM (Table 104). The incidence of hypertension in those women with GDM that so far have no evidence of overt diabetes was 12.9%. Hypertension was found in 44.8% of those women with overt diabetes, with no significant difference between hispanics and blacks. The incidence of arterial hypertension was lower in patients treated with diet (36.3%) than in those taking oral hypoglycemic agents (42.1%) or in-

Table 106. Incidence of obesity and overt diabetes

Treatment	#	Normal weight	Obesity
Non-diabetics	31	16 (51.6)	15 (48.4)
Total diabetics	58	15 (25.8)	43 (74.2)
Diet	11	5 (45.4)	6 (54.6)
Oral agents	19	7 (33.8)	12 (66.2)
Insulin	28	3 (10.7)	25 (89.3)

Diet vs. oral p N.S.; Diet vs. insulin p < 0.05.
Oral vs. insulin p < 0.05.
Non-diabetics vs. oral p N.S.; vs. diet p N.S.; vs. insulin p < 0.01.

Table 107. Incidence of complications in gestational diabetic women developing overt diabetes

Treatment	#	Hypertension	CVA	Heart	Dialysis
Diet	11	4 (36.3)	—	1	—
Oral agents	19	8 (42.1)	2	—	—
Insulin	8	14 (50.0)	3	3	2
	58	26 (44.8)[a]	5 (8.6)	4 (6.8)	2 (3.4)
Non-diabetic	31	4 (12.9)	—	—	—

[a] p < 0.05.

sulin therapy (50%) (Table 105). Due to the small number of patients in each group, there was no statistical significant differences between the groups.

The incidence of obestiy in the non-diabetic group was 48.4%, while it was 74.2% in the overt DM (Table 106). More obesity was found in women on insulin therapy (89.3%), than in the oral agent group (66.2%) or in the group treated with diet alone (54.6%). The difference was statistically significant when insulin treated women are compared to the other two treatment groups (p < 0.05); it was also significant when they are compared to the non-diabetic control group (p < 0.01).

In addition to arterial hypertension the incidence of cerebral vascular accidents (CVA), heart attacks and renal insufficiency requiring dialysis was significant in patients with overt diabetes as compared to non-diabetic women with a previous history of GDM (Table 107).

Discussion

Gestational Diabetes Mellitus (GDM) defined once as diabetes first diagnosed during pregnancy, with normalization of glucose intolerance following delivery (WHITE, 1949), does not appear to be a benign, transitory entity. The present study presents further evidence of the importance in recognizing this entity and the potential value of early intervention in preventing late diabetic complications. We had added a small group of 5 women, in whom only one abnormal value was abnormal in the GTT. Striking is the fact that 3 of them developed overt diabetes, and 2 of them were on insulin therapy. Long-term follow-up studies are needed in order to better delineate normal values for the GTT in pregnancy. O'SULLIVAN (1984) recently summarized his experience with a group of 615 gestational diabetic women studied 22 to 28 years later. All but 11 of the 615 women had a normal GTT 6 weeks postpartum. The total incidence of diabetes mellitus was almost 50%, and the estimated incidence, assuming all patients had been followed a full 24 years was 73%. The incidence of systolic hypertension, proteinuria, hyperlipidemia and abnormal resting electrocardiograms was significant as compared to the control group. Mortality rate was 8.1% as compared to 5.2% of control women. O'SULLIVAN concluded from his study that "the diagnosis made years earlier during pregnancy were far from innocuous and even at that early stage have important implications for preventive medicine". METZGER et al. (1985) studied a group of 113 women with GDM within one year following delivery. The incidence of abnormal postpartum GTT was 38% in those patients with normal fasting serum glucose but abnormal GTT (Class A 1), 67% in those with slight elevation in the fasting serum glucose (Class A 2), and it was 95% in women with significant pregnancy fasting hyperglycemia (Class B 1). Overt DM occurred in 23%, 43%, and 86% respectively. They also found a significant group of GDM women with positive cytoplasmic islet cell antibodies, especially in women with fasting hyperglycemia. In 8 out of 9 of such patients, abnormal GTT developed one year after delivery, 2 requiring insulin therapy. The presence of HLA-DR 3 and/or -DR 4 antigens was not predictive of the status of GTT during the first year postpartum. Presence of islet cell antibodies have been reported in the serum of 20 patients out of 52 with GDM by RUBINSTEIN et al. (1981). The significance of islet cell antibodies in Gestational Diabetic women is unclear, since they have been detected in relatives of patients with Type I diabetes (SRIKANTA et al., 1985), and appear to predict the later development of Type I

or juvenile diabetes. The above studies have to be confirmed using a more specific assay for islet cell antibodies.

STOWERS et al. (1985) reported their experience in 112 women who *all* had an abnormal result of an Intravenous GTT (IVGTT) after pregnancy, performed because of signs of potential diabetes. Patients were advised to reduce the intake of refined carbohydrates and to lose weight when indicated. In about 1/3 of the patients 100 mg or more of chlorpropamide were also prescribed. Of 78 subjects treated with diet only and followed up for a mean period of nearly 13 years, 6.4% developed overt diabetes and 29% of them had an abnormal glucose tolerance test. The difference between the low incidence of overt diabetes mellitus found in this series as compared to the other ones, could be related to diet advice and close follow-up given by the Aberdeen group. Chlorpropamide was not more effective than diet alone in preventing the development of DM, but age and initial value of the FSG (in the postpartum period) were significant factors. Obesity was not a significant factor in the development of overt diabetes.

Obesity as an independent risk factor was studied by O'SULLIVAN (1982). He compared a group of obese GDM women with a noneobese group 10 to 16 years after their pregnancy. The incidence of diabetes was significantly higher for overweight subjects (46.6%) than for those of normal weight (25.6%). However when the same comparison was made between overweight and nonoverweight women with normal pregnancy GTT, the incidence of diabetes was 4.5% v. 1.9%, a difference that was not significant. He concluded that overweight has no substantial predictive value for diabetes mellitus, unless an additional factor such as gestational diabetes mellitus is present.

In the present study, the high incidence of overt diabetes years after the original diagnosis of GDM is confirmed. The high incidence may be due to the fact that 73% of the women were of hispanic extraction, since the incidence of Type II diabetes in this ethnic population is 2 or 3 times higher than caucasians (GARDNER et al., 1984) and the prevalence of GDM also appears to be increased (MESTMAN et al., 1972; MESTMAN, 1980) as compared to other ethnic groups. Obesity was higher, as expected in those women who developed diabetes, as compared to those non-diabetics. More obese diabetic women required oral hypoglycemic agents or insulin for diabetes management as compared to those diabetics of normal body weight.

Arterial hypertension was significant in diabetic patients independent of race, 43.6% as compared to 12.9% of the non-diabetic group. Not only hypertension was prevalent in these group, but the incidence of other complications such as cerebrovascular accidents (CVA), myocardial infarction and renal insufficiency requiring dialysis was significant.

Early recognition of carbohydrate metabolism abnormalities in patients with a history of GDM may have a significant role in the management of this medical entity. Recent studies in women with a history of GDM, has shown deficiencies in insulin release to the administration of intravenous glucose, and tissue responsiveness to insulin (TÜRNER et al., 1979; WARD et al., 1985, 1985 a; CATALANO et al., 1986). These subclinical defects, present in postpartum women at the time of normal routine glucose tolerance tests, are more pronounced in obese women with previous gestational diabetics than on lean ones.

Conclusions

1. Gestational Diabetes Mellitus is a potential serious disease.

2. In a high percent of patients it is the first manifestation of clinical diabetes, mainly Type II.

3. GDM may be followed by a long period of asymptomatic hyperglycemia.

4. Every patient with fasting hyperglycemia during gestation developed overt diabetes later on in life.

5. The incidence of complications such as hypertension, cerebral vascular accident, myocardial infarct and renal failure is significant.

6. Professional and patient education, periodic follow-up, and weight control are strongly recommended as the first step in the care of patients with GDM.

7. Further prospective studies in GDM women and their offsprings, including but not limited to, determination of islet cell antibodies, HLA-antigens, prophylactic insulin therapy during pregnancy, are needed in order to unveil the natural history of this important medical entity.

Summary

1. A group of 89 women were studied 12 to 18 years after the diagnosis of gestational diabetes mellitus.

2. The incidence of overt diabetes was 65.2%.

3. Of 67 GDM women, with normal FSG (Class A), 38 (56.7%) developed overt diabetes. Of these 11 (28.9%) were on diet, 15 (39.4%) were on oral hypoglycemic agents and 12 (31.6%) were on insulin therapy.

4. Of 17 women with elevated FSG in pregnancy (Class A 2, B) all of them had developed overt diabetes. Fourteen were on in-

sulin therapy (82.3%) and 3 on oral hypo-glycemic agents.

5. The incidence of arterial hypertension was 44.8% in diabetics as compared to 12.9% in those women without evidence of overt diabetes mellitus.

6. The incidence of obesity was 74.2% in overt diabetic women as compared to 48.4% in non-diabetic subjects.

7. Of the 58 women who developed overt diabetes, 5 (8.6%) had had cerebral vascular accidents, 4 (6.8%) had suffered myocardial infarcts, and 2 (3.4%) were on dialysis therapy.

21

The Emergence of Diabetes and Impaired Glucose Tolerance in Women Who Had Gestational Diabetes

J. N. Oats, N. A. Beischer, and *P. T. Grant*

Department of Obstetrics and Gynaecology, Mercy Maternity Hospital, University of Melbourne, Australia

Introduction

Diabetes mellitus is a common medical disorder that leads to significant morbidity and mortality. In the Australian community it is estimated that 3.4% of the population over 24 years of age suffer from the disease (GLATTHAAR et al., 1985). A number of factors, particularly hormonal, act during pregnancy to increase blood sugar levels but pregnant women with normal pancreatic function can maintain these levels within the normal range. The small proportion of women who are unable to do this develop gestational diabetes. Following the removal of the fetus and placenta this glucose intolerance usually disappears, but in some patients it re-emerges at a later date (KÜHL et al., 1984).

The antenatal period is an ideal time to screen for disorders such as diabetes and it has been the policy of the Mercy Maternity Hospital, Melbourne to aim to detect abnormal glucose tolerance in all women attending the hospital. Following the report from Boston that up to 60% of women diagnosed as having gestational diabetes eventually developed diabetes (O'SULLIVAN, 1979) we felt it was imperative to endeavour to trace and retest as many of those women with gestational diabetes as we could. This paper updates the first report from this study published in the Australian and New Zealand Journal of Obstetrics and Gynaecology (GRANT et al., 1986) particularly to conform with the amended WHO recommended definitions of diabetes and impaired glucose tolerance [Diabetes mellitus. Report of a WHO Study Group, Geneva: WHO, 1985 (Technical Report Series 727)].

Materials and methods

Since the opening of the Mercy Maternity Hospital in 1971 it has been the policy to screen all patients for glucose intolerance in the third trimester of pregnancy; that is they were not tested because of perceived indications.

The glucose tolerance test (GTT) was usually performed between 32 and 34 weeks' gestation. Patients continued their normal diet and were instructed to fast from 10 p.m. on the evening before their test. At 9 a.m. a 50 g glucose load was given after the fasting sample was taken and further samples were taken at hourly intervals. Capillary blood was obtained by finger prick, collected into a heparinized tube and centrifuged immediately. Plasma glucose was assayed by the glucose oxidase method using a Beckman

Table 108. WHO diagnostic values for the oral glucose tolerance test glucose concentration, mmol/litre (mg/dl)

	Whole blood		Plasma	
	Venous	Capillary	Venous	Capillary
Diabetes mellitus				
Fasting value and/or	$\geqslant 6.7 \ (\geqslant 120)$	$\geqslant 6.7 \ (\geqslant 120)$	$\geqslant 7.8 \ (\geqslant 140)$	$\geqslant 7.8 \ (\geqslant 140)$
2-hours after glucose load	$\geqslant 10.0 \ (\geqslant 180)$	$\geqslant 11.1 \ (\geqslant 200)$	$\geqslant 11.1 \ (\geqslant 200)$	$\geqslant 12.2 \ (\geqslant 220)$
Impaired glucose tolerance				
Fasting value and	$< 6.7 \ (< 120)$	$< 6.7 \ (< 120)$	$< 7.8 \ (< 140)$	$< 7.8 \ (< 140)$
2-hours after glucose load	6.7–10.0 (120–180)	7.8–11.1 (140–200)	7.8–11.1 (140–200)	8.9–12.2 (160–220)

glucose analyser (Beckman Instruments, Fullerton, California 92634).

Initially the criteria used to diagnose gestational diabetes were a 1-hour capillary plasma glucose level of 10 mmol/1 (180 mg/100 ml) or more together with a minimum 2-hour level of 7.8 mmol/1 (140 mg/100 ml). The incidence of gestational diabetes using these criteria was 0.7%. It was felt that these criteria were too restrictive since reports from other centres quoted incidences of up to 12.3% (MESTMAN, 1980) so new criteria were derived by computer analysis of the first 18,679 patients tested. The combination of a 1-hour level of 9 mmol/1 or more together with a 2-hour level of 7 mmol/1 or more identified the worst perinatal outcome associated with hyperglycaemia and selected 2.5% of the hospital population.

In the 14 year 6 month interval under review, 1971 to June 1985, glucose tolerance tests were performed on 43,294 out of the 63,883 confinements (67.8%) and 1,261 women were diagnosed as having gestational diabetes. The review ceased at June 1985 to allow a minimum interval of 1-year before review.

These 1,261 patients were contacted, asking them to return for a follow-up GTT and to the end of June 1986, 485 (38.5%) had returned for review. To conform with the recommendations of the WHO Expert Committee on Diabetes mellitus all nonpregnant patients were given a 75 g oral glucose load and capillary plasma samples were collected and analysed as described above. The results of these tests were classified using the WHO criteria of normal, impaired glucose tolerance and diabetes (1985). These criteria are summarized in Table 108.

At the review test further clinical information was obtained, in particular details of family history of diabetes, subsequent obstetric medical and surgical history, and symptoms of hyperglycaemia. Body mass index was calculated using the formula weight in kg/(height in m)^2 (BRAY, 1985).

Statistical analysis was performed by means of X^2, Fisher's exact probability and Kolmogorov-Smirnov test, where appropriate.

Results

At the time of their last review 44 patients (9%) were diabetic using the WHO criteria and 82 (17%) had impaired glucose tolerance (Table 109). Fourteen of the 44 (32%) diabetic patients were known to be diabetic at the time of recall and were receiving appropriate therapy. Only 8 of the remaining 30 patients (27%) had 1 or more of the classic symptoms of diabetes (polyuria, polydipsia, lethargy) and the presence of these

Table 109. Pattern of glucose tolerance at follow-up test according to severity of gestational diabetes during pregnancy

	Total tested	Follow-up glucose tolerance test					
		Normal		Impaired		Diabetic	
		No.	%	No.	%	No.	%
Old criteria[a]	136	81	60	26	19	29	21
New criteria[b]	349	278	80	56	16	15	4[c]
Total	485	359	74	82	17	44	9

[a] 1-hour ⩾ 10.0 mmol/1 + 2-hour ⩾ 7.8 mmol/1.
[b] 1-hour ⩾ 9.0 mmol/1 + 2-hour ⩾ 7.0 mmol/1.
[c] $p < 0.001$.

bore little relationship to the level of hyperglycaemia; 5 with fasting levels over 9.0 mmol/l were asymptomatic and 4 with fasting levels between 6 and 7 mmol/l were symptomatic.

Incidence according to severity of gestational diabetes

When the incidence of abnormal glucose tolerance at the follow-up test was examined in relation to the severity of gestational diabetes, as expected those diagnosed using the original criteria (1-hour $\geq$ 10 mmol/l plus 2-hour $\geq$ 7.8 mmol/l) had a five fold higher incidence of diabetes (21%) than those diagnosed using the new criteria (1-hour 9 mmol/l plus 2-hour $\geq$ 7 mmol/l) $-$ 4% (p < 0.001) (Table 109). By contrast the incidence of impaired glucose tolerance did not differ significantly according to the severity of the gestational diabetes. It is noteworthy that 34% (15 of 44) of diabetics and 68% (56 of 82) of those with impaired tolerance were identified using the new criteria and a number of patients with the highest blood glucose levels at the time of diagnosis were among those who were only classified as gestational diabetics using the new criteria.

Family history

A family history of diabetes in either the immediate family (parent or sibling) or distant (grandparent, uncle, aunt or cousin) was sought; 103 (21%) had an immediate history and 86 (18%) had a distant history (Table 110). When these incidences were compared with the results of the review glucose tolerance test there was a significantly increased incidence of an immediate family history of diabetes in those found to be diabetic (17/44, 39%) compared with those who had normal glucose tolerance (66/359, 18%) (p < 0.05). A distant family history was not a significant predictive factor for those likely to have an abnormal glucose tolerance test.

Subsequent pregnancies (Fig. 51)

One hundred and twenty of the 210 (57%) who had subsequent pregnancies had glucose tolerance tests performed in these later pregnancies. This surprisingly low figure of reassessment is in part explained by the fact that, as discussed above, the criteria for diagnosing gestational diabetes were redefined in 1980 so that the follow-up programme included a large number of women who were not classified during the index pregnancy as having gestational diabetes. In the subsequent pregnancy, 75 (63%) reverted to normal glucose tolerance, 43 (36%) were again gestational diabetics and 2 (2%) were diabetic by WHO criteria (Fig. 51). When these findings were examined according to glucose tolerance at follow-up, 65 of the 75 (87%) who remained normal in the subsequent tested pregnancy were normal, and the remaining 10 (13%) had impaired tolerance;

Table 110. Family history of diabetes of 485 women with gestational diabetes

| Family history | Total | | Follow-up glucose tolerance test | | | | | |
| | | | Normal | | Impaired | | Diabetic | |
	No.	%	No.	%	No.	%	No.	%
Immediate	103	21	66	64	20	19	17	17
Distant	86	18	62	72	16	19	8	9
None	296	61	231	78	46	16	19	6
Total	485		359	74	82	17	44	9

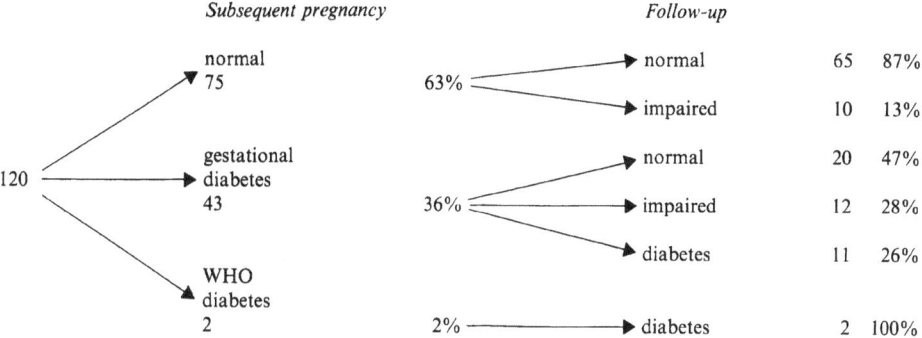

Fig. 51. Patterns of glucose tolerance in 120 gestational diabetes retested in a subsequent pregnancy and their test results at follow-up

Table 111. Incidence of abnormal glucose tolerance at follow-up according to results of the postnatal glucose tolerance test

Postnatal glucose tolerance test			Follow-up glucose tolerance test					
			Normal		Impaired		Diabetic	
	No.	%	No.	%	No.	%	No.	%
Normal	51	65	34	67	10	20	7	14
Abnormal per mercy maternity criteria	19	24	9	47	5	26	5	26
Impaired glucose tolerance[a]	5	6	0	0	2	40	3	60
Diabetes[a]	4	5	0	0	0	0	4	100
Total	79	100	43	54	17	22	19	24

[a] WHO criteria.

20 of the 45 (44%) who remained gestational diabetics were normal, 12 (27%) had impaired tolerance and 13 (29%) were diabetic. These differences were statistically significant ($X^2 = 31.47$, $p < 0.001$).

Postnatal tests

A glucose tolerance test had been performed in the postnatal period in 79 of the 485 (16%) patients. The results of these tests with the subsequent follow-up test results are shown in Table 111. The less strict Mercy Maternity pregnancy criteria (9 mmol/l or more at 1-hour plus 7 mmol/l or more at 2-hour) identified 10 women who were later found to have abnormal tolerance – 5 impaired and 5 diabetic. Thirty-four of the 51 women (67%) who had normal postnatal tests had normal glucose tolerance at follow-up compared with 9 of 28 (32%) who had an abnormal postnatal test either by Mercy Maternity or WHO criteria ($X^2 = 10.59$, $p < 0.01$). Although the postnatal test is a good discriminator of those at higher risk of developing abnormal glucose tolerance, 34% with a normal postnatal test were later found to have abnormal glucose tolerance – these included 7 of the 19 patients (37%) ultimately found to have diabetes.

Age at diagnosis (Table 112)

Diabetes, particularly the noninsulin dependant variety, is age related, however, this was

Table 112. Age at diagnosis of abnormal glucose tolerance at review test using WHO criteria

Age (years)	Total	Follow-up glucose tolerance test					
		Normal		Impaired		Diabetic	
		No.	%	No.	%	No.	%
20–24	6	5	83	0	0	1	17
25–29	69	57	83	5	7	7	10
30–34	168	125	74	32	19	11	7
35–39	148	115	78	21	14	12	8
40–44	71	44	62	16	23	11	15
45 or more	23	13	57	8	35	2	9
Total	485	359	74	82	17	44	9

Table 113. Interval from index pregnancy to diagnosis of abnormal glucose tolerance at review test

Interval (years)	Total	Follow-up glucose tolerance test					
		Normal		Impaired		Diabetic	
		No.	%	No.	%	No.	%
0–2	72	51	71	12	17	9	13
3–4	113	81	72	21	19	11	10
5–6	134	109	81	17	13	8	6
7–8	84	61	73	15	18	8	10
9–10	53	36	68	10	19	7	13
11–12	21	14	67	6	29	1	5
13–14	8	7	88	1	13	0	0
Total	485	359	74	82	17	44	9

not demonstrated in this study. There was an increasing incidence of impaired tolerance with advancing years but this did not achieve statistical significance.

Interval from index pregnancy to diagnosis (Table 113)

As the follow-up study started in 1980, 9 years after the commencement of the routine screening in pregnancy programme, the time interval from the index pregnancy to diagnosis at follow-up shown in Table 113 may not accurately reflect the true lapse in time before glucose intolerance emerged. Forty-

five percent of the diabetics (20 of 44) were identified within 4 years and 9 (45%) of these patients were not diagnosed until identified by the follow-up programme. The incidence of impaired glucose tolerance rose with the passage of time, but the differences were not statistically significant.

Body mass index (Table 114)

The body mass index was calculated from the patient's weight and height measured at the time of the latest follow-up test (Table 114). Forty-five percent of the dia-

Table 114. Body mass index calculated at the time of the review test

| | Total | | Follow-up glucose tolerance test | | | | | |
| | | | Normal | | Impaired | | Diabetic | |
B.M.I.	No.	%	No.	%	No.	%	No.	%
16–17 (underweight)	4	1	3	75	1	25	0	0
18–23 (acceptable)	237	49	197	83	33	14	7	3
24–29 (overweight)	171	35	123	72	32	19	16	9
30 or more (obese)	69	14	34	49	16	23	19	28
NR	4	1	2	50	0	0	2	50
Total	485		359	74	82	17	44	9

B.M.I. weight in kg/(height in m)^2.
NR weight or height not recorded.

betics were obese (BMI > 29) which was significantly more than women with impaired (20%) and normal (10%) tolerance ($X^2 = 47.25$, p < 0.001). Furthermore only 17% of the diabetics were not overweight or obese.

Discussion

The largest follow-up study of women with gestational diabetes is that of O'SULLIVAN and colleagues from Boston, U.S.A. (O'SULLIVAN, 1984). They have followed 615 women together with 328 "negative controls" (women who had a normal glucose tolerance test in pregnancy) for 22 to 28 years. Their reported incidence of diabetes is currently 49.9% in those who had gestational diabetes and 7.0% in the controls. The cumulative incidence of diabetes in their study rose from approximately 16%, 5 years after diagnosis, to 40% at 15 years and reached 72% at 24 years.

Comparison of the Boston results with the findings from the present study are hampered by a number of factors. The major difficulty is the different criteria used to diagnose gestational diabetes and to classify the nonpregnant tests. To diagnose gestational diabetes O'SULLIVAN required 2 or more venous plasma glucose concentrations to equal or exceed the following − fasting 5.8 mmol/l, 1-hour 10.6 mmol/l, 2-hour

9.2 mmol/l and 3-hour 8.1 mmol/l after a 100 g oral glucose load. For nonpregnant patients they used the USPHS criteria as above except that 3 or more, or the fasting and 3-hour, combinations were required to diagnose diabetes. In our study capillary rather that venous samples have been used which results in plasma glucose levels that are 7 to 8% higher. The other principal difference was the means by which the patients were selected for the initial screening test for gestational diabetes. As discussed earlier it has been the policy at the Mercy Maternity Hospital to screen *all* women for glucose intolerance between 30 and 34 weeks' gestation. O'SULLIVAN performed a screening test at the first visit using a 50 g oral load and then a 100 g oral glucose tolerance test was done if the venous whole blood sugar level was 130 mg/100 ml or more (7.2 mmol/l) 1-hour later. If this was negative and the patient had 1 or more of the following indications − positive family history of diabetes, previous infant with birthweight greater than 4.1 kg, a previous stillbirth or neonatal death, a major congenital abnormality, a premature delivery, preeclampsia in 2 or more pregnancies – the screening test was repeated in the 2 subsequent trimesters (DANDROW and O'SULLIVAN, 1966).

Glucose intolerance however increases during pregnancy and since MACAFEE and BEISCHER (1974) have shown that of all the indications for screening for glucose intolerance only age above 30 years has a significantly increased pick-up rate that is greater than that obtained from screening all the pregnant population, it is possible that the O'SULLIVAN series has missed some gestational diabetics.

Despite these differences a number of similarities in the findings were present. Factors related to the development of diabetes in the Boston study were obesity and age, whereas perinatal mortality, macrosomia and a family history were not significant factors. In our study the only difference was that age was not an associated factor, however this is probably due to the shorter duration of the study since to date only 33 or 7% of the study group are aged 45 years or more.

The most significant predictive factor for emergence of diabetes found in this study was glucose intolerance in a subsequent pregnancy (Fig. 51). If the test was abnormal in a subsequent pregnancy 56% had abnormal glucose tolerance at follow-up compared with 13% of those who had a normal test in a subsequent pregnancy.

The postnatal test also proved to be a useful discriminator, since all 9 patients who had an abnormal test 4–6 weeks after delivery by WHO criteria were abnormal at follow-up compared with 39% (27 of 70) of the others (Table 111). The fact that 12 of the 70 (17%) with normal postnatal tests using WHO criteria became diabetic and a further 15 (21%) had impaired tolerance however emphasizes the necessity to follow-up all women with gestational diabetes. The incidence of 11% with WHO classified intolerance and diabetes at the postnatal test is considerably higher than the 2% from O'SULLIVAN's study; again this probably reflects the differing selection criteria and suggests that the Mercy Maternity Hospital criteria are realistic ones.

We were concerned at the low rate of tests performed at the postnatal visit, only 16%, and have since instituted a more stringent programme to retest women with gestational diabetes at 4–6 weeks after delivery.

MESTMAN et al. (1972) followed 360 patients who had glucose tolerance tests in pregnancy for 5 years. They reported that of those with a positive oral glucose tolerance test in pregnancy, 13% became overt diabetics and a further 33% had abnormal tests by 5 years after delivery.

HADDEN (1979), however has concluded from his well-known studies of an Irish population that women who had glucose intolerance in pregnancy were *not* at increased risk of developing glucose intolerance later in life. He retested 234 women 10 years after delivery who were traced from 625 mothers who had a positive 50 g oral glucose load screening test, and found that only 6 were abnormal and 1 was diabetic.

It is obviously important to study age and weight matched control patients who had normal glucose tolerance during pregnancy and these have now been added to our follow-up programme.

Summary

Four hundred and eighty-five women who had gestational diabetes have been retested at intervals from 1 to 15 years following diagnosis; 44 (9.1%) were found to be diabetic and 82 (16.9%) had impaired glucose tolerance using the WHO criteria. An abnormal glucose tolerance test in the puerperium and obesity at the time of retesting had significant associations with abnormal glucose tolerance at follow-up. However, the best predictive factor of the likelihood of the development of significant hyperglycaemia was the recurrence of gestational diabetes in a subsequent pregnancy, since 29% of these women were diabetic and a further 27% had impaired glucose tolerance at the time of follow-up. These findings indicate that the criteria used for the diagnosis of gestational diabetes at the Mercy Maternity Hospital, Melbourne (1-hour ≥ 9 mmol/1 together with a 2-hour ≥ 7 mmol/1) are appropriate for an Australian population.

References

Aantaa K, Forss M (1980) Growth of the fetal biparietal diameter in different types of pregnancies. Radiology 137: 167

Abell DA, Beischer NA, Wood C (1976) Routine testing for gestational diabetes, pregnancy hypoglycemia and fetal growth retardation, and results of treatment. J Perinat Med 4: 197

Abraham EC, Perry RE, Stallings M (1983) Application of affinity chromatography for separation and quantitation of glycosylated hemoglobins. J Lab Clin Med 102: 189

Abraham EC, Stallings M, Abraham A, Elseweidy MM (1983) Affinity chomatographic quantitation of glycosylated hemoglobin in newborn infants. Hemoglobin 7: 449

Abramovici A, Sporn J, Prager R, Shaltiel A, Laron Z, Liban E (1978) Glycogen metabolism in the placenta of streptozotocin diabetic rats. Horm Metab Res 10: 195

Abramovici A, Svejcar J (1982) Histochemical and quantitative biochemical changes in mucopolysaccharides of the term placenta of the diabetic rat. Placenta 3: 81

Adam PAJ, Räihä N, Rahiala EL, Kekomäki M (1975) Oxidation of glucose and D-beta-OH-butyrate by the early human fetal brain. Acta Paediatr Scand 64, 17

Adashi EY, Pinto H, Tyson JE (1979) Impact of maternal euglycemia on fetal outcome in diabetic pregnancy. Am J Obstet Gynecol 133: 268

Aerts L, Van Assche FA (1977) Rat foetal endocrine pancreas in experimental diabetes. J Endocrinol 73: 339

Aerts L, Van Assche FA (1979) Is gestational diabetes an acquired condition? J Develop Physiol 1: 219

Aladjem S (1967) Morphologic aspects of the placenta in gestational diabetes seen by phase-contrast microscopy. Am J Obstet Gynecol 99: 341

Albutt EC, Natrais M, Northam BE (1985) Glucose tolerance test and glycosilated haemoglobin measurement for diagnosis of diabetes mellitus and assessment of the criteria of WHO expert comittee on diabetes mellitus 1980. Ann Clin Biochem 22: 67

Alsat E, Bouali Y, Goldstein S, Malassine A, Laudat MH, Cedard L (1982) Characterization of specific low-density lipoprotein binding in human term placental microvillous membranes. Mol Cell Endocrinol 28: 439

Alsat E, Bouali Y, Goldstein S, Malassine A, Berthelier M, Mondon F, Cedard L (1984) Low-density lipoprotein binding sites in the microvillous mem-branes of human placenta at different stages of gestation. Mol Cell Endocrinol 38: 197

Amankwah KS, Prentice RL, Fleury FJ (1977) The incidence of gestational diabetes. Obstet Gynecol 49: 497

Amendt P, Michaelis D, Hildemann W (1976) Clinical and metabolic studies in children of diabetic mothers. Endokrinologie 67: 351

Andersen O, Kühl C (1985) Insulin receptor binding and biological effect in adipocytes from normal women. In: Grech ES (ed) Diabetic pregnancy study group of the european association for the study of diabetes, XVI annual meeting. University of Malta Press, p 21 (abstracts)

Andersen O, Kühl C, Buch I (1986) Insulin receptors in normal pregnant women and women with gestational diabetes. Acta Endocrinol 112: 27

Andersen O, Saurbrey N, Kühl C, Molsted-Pedersen L (1986) Evaluation of 75 g glucose load in oral glucose tolerance test (75-g-oGTT) in normal pregnant and non pregnant women. 1st international Graz symposium: Disturbances of carbohydrate metabolism during pregnancy. In: Weiss PAM (ed) Probl perinat med 15. Maudrich, Wien München Bern, p 48

Archimaut, G, Belizan JM, Ross NA, Althabe O (1974) Glucose concentration in amniotic fluid: its possible significance in diabetic pregnancy. Am J Obstet Gynecol 119: 596

Armstrong GD, Hollenberg MD, Bhaumick B, Balka RM (1982) Comparative studies on human placental insulin and basic somatomedin receptors. J Cell Biochem 20: 283

Artal R, Golde SH, Dorey F, Mc Clellan SN, Gratacos J, Lirette T, Montoro M, Wu PYK, Anderson B, Mestman J (1983) The effect of plasma glucose variability on neonatal outcome in the pregnant diabetic patient. Am J Obstet Gynecol 147: 537

Artal R, Mosley GM, Dorey FJ (1984) Glycohemoglobin as a screening test for gestational diabetes. Am J Obstet Gynecol 148: 412

Auinger W (1978) Diabetes und Schwangerschaft: 1. Mitteilung. Wien Klin Wochenschr 90: 109

Auinger W (1978a) Diabetes und Schwangerschaft: 2. Mitteilung. Veränderungen der Kohlenhydrattoleranz und des Kindesgewichtes unter Diätbehandlung bei latentem Diabetes in der Schwangerschaft. Wien Klin Wochenschr 90: 263

Avruch J, Nemenoff RA, Blackshear PJ, Pierce MW (1982) Insulin stimulated tyrosine phosphorylation of the insulin receptor in detergent extracts of human placental membranes. J Biol Chem 257: 15162

Baciagalupo G, Langner K, Saling E (1984) Glycosylated hemoglobin (HbA 1), glucose tolerance and neonatal outcome in gestational diabetic and nondiabetic mothers. J Perinat Med 12: 137

Bajaj JS (1986) Malnutrition related fibro-calculous pancreatic diabetes. In: Serrano-Rios M, Lefebvre PJ (eds) Diabetes 1985. Elsevier, pp 1055–1060

Bakir TMF (1987) Prevalence of antibody to cytomegalic virus (CMV) in a Saudi Arabian population. Saudi Medical J 8: 40

Baker JR, O'Connor JP, Metcalf PA, Lawson MR, Johnson RN (1983) Clinical usefulness of estimation of serum fructosamine concentration as a screening test for diabetes mellitus. Br Med J 287: 863

Bantle JP, Laine DC, Castle GW, Thomas JW, Hoogwerf BJ, Goetz FC (1983) Postprandial glucose and insulin response to meals containing different carbohydrates in normal and diabetic subjects. New Engl J Med 309: 7

Baranyi E, Tamas Gy, Jr, Dimeny E, Kerenyi Z, Petranyi Gy, Jr, Egyed J, Bekefi D (1981) Basal insulin supplementation a new form of treatment of pregnant diabetics. In: Irsigler K, et al (eds) New approaches to insulin therapy. MTP Press, Falcon House. Lancaster, p 463

Bard H (1974) The effect of placental insufficiency on fetal and adult hemoglobin synthesis. Am J Obstet Gynecol 120: 67

Bard H, Prosmanne J (1985) Relative rates of fetal hemoglobin and adult hemoglobin synthesis in cord blood of infants of insulin-dependent diabetic mothers. Pediatrics 75: 1143

Baxi L, Barad D, Reece EA, Farber R (1984) Use of glucosylated hemoglobin as a screen for macrosomia in gestational diabetes. Obstet Gynecol 64: 347

Beaconsfield P, Ginsburg J, Korinski Z (1985) Glucose metabolism via the pentose phosphate pathway relative to cell replication and immunological response. Nature 205: 50

Beck-Nielsen H, Kühl C, Pedersen O, Bjerre-Christensen C, Klebe JG (1979) Decreased insulin binding to monocytes from normal pregnant women. J Clin Endocrinol Metab 49: 810

Beck P, Daughaday WH (1967) Human placental lactogen: Studies of its acute metabolic effects and disposition in normal man. J Clin Invest 46: 103

Beck P, Wells SA (1969) Comparison of the mechanisms underlying carbohydrate intolerance in subclinical diabetic women during pregnancy and during post-partum oral contraceptive steroid treatment. J Clin Endocrinol Metab 29: 807

Beischer N (1982) Das Vorkommen von Proinsulin und C-Peptid bei Gesunden und bei Diabetikerinnen. Fortschr Med 100: 1508

Bellmann O (1978) Zur Regulation des Kohlenhydratstoffwechsels beim Foeten und Neugeborenen — ein Konzept. Gynäkologe 11: 88

Bellmann O (1978 a) Der Einfluß der normalen Schwangerschaft auf den Kohlenhydratstoffwechsel. Gynäkologe 11: 56

Bellmann O (1978 b) Glucose-Toleranz und Insulin-Sekretion in der Schwangerschaft. In: Irsigler K, Regal H, Brändle J (Hrsg) Diabetes-Problem in der Schwangerschaft. Urban und Schwarzenberg, München Wien Baltimore, S 5–22

Bellmann O (1984) Diabetes Früherkennung in der Schwangerschaft. Laboratoriumsblätter 34: 97

Bellmann O, Hartmann E (1975) Influence of pregnancy on the kinetics of insulin. Am J Obstet Gynecol 122: 829

Bellmann O, Schlebusch H, Niesen M, Lang N (1981) Die Beziehungen zwischen „impaired glucose tolerance" und Blutzucker-Tagesprofilen in der Schwangerschaft. Akt Endokrin 2: 83

Benjamin F (1968) Glucose tolerance tests in the early puerperium. Am J Obstet Gynecol 100: 1102

Ben Khalifa F (1986) Diabetes in maternal and child health, the problem in North Africa. IDF Bulletin, pp 31-57-59

Benzie RJ, Doran TA, Harkins JL, Owen VMJ, Porter CJ (1974) Composition of the amniotic fluid and maternal serum in pregnancy. Am J Obstet Gynecol 119: 798

Berger W, Sonnenberg GE (1980) Blutzuckertagesprofile und Hämoglobin A 1 bzw. A 1 C zur Überwachung der Diabetesbehandlung. Schweiz Med Wochenschr 110: 485

Berger M, Chantelau E, Cüppers H, Gösseringer G, Jörgens V, Sonnenberg G, Wasser K (1981) Neuer Schulungskurs. Diabetes-Journal 31: 287

Berger M, Mühlhauser J, Jörgens V (1983) Die Evaluation der Diabetiker-Edukation. Fortschr Med 101: 212

Berglund G, Zetterstrom R (1954) "Infants of diabetic mothers. I.—Fetal hypoxia in maternal diabetes". Acta Paediatr 43: 368

Bergmeyer HU, Berndt E, Schmidt F, Stork H (1974) D-Glucose-Bestimmung mit Hexokinase und Glucose-6-Phosphat-Dehydrogenase. In: Bergmeyer HU (ed) Methoden der enzymatischen Analyse, vol 2, 3. Aufl. Verlag Chemie, Weinheim, S 1241

Berntorp K, Trell E, Thorell J, Hood B (1983) Relation between plasma insulin and blood glucose in a cross-sectional population study of the oral glucose tolerance test. Acta Endocrinol 102: 549

Bishop JS, Larner J (1967) Rapid activation-inactivation of liver uridine diphosphate glucose-glycogen transferase and phosphorylase by insulin and glucagon in vivo. J Biol Chem 242: 1354

Bisse E, Berger W, Fluckinger R (1982) Quantitation of glycosylated hemoglobin. Elimination of labile glycohemoglobin during sample hemolysis at pH 5. Diabetes 31: 630

Bissonnette JA, Black JA, Wickham WK, Acott KM (1981) Glucose uptake into plasma membrane vesicles form the maternal surface of human placenta. J Membrane Biol 58: 75

Björk O, Persson B (1984) Villous structure in different parts of the cotyledon in placentas of insulin-dependent diabetic women. Acta Obstet Scand 63: 37

Blohme G, Karlsson K, Waldenstrom J (1980) Early insulin response in latent gestational diabetes. Acta Med Scand 208: 309

Bloom SR, Johnston DI (1972) Failure of glucagon release in infants of diabetic mothers. Br Med J: 453

Blouquet Y, Senan C, Rosa J (1983) An automatic method for determination of glycosylated hemoglobins using low-pressure liquid chromatography. Applications 275: 41

Boyd JD (1959) Implantation of Ova. In: Eckstein P (ed) Memoirs of the society for endocrinology, no 6. Cambridge University Press, London, p 26

Boyd PA, Scott A, Keeling JW (1986) Quantitative structural studies on placentas from pregnancies complicated by diabetes mellitus. Br J Obstet Gynecol 93: 31

Bracero LA, Baxi LV, Rey HR, Yeh MN (1985) Use of ultrasound in antenatal diagnosis of large-for-gestational age infants in diabetic gravid patients. Am J Obstet Gynecol 152: 43

Brambell FWR (1970) The transmission of passive immunity from mother to young. Elsevier/North-Holland, Amsterdam

Brandau H (1974) Abhängigkeit endokrinologischer Parameter vom Enzymaktivitätsmuster der Placenta. Arch Gynecol 219: 380

Brans YW, Huff RW, Shannon DL, Hunter MA (1982) Maternal diabetes and neonatal macrosomia. I. Postpartum maternal hemoglobin A 1 C levels and neonatal hypoglycemia. Pediatrics 70: 576

Bray GA (1985) Obesity: definition, diagnosis and disadvantages. Med J Aust 142: 2

Brazy JE, Crenshaw MC, Brumley GW (1978) Amniotic fluid cortisol in normal and diabetic pregnant women and its relation to respiratory disease in the neonate. Am J Obstet Gynecol 132: 567

Brenner WE, Edelmann DA, Hendricks CH (1976) A standard of fetal growth for the United States of America. Am J Obstet Gynecol 126: 555

Bromberg YM, Abrahamov A, Salzberger M (1956) The effect of maternal anoxaemia on the foetal hemoglobin of the newborn. J Obstet Gynecol Br Emp 63: 875

Bucalo LR, Cohen RS, Ostrander CR, et al (1984) Pulmonary excretion of carbon monoxide in the human infant as an index of bilirubin production. II c. Evidence for the possible association of cord blood erythropoietin levels and postnatal bilirubin production in infants of mothers with abnormalities of gestational glucose metabolism. Am J Perinatol 2: 177

Buchanan TA, Unterman TG, Metzger BE (1985) The medical management of diabetes in pregnancy. Clin Perinat 12: 625

Bunn HF, Haney DN, Gabbay KH, Gallop PM (1975) Further identification of the nature and linkage of the carbohydrate in hemoglobin A 1 C. Biochem Biophys Res Comm 67: 103

Bunn HF, Gabbay KH, Gallop PM (1978) The glycosylation of hemoglobin: relevance to diabetes mellitus. Science 200: 21

Burkart W, Dame WR, Ruppin E, Schneider HPG (1984) Die Bedeutung von Hormonen im Fruchtwasser. I. Insulin und C-Peptid. Geburtsh Frauenheilk 44: 781

Burkart W, Braulke I, Hanker JP, Schneider HPG (1986) Complications and fetal outcome in diabetic pregnancies—164 cases reviewed. J Perinat Med 14: 293

Burke BJ, Sheriff RJ, Savage PE, Dixon HG (1979) Diabetic twin pregnancy: An unequal result. Lancet i: 1372

Burstein R, Berus AW, Hirata W, Blumenthal HT (1963) A comparative histo- and immunopathological study of the placenta in diabetes mellitus and in erythroblastosis fetalis. Am J Obstet Gynecol 86: 66

Burt RL, Davidson IWF (1974) Insulin half-life and utilization in normal pregnancy. Obstet Gynecol 43: 161

Buschard K, Busch I, Molsted-Petersen L, Hougaard P, Kuhl C (1987) Increased incidence of true type 1 diabetes acquired during pregnancy. Br J Med 294: 275

Cachon AM, Ghaddab M, Ketzis A, et al (1982) Fast fluctuations of glycosylated hemoglobins. I. Implications for the preparation and storage of samples for hemoglobin A 1 C determinations. Clin Chim Acta 121: 125

Calandra C, Abell DA, Beischer NA (1981) Maternal obesity in pregnancy. Obstet Gynecol 57: 8

Cameron BF (1979) Hemoglobin A 1 C (glycohemoglobin) in diabetic pregnancy: an indicator of glucose control and fetal size. Br J Obstet Gynaecol 73: 672

Carpenter MW, Coustan DR (1982) Criteria for screening tests for gestational diabetes. Am J Obstet Gynecol 144: 768

Carpenter MW, Coustan DR (1986) Screening for gestational diabetes. In: Jovanovic L, Peterson CM, Fuhrmann K (eds) Diabetes and pregnancy: teratology, toxicity and treatment, chapter 13. Praeger, p 245

Carson BS, Phillips AF, Simmons MA, et al (1980) Effects of a sustained insulin infusion upon glucose uptake and oxygenation of the ovine fetus. Pediatr Res 14: 147

Casper DJ, Benjamin F (1970) Immunoreactive insulin in amniotic fluid. Obstet Gynecol 35: 389

Cassady G, Blake M, Bailey P, Younger B, Sumners J (1977) Amniotic fluid glucose in pregnancies complicated by diabetes. Am J Obstet Gynecol 127: 21

Castagnola M, Caradonna P, Salvi ML, et al (1983) The chromatographic separation of glycosylated haemoglobins: A comparison between macro- and micromethods. J Clin Chem Clin Biochem 21: 223

Catalano PM, Bernstein IM, Wolfe RR, et al (1986)

Subclinical abnormalities of glucose metabolism in subjects with previous gestational diabetes. Am J Obstet Gynecol 155: 1255

Cerasi E, Grill V (1977) Die Rolle des zyklischen AMP bei der Kontrolle der Insulinsekretion. Münch Med Wochenschr 119: 1229

Challier JC, Hanguel S, Desmaiziers V (1986) Effect of insulin on glucose uptake and metabolism in the human placenta. Endocrin Metab 62: 803

Charache S, Catalano P, Burns S, et al (1985) Pregnancy in carriers of high-affinity hemoglobins. Blood 65: 713

Cheney·C, Shragg P, Hollingsworth D (1985) Demonstration of heterogeneity in gestational diabetes by a 400-kcal breakfast meal tolerance test. Obstet Gynecol 65: 17

Christensen HN, Streicher JA (1948) Association between rapid growth and elevated cell concentrations of amino acids. I. In fetal tissues. Biol Chem 175: 95

Churchill JA, Berendes HW, Nemore J (1969) Neuropsychological deficits in children of diabetic mothers. A report from the collaborative study of cerebral palsy. Am J Obstet Gynecol 105: 257

Cimbala MA, Lamers WH, Nelson K, Nonahan JE, Yoo-Warren H, Hanson RW (1982) Rapid changes in the concentration of phosphoenolpyruvate carboxykinase mRNA in rat liver and kidney. J Biol Chem 257: 7629

Cirkel U, Burkart W, Stähler E, Buchholz R (1986) Effects of alloxan-induced diabetes mellitus on the metabolism of the rat placenta. Arch Gynecol 237: 155

Clavero JA, Botella Llusia J (1963) Measurement of the villous surface in normal and pathologic placentas. Am J Obstet Gynecol 80: 234

Coen RW, Porreco R, Cousins L, Sandler JA (1980) Postpartum glycosylated hemoglobin levels in mothers of large-for-gestational age infants. Am J Obstet Gynecol 136: 380

Coetzee EJ, Jackson WPU, Berman PA (1980) Ketonuria in pregnancy—with special reference to calorie restricted food intake in obese diabetics. Diabetes 29: 177

Compagnucci P, Crechini MG, Bolli G, et al (1983) Hyperglycemia alters the physico-chemical properties of proteins in erythrocyte membranes of diabetic patients. Horm Metabol Res 15: 263

Cornblath M (1975) The effect of hi fat diet during pregnancy on fetal brain metabolism. Int symp on The adipose child, Herzliya 1975

Cornblath M, Schwartz R (1976) Disorders of carbohydrate metabolism in infancy. WB Saunders, Philadelphia, p 115

Coulsten M, Hollenbeck CB, Liu GC, Williams RA, Starich GH, Mazzaferri EL, Reaven GM (1984) Effect of source of dietary carbohydrates on plasma glucose, insulin and gastric inhibitory polypeptide responses to test meals in subjects with non insulin

dependent diabetes mellitus. Am J Clin Nutrit 40: 965

Cousins L, Dattel J, Hollingsworth DR, Zettner A (1984) Glycosylated hemoglobin as a screening test for carbohydrate intolerance in pregnancy. Am J Obstet Gynecol 150: 455

Cousins L, Dattel B, Hollingsworth D, Hulbert D, Zettner A (1985) Screening for carbohydrate intolerance in pregnancy: A comparison of two tests and reassessment of a common approach. Am J Obstet Gynecol 153: 381

Coustan DR, Lewis SB (1978) Insulin therapy for gestational diabetes. Obstet Gynec 51: 306

Coustan DR, Imarah J (1984) Prophylactic insulin treatment of gestational diabetes reduces the incidence of macrosomia, operative delivery and birth trauma. Am J Obstet Gynecol 150: 836

Craighead JE (1985) Viral diabetes. In: Volk BW, Arquilla ER (eds) The diabetic pancreas, 2nd edn. Plenum Medical Book Company, New York London, pp 439–466

Crapo PA, Reaven RD, Olefsky J, Alto P (1976) Plasma glucose and insulin response to orally administered simple and complex carbohydrates. Diabetes 25: 741

Cremer HD, Aign W, Elmadfa I, Muskat E, Schäfer H (1983) Die große Nährwert Tabelle. Die Kalorien/Joule- und Nährstoffgehalte unserer Lebensmittel. Gräfe und Unzer, München

Cruikshank DP, Pitkin RM, Varner MW (1983) Calcium metabolism in diabetic mother, fetus, and newborn infant. Am J Obstet Gynecol 145: 1010

Curtiss LK, Witztum JL (1983) A novel method for generating region-specific monoclonal antibodies to modified proteins. Application to the identification of human glucosylated low density lipoproteins. J Clin Invest 72: 1427

Dahl-Jorgensen K, Larsen AE (1982) HbA$_1$ determination by agar gel electrophoresis after elimination of labile HbA$_1$: A comparison with ion-exchange chromatography. Scand J Clin Laborat Invest 42: 27

Dandrow RV, O'Sullivan JB (1966) Obstetric hazards of gestational diabetes. Am J Obstet Gynecol 96: 1144

Daneman D, Luley N, Becker DJ (1982) Diurnal glucose-dependent fluctuations in glycosylated hemoglobin levels in insulin-dependent diabetes. Metab Clin Exper 31: 989

Dankmeijer HF (1981) Diabetes mellitus. A guide for the general practitioner. Boehringer Ingelheim GmbH (Hrsg)

Davidsen O (1974) Hemoglobin F in newborn infants of diabetic mothers. Acta Endocrinol 182: 73

Daweke H, Hüter KA (1970) Diabetes und Gravidität. Verh Dtsch Ges Inn Med 76: 341

Demers LM, Gabbe SG, Villee CA, Greep RO (1972) The effect of insulin on human placental glycogenesis. Endocrinology 91: 270

Dempe A, Michaelis D, Heinke P, Franke D, Seitz W,

Barthel D, Bauch K (1978) Häufigkeit des Gestationsdiabetes sowie der Veränderung der Insulinsekretion bei Schwangeren. 1. Mitteilung: Untersuchungen von diabetesverdächtigen Schwangeren mittels des Glucoseinfusionstests (GIT). Zbl Gynäkol 100: 1559

De Muylder X (1984) Perinatal complications of gestational diabetes: the influence of the timing of the diagnosis. Europ J Obstet Gynecol Reprod Biol 18: 35

De Pirro R, Forte F, Bertoli A, Greco AV, Lauro R (1981) Changes in insulin receptors during oral contraception. J Clin Endocrinol Metab 52: 29

De Prins R, Van Asche A, Milner RDG (1983) C-peptide levels in amniotic fluid in experimental fetal growth retardation. Biol Neonate 43: 181

D'Ercole AJ, Applewhite GT, Underwood LE (1980) Evidence that somatomedin is synthesized by multiple tissues in the fetus. Develop Biol 75: 315

De Simone J, Biel M, Heller P (1982) Maintenance of fetal hemoglobin (HbF) elevations in the baboon by prolonged erythropoietic stress. Blood 60: 519

Desoye G, Hofmann HH, Weiss PAM (1986) Insulin regulates by its receptor the placental concentrations of glycogen, and N-acetyl-neuraminic acid. Placenta 7: 463

Desoye G, Hoffmann H, Weiss PAM (1986 a) The affinity of the placental insulin receptor is modulated by the membrane lipid composition. IIIrd intern symp insulin receptors insulin action. Abstract of reports, p 109

Desoye G, Schweditsch MD, Pfeiffer KP, Zechner R, Kostner GM (1987) Correlation of hormones with lipid and lipoprotein levels during normal pregnancy and postpartum. J Clin Endocrinol Metab 64; 704

Desoye G PAM (1987a) Influence of the lipid environment on insulin binding to placental membranes from normal and diabetic mothers. Troph Res 2: 29

Diabetes mellitus. Report of a WHO study group (1985) WHO, Geneva (Technical Report Series 727)

Diamant YZ, Mayorek N, Neumann S, Shafrir E (1975) Enzymes of glucose and fatty acid metabolism in early and term human placenta. Am J Obstet Gynecol 121: 58

Diamant YZ, Metzger BE, Freinkel N, Shafrir E (1982) Placental lipid and glycogen content in human and experimental diabetes mellitus. Am J Obstet Gynecol 145: 5

Diamant YZ, Kisselevitz R (1983) Metabolic changes in human placental tissue in diabetes mellitus. Troph Res 1: 209

Diamant YZ, Kisselevitz R, Shafrir E (1984) Changes in activity of enzymes related to glycolysis, gluconeogenesis and lipogenesis in placentae from diabetic women. Placenta 5: 55

Dimitriadis G, Gryer P, Gerich J (1985) Prolonged hyperglycemia during infusion of glucose and somatostatin impairs pancreatic A- and B cell responses to decrements of plasma glucose in normal man: evidence of induction of altered sensitivity to glucose. Diabetologia 28: 63

Di-Pietro DL, Gutierrez-Correa J, Thaidigsman JH (1967) Glucose metabolism by human placental villi. Biochem J 103: 246

Ditzel J (1976) Oxygen transport impairment in diabetes. Diabetes 25 [Suppl] 2: 832

Draisey TF, Gagneja GL, Thibert RJ (1977) Pulmonary surfactant and amniotic fluid insulin. Obstet Gynecol 50: 197

Drazancic A, Kuvacic J (1974) Amniotic fluid glucose concentration. Am J Obstet Gynecol 120: 40

Duncan RW, Gilham PT (1975) Isolation of transfer RNA isoacceptors by chromatography on dihydroxy-boryl-substituted cellulose, polyacrylamide, and glass. Anal Biochem 66: 532

Dunn PJ, Cole RA, Soeldner JS, Gleason RE, Kwa E, Firoozabadi H, Younger D, Graham CA (1979) Temporal relationship of glycosylated hemoglobin concentrations to glucose control in diabetics. Diabetologia 17: 213

Dunn PM (1979) Low birthweight. Incidence, aetiology and prevention. In: Hugh Philpott R (ed) Maternity services in the developing world; what the community needs. Proceedings of the seventh group of the royal college of obstetricians and gynecologists, pp 228–232

Duran-Garcia S, Nieto JG, Cabello AM (1979) Effect of gestational diabetes on insulin receptors in human placenta. Diabetologia 16: 87

Ebrahim G (1979) Influence of maternal nutrition on pregnancy outcome. In: Hugh Philpott R (ed) Maternity services in the developing world; what the community needs. Proceedings of the seventh group of the royal college of obstetricians and gynecologists, pp 228–232

Edlinger B (1986) System einer direkten Vergleichsmöglichkeit von Kohlenhydrat-induzierten Blutglucosespiegeln in Hinblick auf die Resorptionsdynamik verschiedener Kohlenhydrate bei verschiedenen Personen. Inauguraldissertation zur Erlangung der Doktorwürde an der Naturwissenschaftlichen Fakultät der Karl-Franzens-Universität in Graz

Elliot JP (1983) Magnesium sulfate as a tocolytic agent. Am J Obstet Gynecol 147: 277

Elliot JP, Garite TJ, Freeman RK, Mc Quown DS, Patel JM (1982) Ultrasonic prediction of fetal macrosomia in diabetic patients. Obstet Gynecol 60: 159

Emerson K (1975) Maternal energy as a guide to diet in pregnancy. In: Camerini-Davalos RA, Cole HS (eds) Early diabetes in early life. Academic Press, New York, p 435

Emmerich P, Godel E, Dempe A (1978) Morphologie der Plazenta bei praemanifestem mütterlichem Diabetes mellitus. Z Geburtsh Perinat 182: 79

Enders RH, Jodol RM, Donohue TM, Smith CH (1976)

Placental amino acid uptake. III. Transport systems for neutral amino acids. Am J Physiol 230: 706

Espinosa De Los Monteros AM, Driscoll SG, Steinke J (1970) Insulin release from isolated human fetal pancreatic islets. Science 168: 1111

Essex NL, Pyke DA, Watkins PJ, Brundenell JM, Gamsu HR (1973) Diabetic pregnancy. Br Med J 4: 89

Evans JS, Gerritsen GC, Mann KM, Owen SP (1965) Anti-tumor and hyperglycemie activity of streptozotocin (NSC 37917) and its cofactor (U 15774). Cancer Chemotherapy Rep 48: 1

Exon PD, Dixon K (1974) Insulin antibodies in diabetic pregnancy. Lancet ii: 126

Faber JJ, Thornburg KL (1983) Placental physiology. Structure and function of fetomaternal exchange. Raven Press, New York

Fadel HE, Hammoud SD, Huff TA, Harp RJ (1979) Glycosylated hemoglobin in normal pregnancy and gestational diabetes mellitus. Obstet Gynecol 54: 322

Fadel HE, Reynolds A, Stallings M, Abraham EC (1981) Minor (glycosylated) hemoglobins in cord blood of infants of normal and diabetic mothers. Am J Obstet Gynecol 239: 397

Fadel HE, Hammond SD (1982) Diabetes mellitus and pregnancy: management and results. J Reprod Med 27: 56

Fajans S (1987) Maturity onset diabetes in young black americans. New Engl J Med 317: 380

Farquhar JW (1969) Prognosis for babies born to diabetic mothers in Edinburgh. Arch Dis Child 44: 36

Farquhar JW (1976) The infant of diabetic mother. Clin Endocrinol Metab 5: 237

Faulk WP, Galbraith RM, Keane, M (1980) Immunological consideration of the feto-placental unit in maternal diabetes. In: Irvine J (ed) Immunology of diabetes. Teviot, Edinburgh, p 309

Faul WP, Jarret R, Keane M, Johnson PM, Boackle RJ (1980) Immunological studies of human placenta: complement components in immature and mature chorionic villi. Clinic Exp Immunol 40: 299

Feige A, Kunzel W, Mitzkat HJ (1977) Fetal and maternal blood glucose, insulin and acid base observations following maternal glucose infusion. J Perinat Med 5: 84

Feige A, Kellermann W, Mitzkat HJ, Lehmann V (1979) Glukosekinetik im Fruchtwasser sub partu während intravenöser Glukoseinfusion an die Mutter. Z Geburtsh Perinat 183: 45

Feige A, Nössner U (1985) Das Verhalten des glycosylierten Hämoglobins (HbA 1) in normaler und pathologischer Schwangerschaft. Z Geburtsh Perinat 189: 13

Feldman F, Rubin H, Nawabi I, Abbondante AM, Posner NA, Stefanyshyn NI (1984) Glucosylated fetal hemoglobin. Diabetes 33: 81

Felig P, Lynch V (1970) Starvation in human pregnancy: Hypogylcemia, hyperinsulinemia and hyperketonemia. Science 170: 990

Finne Ph (1964) Erythropoietin level in the amniotic fluid, particularly in Rh-immunized pregnancies. Acta Paediatr Scand 53: 269

Finne Ph (1968) Erythropoietin production in fetal hypoxia and in anemic-uremic patients. Ann NY Acad Sci 149: 497

Firpo A, Jovanovic L, Peterson CM (1981) Basement membrane thickening of placenta from pregnancies complicated by diabetes. Diabetes 30 [Suppl] 1: 140

Fischer U, Horkey Z (1966) Vorläufige Untersuchungen zum Glykogengehalt sowie Sauerstoff- und Glukoseverbrauch der Plazenta in vitro bei Diabetes mellitus. Zentralbl Gynaekol 88: 1427

Fischl F, Binstorfer E, Reinold E (1981) Neugeborene über 4000 Gramm. Wien Med Wochenschr 131: 471

Fitzgerald MG, Malins JM, O'Sullivan DJ (1961) Prevalence of diabetes in women thirteen years after bearing a big baby. Lancet i: 1250

Fitzgibbons JR, Koler RD, Jones RT (1976) Red cell age-related changes of hemoglobins A 1 a + b and A 1 c in normal and diabetic subjects. J Clin Invest 58: 820

Foglia VG, Chieri RA, Mirta Cattaneo De Peralta Ramos (1970) Mechanisms of disturbance during pregnancy in the diabetic female rat. Horm Metab Res 2: 76

Forrest JM, Menser MA, Burgess JA (1971) High frequency of diabetes mellitus in young adults with congenital rubella. Lancet ii: 332

Fox H (1978) Diabetes mellitus. In: Pathology of the placenta. WB Saunders, London, p 223

Fox H (1978 a) The pathology of the human placenta. In: Major problems in pathology, vol 7. WB Saunders, London

Fraser RB, Ford FA, Milner RDG (1983) A controlled trial of a high dietary fibre intake in pregnancy-effects on plasma glucose and insulin levels. Diabetologia 25: 238

Fredholm BB, Lunell NO, Persson B, Wagner W (1978) Actions of Salbutamol in late pregnancy: Plasma cyclic AMP, insulin and C-peptide, carbohydrate and lipid metabolites in diabetic and nondiabetic women. Diabetologia 14: 235

Freinkel N (1965) Effects of the conceptus on maternal metabolism during pregnancy. In: On the nature and treatment of diabetes. Amsterdam, Excerpta Medica Foundation, p 679

Freinkel N, Metzger BE (1979) Pregnancy as a tissue culture experience: the critical implications of maternal metabolism for fetal development. In: Pregnancy metabolism, diabetes and the fetus. Excerpta Medica, Amsterdam [Ciba Foundation Symposium 63 (new series), p 3]

Freinkel N (1980) Of pregnancy and progeny. Diabetes 29: 1023

Freinkel N, Josimovich J (1980) American Diabetes Association workshop-conference on gestational diabetes: Summary and recommendations. Diabetes Care 3: 499

Freinkel N, Metzger BE, Phelps RL, Dooley SL, Ogata ES, Radvany RM (1985) Heterogeneity of maternal age, weight, insulin secretion, HLA antigens, and islet cell antibodies and the impact of maternal metabolism on pancreatic B-cell and somatic development in the offspring. Diabetes 34 [Suppl] 2: 1

Freinkel N, Hoet JJ (1986) Guest editors' introduction: Steering a future course for pregnancies complicated by diabetes. IDF Bulletin: 55

Freychet P, Roth J, Neville DM (1971) Monoiodoinsulin: Demonstration of its biological acitvity and binding to fat cells and liver membranes. Biochem Biophys Res Comm 43: 400

Fuhrmann K (1982) Diabetic control and outcome in the pregnant patient. In: Petersen CM (ed) Diabetes management in the 80's. Praeger, New York

Fuhrmann K (1982) Gestationsdiabetes — Endokrinmetabolische Veränderungen, Diagnostik, Reproduzierbarkeit und prädisponierende Faktoren. Inaugural-Dissertation, Ernst-Moritz-Arndt-Universität Greifswald

Fuhrmann K, Semmler K, Reiher H (1980) Das Verhalten des HPL (Humanes plazentares Laktogen) unter Glukoseinfusionsbelastung in der Spätschwangerschaft. Zbl Gynäkol 102: 1031

Fuhrmann K, Keilacker (1983) Insulinantikörper im Fruchtwasser. Persönliche Mitteilung

Fujita H, Fujino-Kurihara (1986) Fine structural aspects of the degenerating process of pancreatic islets of NOD mice. In: Tarui S, Tochino Y, Nonaka K (eds) Insuletis and type 1 diabetes. Academic Press Japan, pp 51–59

Gabbay KH (1982) Glycosylated hemoglobin and diabetes mellitus. Med Clin N Am 66: 1309

Gabbay KH, Hasty K, Breslow JL, Ellison RC, Bunn HF, Gallop PM (1977) Glycosylated hemoglobins and long-term blood glucose control in diabetes mellitus. J Clin Endocrin Met 44: 859

Gabbe SG (1979) Application of a scientific rationale to the management of the pregnant diabetic. In: Merkatz IR, Adams PJ (eds) The diabetic pregnancy. Grune and Stratton, New York

Gabbe SG, Demers LM, Greep RO, Villee CA (1972) Placental glycogen metabolism in diabetes mellitus. Diabetes 21: 1185

Gabbe SG, Demers LM, Greep RO, Villee CA (1972 a) The effects of hypoxia on placental glycogen metabolism. Am J Obstet Gynecol 114: 540

Gabbe SG, Qilligan EJ (1977) Fetal carbohydrate metabolism: Its clinical importance. Am J Obstet Gynecol 127: 92

Gabbe SG, Lowensohn RJ, Mestman JH, Freeman RK, Goebelsmann U (1977) Lecithin/sphingomyelin ratio in pregnancies complicated by diabetes mellitus. Am J Obstet Gynecol 128: 757

Gabbe SG, Mestman JH, Freeman RK, Anderson GV, Lowensohn RL (1977) Management and outcome of class A diabetes mellitus. Am J Obstet Gynecol 127: 465

Gabbe SG, Mestman JH, Freeman RK, Goebelsmann UT, Lowensohn RJ, Nochimson D, Cetrulo C, Quilligan EJ (1977) Management and outcome of pregnancy in diabetes mellitus, class B to R. Am J Obstet Gynecol 129: 723

Galbraith GMP, Galbraith RM, Paulsen EP (1981) Placental immunopathology in gestational diabetes. Placenta [Suppl] 3: 183

Galbraith RM (1978) Diabetes and pregnancy. In: Immunological aspects of diabetes mellitus. CRC Press, Boca Raton, p 65

Galbraith RM, Faulk WP (1979) Immunologic considerations of the maternofetal relationship in diabetes mellitus. In: Merkatz IR, Adam PAJ (eds) The diabetic pregnancy: a perinatal perspective. Grune and Stratton, New York, p 111

Galbraith RM, Fox H, Hsi B, Galbraith GMP, Bray RS, Faulk WP (1980) The human materno-fetal relationship in malaria. II. Histological, ultrastructural and immunopathological studies of the placenta. Trans Royal Soc Trop Med Hyg 74: 61

Gambhir KK, Archer JA, Bradley CJ (1978) Characteristics of human erythrocyte insulin receptor. Diabetes 27: 701

Gamlen TR, Ayusley-Green A, Irvine WJ, McCallum CJ (1977) Immunological studies in the neonate of a mother with Addison's disease and diabetes mellitus. Clin Exp Immunol 28: 192

Gardner LI, Stern MP, Haffner SM, et al (1984) Prevalence of diabetes in Mexican Americans. Relationship to percent of gene pool derived from native american sources. Diabetes 33: 86

Garlick RL, Mazer JS, Higgins PJ, Bunn HF (1983) Characterization of glycosylated hemoglobins. Relevance to monitoring of diabetic control and analysis of other proteins. J Clin Invest 71: 1062

Giedion A, Haeflinger H, Dangel P (1973) Acute pulmonary X-ray changes in hyaline membrane disease treated with artificial ventilation and positive endexpiratory pressure (PEEP). Pediat Radiol 1: 154

Gillmer MDG, Mazibuko D (1979) Pyridoxine treatment of chemical diabetes in pregnancy. Am J Obstet Gynecol 133: 499

Gillmer MDG, Oakley NW, Beard RW, Nithyananthan R, Cawston M (1980) Screening for diabetes during pregnancy. Br J Obstet Gynec 87: 377

Ginsburg J, Jeacock NN (1966) Some aspects of human carbohydrate metabolism in human diabetes. J Obstet Gynaec Br Cwlth 73: 452

Glatthaar C, Welborn TA, Stenhouse NS, Carcia-Webb P (1985) Diabetes and impaired glucose tolerance: a prevalence estimate based on the Busselton 1981 survey. Med J Aust 143: 436

Goldberg JD, Franklin B, Lasser D, Jornsay DL, Hausknecht RU, Ginsberg-Fellner F, Berkowitz RL (1986) Gestational diabetes: impact of home glucose monitoring on neonatal birth weight. Am J Obstet Gynecol 154: 546

Golditch IM, Kirkman K (1978) The large fetus. Management and outcome. Obstet Gynecol 52: 26

Goldstein A, Elliott J, Lederman S, Worcester B, Russell P, Linzey EM (1983) Economic effects of self-monitoring of blood glucose concentrations by women with insulin dependent diabetes during pregnancy. J Reprod Med 126: 449

Golz N, Klausmeyer A, Mast H (1986) Glycohemoglobin as a screening test for gestational diabetes in mothers of large-for-gestational-age babies (LAG)? In: Weiss PAM (ed) Abstracts of free communications. 1. Internationales Grazer Symposium Kohlenhydratstoffwechselstörungen in der Schwangerschaft. Maudrich, Wien München Bern (Probl perinat Med 15, p 195)

Gouedard HA, Menez JF, Meskar A, Caroff J, Lucas D, Legendre JM (1984) Perinatal assessment of glycaemic control in newborn infants of diabetic mothers. Diabetologica 27: 553

Gould BJ, Hall PM, Cook JGH (1982) Measurement of glycosylated haemoglobins using an affinity chromatography method. Clin Chim Acta 125: 41

Graefenstein K, Duchna W (1981) Zur Abhängigkeit des HbA 1 (Glykohämoglobin) von Lebensalter, Schwangerschaft und Glukosetoleranz. Z Ges Inn Med 36: 920

Gragnoli G, Tanganelli I, Signorini AM, et al (1982) Non-enzymatic glycosylation of serum proteins as in indicator of diabetic control. Acta Diabetol Lat 19: 161

Grandjean H, Sarramon MF, De Monzon J, Reme JM, Pontonier G (1980) Detection of gestational diabetes by means of ultrasonic diagnosis of excessive fetal growth. Am J Obstet Gynecol 138: 790

Granner D, Andreone T, Sasaki K, Beale E (1983) Inhibition of transcription of the phosphoenolpyruvate carboxykinase gene by insulin. Nature 305: 549

Grant PT, Oats JN, Beischer NA (1986) The long-term follow-up of women with gestational diabetes. ANZ J Obstet Gynaecol 26: 17

Gratacos JA, Neufeld N, Kumar D, Artal R, Paul RH, Mestman J (1981) Monocyte insulin binding studies in normal and diabetic pregnancies. Am J Obstet Gynecol 141: 611

Gross T, Sokol RJ, King KC (1980) Obesity in pregnancy: risk and outcome. Obstet Gynecol 56: 446

Guibaud S, Bonnet M, Khalil F, Combet A, Toulon JM, Dumont M (1978) Glucose concentration in amniotic fluid from anencephalic pregnancies (letter). Lancet i: 661

Gyves MT, Schulman PK, Merkatz IR (1980) Results of individualized intervention in gestational diabetes. Diabetes Care 3: 495

Habicht JP, Lechtig A, Yarbrough CH, Klein RE (1974) Maternal nutrition, birthweight and infant mortality. In: Size at birth. Associated Sci Publ, (Ciba Foundation Symposium, pp 354–377)

Hadden DR (1979) Asymptomatic diabetes in pregnancy. In: Sutherland HW, Stowers JM (eds) Carbohydrate metabolism in pregnancy and the newborn. Springer, Berlin Heidelberg New York, p 408

Hadden DR (1980) Screening for abnormalities of carbohydrate metabolism in pregnancy 1966–1967: the Belfast experience. Diabetes Care 3: 440

Hafez ESE (1964) Uterine and placental enzymes. Acta Endocrinol 46: 217

Hall PM, Cawdell GM, Cook JGH, Gould BJ (1983) Measurement of glycosylated hemoglobins and glycosylated plasma proteins in maternal and cord blood using an affinity chromatography method. Diabetologia 25: 477

Hammons GT, Junger K, McDonald JM, Ladenson JH (1982) Evaluation of three minicolumn procedures for measuring hemoglobin A 1. Clin Chem 28: 1775

Hansmann M (1976) Ultraschallbiometrie im II. und III. Trimester der Schwangerschaft. Gynäkologe 9: 133

Haour F, Bertrand J (1974) Insulin receptors in the plasma membranes of human placenta. J Clin Endocrinol Metab 38: 334

Harrison LC, Billington T, Clark S, Nichols R, East I, Martin FIR (1977) Decreased binding of insulin by receptors on placental membranes from diabetic mothers. J Clin Endocrinol Metab 44: 206

Harrison LC, Itin A (1980) Purification of the insulin receptor from human placenta by chromatography on immobilized wheat germ lectin and receptor antibodies. J Biol Chem 255: 12066

Haust MD (1981) Maternal diabetes mellitus—effects on the fetus and placenta. Monographs in Pathology 22: 201

Haycock GB, Schwartz GJ, Wisotsky DH (1978) Geometric method for measuring body surface area: a height-weight formula validated in infants, children and adults. J Pediatr 93: 62

Heijkenskjold F (1958) Lipoid content of placentae from normal and diabetic mothers. Acta Obstet Gynec Scand 37: 386

Heijkenskjold F, Gemzell CA (1957) Glycogen content in the placenta of diabetic mothers. Acta Paediatr 46: 74

Heilmann L (1981) Hämorheologische Untersuchungen in der Schwangerschaft. Perimed, Erlangen

Heisig N (1975) Diabetes und Schwangerschaft. G Thieme, Stuttgart

Herold DA, Boyd JC, Bruns DE, et al (1983) Measurement of glycosylated hemoglobins using boronate affinity chromatography. Annals Clin Lab Sci 13: 482

Hill DE (1978) Effect of insulin on fetal growth. Semin Perinat 2: 319

Hill DJ, Milner RDG (1980) Increased somatomedin and cartilage metabolic activity in rabbit fetuses injected with insulin in utero. Diabetologia 19: 143

Hinckers HJ (1978) Kindsgeschlecht und Glukosetoleranz: Untersuchungen mittels des kontinuierlich registrierten i.v. Glukosetoleranztests (i.v.-Glucogramm). Gynäkologe 11: 99

Hochberg Z, Perlman R, Benderli A, Brandes JM (1982) The effect of glucose 2-deoxy-D-glucose and insulin on estradiol secretion by cultured human trophoblast. Biochem Biophys Res Commun 108: 102

Hodge JE (1955) The Amadori rearrangement. Adv Carbohydr Chem 10: 169

Hoet JJ (1977) Diabetes und Schwangerschaft. Münch Med Wochenschr 119: 659

Hoet JJ, Person B, Van Assche FA (1985) Diabetes and pregnancy. In: Shearman RS (ed) Clinical reproductive endocrinology. Churchill/Livingstone, Edinburgh, pp 229–327

Hoet JJ, Reusens-Billen B (1986) Relationship of pancreatic morphologic changes to the onset of diabetes. In: Krall LP (ed) Worldbook of diabetes in practice, vol 2. Elsevier, Amsterdam New York Oxford, pp 33–37

Hoet JJ, Reusens B, Remacle C (1986) Effect du diabete maternel sur le systeme entero-pancreatique du foetus du rat. Bull l'Acad Roy Med Belgique 141: 370

Hoet JJ, Remacle C (1987a) Organisation of the pancreatic islets with special reference to diabetes. In: De Groot LJ, Rubenstein AH (eds) Endocrinology; the pancreatic islets, 2nd edn. Grune & Stratton, Florida (in press)

Hoet JJ, Reusens B, Remacle C (1987b) Lessons from the diabetic pancreas. Horm Metab Res 19 (in press)

Hofmann HMH, Weiss PAM (1986) Insulintherapie des Gestationsdiabetes. In: Weiss PAM (ed) 1. Internationales Grazer Symposium: Kohlenhydratstoffwechselstörungen und Schwangerschaft. Maudrich, Wien München Bern 1987 (Probl perinat med 15, p 103)

Hohenauer L (1980) Intrauterine Wachstumskurven für den Deutschen Sprachraum. Z Geburtsh Perinat 184: 167

Holländer HJ (1976) Das Ultraschallbild des Feten im zweiten und dritten Trimester der Schwangerschaft. Gynäkologe 9: 123

Hollingsworth DR (1983) Alterations of maternal metabolism in normal and diabetic pregnancies: Differences in insulin dependent and gestational diabetes. Am J Obstet Gynecol 146: 417

Hollingsworth DR (1984) Pregnancy, diabetes and birth. A management guide, 1st edn. Williams and Wilkins, Baltimore

Hollingsworth DR, Grundy SM (1982) Pregnancy-associated hypertriglyceridemia in normal and diabetic women. Differences in insulin dependent, non-insulin-dependent, and gestational diabetes. Diabetes 31: 1092

Hommel G, Muck BR (1977) Ein diskriminanzanalytischer Ansatz zur Beurteilung der Glucosetoleranz Schwangerer anhand von Seruminsulin-Verlaufsbeobachtungen. Arch Gynäk 223: 315

Horky Z (1965) Die Reifungsstörungen der Placenta bei Diabetes mellitus. Zentralbl Gyn 45: 1555

Hornnes PJ, Kühl C (1980) Endocrine pancreatic sensitivity to glucose in women with gestational diabetes. Obstet Gynecol 62: 305

Hornnes PJ, Kühl C (1980) Plasma insulin and glucagon responses to isoglycemic stimulation in normal pregnancy and post partum. Obstet Gynecol 55: 425

Hornnes PJ, Kühl C, Lauritsen KB (1981) Gastro-enteropancreatic hormones in normal pregnancy: Response to a protein rich meal. Eur J Clin Invest 11: 345

Hornnes PJ, Kühl C, Lauritsen KB (1981) Gastrointestinal insulinotropic hormones in normal and gestational diabetic pregnancy: Response to oral glucose. Diabetes 30: 504

Hornnes PJ, Kühl C, Lauritsen KB (1982) Gastro-enteropancreatic hormones in gestational diabetes: response to a protein rich meal. Horm Metab Res 14: 335

Hornnes PJ, Kühl C (1984) Cortisol and glucose tolerance in normal pregnancy. Diabet Metabol 10: 1

Hosoya N, Hagerman D, Villee C (1960) Stimulation of fatty acid synthesis by estradiol in vitro. Biochem J 76: 297

Houghton DJ, Shackleton P, Obikekwe BC, Charol T (1984) Relationship of maternal and fetal levels of human placental lactogen to the weight and sex of the fetus. Placenta 5: 455

Huell D, Elphick MC (1979) Evidence for fatty acid transfer across the human placenta. In: Blarol WB (ed) Pregnancy, metabolism diabetes and fetus, Ciba Foundation Series 63: 75

Hubert C, Mondon F, Cedard L (1981) Distribution, quantification and biological activity of messenger RNA coding for human chorionic somatomammotropin during normal pregnancy. Mol Cell Endocrin 24: 339

Huisman W, Kuijkjken JPAA, Tan-Tjiong HL, et al (1982) Unstable glycosylated hemoglobin in patients with diabetes mellitus. Clin Chim Acta 118: 303

Hull D (1975) Storage and supply of fatty acids before and after birth. Br Med Bull 31: 32

Hulstaert CE, Torrings JL, Koudstall J, Hardonk MJ, Molenaar I (1973) The characteristic distribution of alkaline phosphatase in the full term human placenta. Gynecol Invest 4: 23

Hultquist GT, Olding L (1975) Pancreatic islet fibrosis in young infants of diabetic mothers. Lancet ii: 1015

Hytten FE (1982) Körpermasse und Körpergewicht Erwachsener. In: Wissenschaftliche Tabellen Geigy. Teilband Somatometrie und Biochemie, 8. Aufl. Ciba-Geigy, Basel

Ilan J, Pierce DR, Hochberg AA, Folman R, Gyves MT (1984) Increased rates of polypeptide chain elongation in placental explants from human diabetics. Proc Natl Acad Sci USA 81: 1366

Illsley NP, Wootion R, Penfold P, Hall S, Duffy S (1985) Lactate transfer across the perfused human placenta. Placenta 7: 209

Irsigler K (1978) Was ist eine gute Diabeteseinstellung?

In: Irsigler K, et al (Hrsg) Diabetes-Probleme in der Schwangerschaft. Urban und Schwarzenberg, München Wien Baltimore, S 53

Irsigler K, Bali Ch, Artner J (1982) Home monitoring and open-loop systems in pregnancy. In: Peterson CM (ed) Diabetes management in the '80s. Praeger, p 268

Irsigler K, Weiss PAM, Leodolter S, Bali C, Rost I, Pilz I, Hofmann HM, Chwatal K, Willinger C, Schmid B, Kirisits M (1986) Impacts of glucose tolerance in pregnancy on neonatal pathology. Results from a prospective study Graz-Vienna. In: Weiss PAM (ed) 1st international Graz symposium: Disturbance of carbohydrate metabolism during pregnancy. Maudrich, Wien München Bern (Probl perinat med 15, p 85)

Jackson WPU, Woolf N (1958) Maternal prediabetes as a cause of the unexplained stillbirth. Diabetes 7: 446

Javid J, Pettis PK, Koenig RJ, Cerami A (1978) Immunological characterization and quantification of HbA 1 c. Haematology 38: 329

Jelliffe DB (1966) In: Wissenschaftliche Tabellen Geigy (1982). Teilband Somatometrie und Biochemie, 8. Aufl. Ciba Geigy, Basel

Johnson LW, Smith LH (1985) Glucose transport across the basal plasma membrane of human placental syncytiotrophoblast. Biochem Biophys Acta 815: 44

Johnson RN, Metcalf PA, Baker JR (1982) Fructosamine: a new approach to the estimation of serum glycosylprotein. An index of diabetic control. Clin Chim Acta 127: 87

Jones CJP, Fox H (1976) Placental changes in gestational diabetes. An ultrastructural study. Obstet Gynecol 48: 274

Jones CJP, Fox H (1976) An ultrastructural study of distribution of acid and alkaline phosphatase in placentae from normal and complicated pregnancies. J Pathol 118: 143

Jones IR, Owens DR, Williams S, et al (1983) Glycosylated serum albumin: An intermediate index of diabetic control. Diabetes Care 6: 501

Jovanovic L, Peterson CM, Saxena BB (1980) Feasibility of maintaining normal glucose profiles in insulin-dependent pregnant diabetic women. Am J Med 68: 105

Jovanovic L, Braun CB, Druzin ML, Peterson CM (1982) The management of diabetes and pregnancy. In: Peterson CM (ed) Diabetes management in the 80's. Praeger, New York, p 248

Jovanovic L, Druzin M, Peterson CM (1982) Effect of euglycemia on the outcome of pregnancy in insulin-dependent diabetic women as compared with normal control subjects. Am J Med 71: 921

Jovanovic L, Peterson CM (1982) Optimal insulin delivery for the pregnant diabetic patient. Diabetes Care 5 [Suppl] 1: 24

Jovanovic L, Peterson CM (1985) Screening for ges-

tational diabetes. Optimum timing and criteria for retesting. Diabetes 34 [Suppl] 2: 21

Juhan-Vagua I, Vague P (1982) Properties of erythrocytes and platelets and the degree of diabetic control. Nouv Rev Franc Hematolog 24: 191 (in French)

Jury DR, Baker JR, Bunn PJ (1983) Clinical importance of the reversible fraction of haemoglobin A 1 c in type 2 (non-insulin-dependent) diabetes. Diabetologia 25: 313

Kahn CR (1982) Insulin receptors and syndromes of insulin resistance. Diabetes Care 5 [Suppl] 1: 98

Kahn CR (1985) Insulin receptors. Diab Ann 1: 446

Kainer F, Rollet H, Hofmann HMH, Weiss PAM, Haas JG (1986) Orale Glukosebelastung (oGTT) mit 50 g, 1 g/kg Körpergewicht und 100 g. Pilot-Studie. In: Weiss PAM (ed) 1. Internationales Grazer Symposium: Kohlenhydratstoffwechselstörungen und Schwangerschaft. Maudrich, Wien München Bern (Probl perinat med, p 57)

Kalhan SC, Savin SM, Adam PAJ (1977) Attenuated glucose production rate in newborn infants of insulin dependent diabetic mothers. N Engl J Med 296: 375

Kalhan SC, D'Angelo LJ, Savin SRM, Adam PAJ (1979) Glucose production in pregnant woman at term gestation. J Clin Invest 63: 388

Kalkhoff RK (1975) Effects of oral contraceptive agents on carbohydrate metabolism. J Steroid Biochem 6: 949

Kalkhoff RK (1985) Therapeutic results of insulin therapy in gestational diabetes mellitus. Diabetes 34 [Suppl] 2: 97

Kanada T, Otsuji S (1983) Lower levels of erythrocyte membrane fluidity in diabetic patients. A spin label study. Diabetes 37: 585

Karlsson K, Kjellmer I (1972) The outcome of diabetic pregnancies in relation to the mother's blood sugar level. Am J Obstet Gynecol 112: 213

Karp W, Sprecher H, Roberson A (1973) Human placental phospholipid synthesis. Biol Neonate 22: 398

Karpellus E, Bichler A (1984) Diabetes-Screening im Wochenbett: oraler Glukosetoleranztest oder glykosyliertes Hämoglobin A 1? Gynäk Rdsch 24 [Suppl] 2: 94

Kaulhausen H (1980) Klinik und Therapie der Gestose (Praeeklampsie). Nieren- und Hochdruckkrankheiten 9: 65

Kauppila A, Tuimala R, Ylikorkala O, Haapalathi J, Karppanen H, Vinikka L (1978) Effects of ritodrine and isoxsuprine with and without dexamethasone during late pregnancy. Obstet Gynecol 51: 288

Kennedy L, Mehl TD, Elder E, et al (1982) Nonenzymatic glycosylation of serum and plasma proteins. Diabetes 31 [Suppl] 3: 52

Kennedy L, Baynes JW (1984) Non-enzymatic glycosylation and the chronic complications of diabetes: an overview. Diabetologia 26: 93

Kerner JA, Stevenson DK, Hattner JA, Cohen RS, Schwartz HC, Sunshine HC (1982) Evidence for the possible relationship of neonatal skinfold thickness to maternal glucose metabolism during the third trimester. J Pediat Gastroenterol Nutr 1: 59

Kervran A, Guillaume M, Jost A (1978) The endocrine pancreas of the fetus from diabetic pregnant rat. Diabetologia 15: 387

Khouzami VA, Ginsburg DS, Daikoku NH, Johnson JWC (1981) The glucose tolerance test as a means of identifying intrauterine growth retardation. Am J Obstet Gynecol 139: 423

Kim YJ, Felig P (1972) Maternal and amniotic fluid substrate levels during caloric deprivation in human pregnancy. Metabolism 21: 507

Kitzmiller JL, Tanenberg RJ, Acki TT, Tabatabaii A, Gleason R, Jewett JF, Hare JW, Soeldner JS (1980) Pancreatic alpha cell response to alanine during and after normal and diabetic pregnancies. Obstet Gynecol 56: 440

Kitzmiller JL, Cloherty JP, Younger MD, Tabatabaii A, Rothchild SB, Sosenko I, Epstein MF, Singh S, Neff RK (1978) Diabetic pregnancy and perinatal morbidity. Am J Obstet Gynecol 131: 560

Kivinen S, Ylikorkala O, Puukka M (1979) Prolactin response to thyrotropin-releasing hormone in normal and complicated late pregnancies. Obstet Gynecol 54: 695

Kjaergaard JJ, Ditzel J (1979) Hemoglobin A 1 c as an index of long term blood glucose regulation in diabetic pregnancy. Diabetes 28: 694

Kleinberger G (1981) Parenterale Ernährung bei Diabetes mellitus. In: Lochs H, Grünet A, Druml W (Hrsg) Klinische Ernährung 5: Aktuelle Probleme der klinischen Ernährung. W Zuckschwerdt, München, S 98

Kleine U (1967) Studies on lipid metabolism of villi of mature human placenta. Clin Chim Acta 17: 95

Klenk DC, Hermanson GT, Krohn RI, et al (1982) Determination of glycosylated hemoglobin by affinity chromatography: Comparison with colorimetric and ion-exchange methods, and effects of common interferences. Clin Chem 28: 2088

Knopp RH, Warth MR, Carrol CJ (1973) Lipid metabolism in pregnancy. I. Changes in lipoprotein triglyceride and cholesterol in normal pregnancy and the effects of diabetes mellitus. J Reprod Med 10: 95

Knopp RH, Montes A, Warth MR (1978) Carbohydrate and lipid metabolism. In: Laboratory indices of nutritional status in pregnancy. National Academy of Sciences, Washington, DC, p 35

Knopp RH, Chapman M, Bergelin R, Wahl PW, Warth MR, Irvine S (1980) Relationships of lipoprotein lipids to mild fasting hyperglycemia and diabetes in pregnancy. Diabetes Care 3: 416

Kohlhoff R, Roth P (1986) Untersuchungen der Glukosetoleranz (1,75 g/kg — oGTT) bei Nachkommen diabetischer Mütter im 1.—3. Lebensjahr. Kinderärztl Praxis 54: 613

Korp W, Rogovits N (1973) Die Bewertung oraler Glu-

cosetoleranzteste in der Schwangerschaft und im Wochenbett. Wien Klin Wochenschr 85: 432

Kono T (1983) Actions of insulin on glucose transport and cAMP phosphodiesterase in fat cells. Involvement of two distinct molecular mechanisms. Rec Prog Horm Res 39: 519

Krishnamoorthy R, Gacon G, Labie D (1977) Isolation and partial characterisation of hemoglobin A 1 b. FEBS Lett 77: 99

Kuberski TT, Bennett PH (1980) Diabetes mellitus as an emerging public health problem on Guam. Diabetes Care 3: 235

Kühl C (1975) Glucose metabolism during and after pregnancy in normal and gestational diabetic women. I. Influence of normal pregnancy on serum glucose and insulin concentration during basal fasting conditions and after a challenge with glucose. Acta Endocrinol (Kbh) 79: 709

Kühl C (1976) Serum proinsulin in normal and gestational diabetic pregnancy. Diabetologia 12: 295

Kühl C, Holst JJ (1976) Plasma glucagon and the insulin: glucagon ratio in gestational diabetes. Diabetes 25: 16

Kühl C (1977) Serum insulin and plasma glucagon in human pregnancy—on the pathogenesis of gestational diabetes. (A review.) Acta Diabetol Lat 14: 1

Kühl C, Hornnes PJ, Jensen SL (1977) Effect of pregnancy on the hepatic extraction of insulin. Diabetologia 13: 411

Kühl C, Hornnes PJ, Faber OK (1981) Hepatic insulin extraction in human pregnancy. Horm Metab Res 13: 71

Kühl C, Hornnes PJ (1982) Insulin and glucagon responses to amino acids in normal and gestational diabetic pregnancy. Diabetologia 23: 182

Kühl C, Andersen GE, Hertel J, Molsted-Pedersen L (1982) Metabolic events in infants of diabetic mothers during first 24 hours after birth. Acta Paediatr Scand 71: 19

Kühl C, Hornnes PJ (1984) Plasma insulin, proinsulin, and pancreatic glucagon in gestational diabetes. In: Melchionda N, et al (eds) Recent advances in obesity and diabetes research. Raven Press, New York, p 129

Kühl C, Hornnes PJ, Andersen O (1985) Etiology and pathophysiology of gestational diabetes mellitus. Diabetes 34 [Suppl] 2: 66

Kühl C, Andersen O (1986) Pathophysiological background for gestational diabetes. In: Weiss PAM (ed) 1st international Graz-Symposium: disturbance of carbohydrate metabolism during pregnancy. Maudrich, Wien München Bern (Probl perinat med 15, p 3)

Lacroix S, Ekoe JM, Assal JPh, Leuenberger PM (1982) Problemes de lecture de tests colorimetriques urinaires chez les diabetiques. Klin Mbl Augenheilk 120: 407

Laga EM, Driscoll SG, Nunro HN (1973) Quantitative

studies of human placenta. I. Morphometry. Biol Neonate 23: 231

Lang N, Bellmann O, Hinckers HJ, Schlebusch H (1978) Diagnostik und klinische Bedeutung des Gestationsdiabetes. Gynäkologe 11: 78

Laureti E, De Galateo A, Gikorgino F (1982) Ricerche quantiatative sul tenore di idrossiprolina, di lipidi e di esosi in placente normali e diabetiche del 3 mese di gravidanza. Boll Soc It Biol Sper 58: 702

Lavin JP, Barden TP, Miodovnik M (1981) Clinical Experience with a screening program for gestational diabetes. Am J Obstet Gynecol 141: 491

Lavin JP, Gimmon Z, Miodovnik M, Meyenfeldt M, Fischer JE (1982) Total parenteral nutrition in a pregnant insulin requiring diabetic. Obstet Gynec 59: 660

Lavin JP (1985) Screening of high-risk and general populations for gestational diabetes. Clinical application and cost analysis. Diabetes 34 [Suppl] 2: 24

Lazarus NR, Penhos JC, Tanese T, Michaels L, Gutman R, Recant L (1970) Studies on the biological activity of porcine proinsulin. J Clin Invest 49: 487

Lechtig A, Yarbrough C, Delgado H, Marborell R, Klein R, Behar M (1975) Effect of moderate maternal malnutrition on the placenta. Am J Obstet Gynecol 123: 191

Lee LPK, Arnott B, Feng M, Hynie I (1982) Comparison of four commercial methods for the determination of fast hemoglobins. Clin Biochem 15: 230

Le-Pape A, Guitton JD, Gutman N, et al (1983) Nonenzymatic glycosylation of collagen in diabetes: Incidence of increased normal platelet aggregation. Haemostasis 13: 36

Le-Pape A, Gutman N, Guitton JD, et al (1983) Nonenzymatic glycosylation increases platelet aggregating potency of collagen from placenta of diabetic human beings. Biochem Biophys Res Commun 111: 602

Lev-Ran A, Goldman JA (1977) Brittle diabetes in pregnancy. Diabetes 26: 926

Lewis SB, Wallin JD, Kuzuya H, Coustan DR, Daane TA, Rubenstein AH (1976) Circadian variation of serum glucose, C-peptide immunoreactivity and free insulin in normal and insulin treated diabetic pregnant subjects. Diabetologia 12: 343

Lewis SB, Murray WK, Murray JD, Wallin DR, Coustan DR, Daane TA, Tredway DR, Navins JP (1976) Improved glucose control in nonhospitalized pregnant diabetic patients. Obstet Gynec 48: 260

Liebhardt M (1971) The electron-microscopic pattern of placental villi in diabetes of the mother. Acta Med Polonica 12: 133

Lin CC, River Ph, Moawad AH, Lowensohn RJ, Blix PM, Abraham M, Rubenstein AH (1981) Prenatal assessment of fetal outcome by amniotic fluid C-peptide levels in pregnant diabetic women. Am J Obstet Gynecol 141: 671

Lin CC, Moawad AH, River P, Blix P, Abraham M, Rubinstein AH (1981 a) Amniotic fluid C-peptide

as an index for intrauterine fetal growth. Am J Obstet Gynecol 139: 390

Lin TM, Halbert SP, Spellacy WN (1978) Pregnancy-associated plasma protein B (PAPP-B) in normal and abnormal pregnancies at term. Br J Obstet Gynaecol 85: 652

Lind T, Harris VG (1976) Changes in the oral glucose tolerance test during the puerperium. Br J Obstet Gynaecol 83: 460

Lind T, Billewicz WZ, Brown G (1973) A serial study of changes occuring in the oral glucose tolerance test during pregnancy. J Obstet Gynaecol Brit Cmwth 80: 1033

Lind T, Bell S, Gilmore E, Huisjes HJ, Schally AV (1977) Insulin disappearance rate in pregnant and non-pregnant women, and in non-pregnant women given GHRIH. Eur J Clin Invest 7: 47

Lind T, Cheyne GA (1979) Effect of normal pregnancy upon the glycosylated hemoglobins. Br J Obstet Gynaecol 73: 210

Lind T, Anderson J (1984) Does random blood glucose sampling outdate testing for glycosuria in the detection of diabetes during pregnancy? Br Med J 289: 1569

Lipshitz J, Vinik AV (1978) The effects of hexoprenalin, a Beta-2-sympathomimetic drug, on maternal glucose, insulin glucagon and free fatty acid level. Am J Obstet Gynecol 130: 761

Little RR, England JD, Wiedmeyer HM, Goldstein DE (1983) Glycosylated hemoglobin measured by affinity chromatography: Microsample collection and room-temperature storage. Clin Chem 29: 1080

Little RR, England JD, Wiedmeyer HM, Goldstein DE (1983) Effects of whole blood storage on results for glycosylated hemoglobin as measured by ion-exchange chromatography, affinity chromatography, and colorimetry. Clin Chem 29: 1113

Liu KS, Wang CY, Mills N, Gyves M, Ilan J (1985) Insulin related genes expressed in human placenta from normal and diabetic pregnancies. Proc Natl Acad Sci USA 82: 3868

Longo LD, Yuen P, Gusseck DJ (1973) Anaerobic, glycogen-dependent transport of amino acids by the placenta. Nature 243: 531

Lorenz U, Rüttgers H, Fux G, Kubli F (1974) Fetal pulmonary surfactant induction by bromhexine metabolite VIII. Am J Obstet Gynecol 119: 1126

Love EJ, Stevenson JAF, Kinch RAH (1964) Evaluation of oral and intravenous glucose tolerance test for the diagnosis of "prediabetes" in the puerperium. Am J Obstet Gynecol 88: 283

Löwenberg E, Jimenerz L, Martinez M, Pommier M (1981) Effects of ambroxol (NA 872) on biochemical fetal lung maturity and prevention of the respiratory distress syndrome. Prog Resp Res 15: 240

Lubchenco LO, Hansemann L, Dressler M, Boyd E (1963) Intrauterine growth as estimated from liveborn birth weight data at 24 to 42 weeks of gestation. Pediatrics 32: 793

Lutterman JA, Beuraad TJ, Van't Laar A (1981) The

relationship between insulin secretion and metabolic stability in type I (insulin dependent) diabetes. Diabetologia 21: 99

Macafee CAJ, Beischer NA (1974) The relative value of the standard indications for performing a glucose tolerance test in pregnancy. Med J Aust 1: 911

Machicao F, Urumow T, Wieland OH (1983) Evidence for phosphorylation of actin by the insulin receptor-associated protein kinase from human placenta. FEBS Lett 163: 76

Machicao F, Wieland OH (1984) Evidence that the insulin receptor associated protein kinase acts as a phosphatidylinositol kinase. FEBS Lett 175: 113

Madan R, Bajaj JS (1986) Pregnancy and diabetes. Problems and perspectives in developing countries. In: Serrano-Rios M, Lefebvre PJ (eds) Diabetes 1985. Elsevier, pp 1067–1071

Madan R, Mohan N (1986) Diabetes in maternal and child health, the asian experience. IDF Bulletin: 31-60-62

Madsen H, Kjaergaard JJ, Ditzel J (1982) Relationship between glycosylation of haemoglobin and the duration of diabetes: a study during the third trimester of pregnancy. Diabetologia 22: 37

Madsen H, Ditzel J (1984) Blood-oxygen transport in first trimester of diabetic pregnancy. Acta Obstet Gynecol Scand 63: 317

Malassine A, Goldstein S, Alsat E, Merger Ch, Cedard L (1984) Ultrastructural localization of low density lipoprotein binding sites on the surface of the syncytial microvillous membranes of the human placenta. IRCS Med Sci 12: 166

Malia AK, Hermanson GT, Krohn RJ, et al (1981) Preparation and use of a boronic acid affinity support for separation and quantitation of glycosylated hemoglobins. Analyt Letters 14: 649

Manda N, Nakayama H, Aoki S, et al (1982) Determination of glucosylated albumin and its clinical significance in diabetes mellitus. J Jap Diab Soc 25: 691

Manners DJ (1974) The structure and metabolism of starch. Essays Biochem 10: 37

Manning FA, Platt LC, Sipos L (1980) Antepartum fetal evaluation: Development of a fetal biophysical profile. Am J Obstet Gynecol 136: 787

Mansani FE, Caltabiano M, Ceruti M, Condemi V (1982) Postpartum glycosylated hemoglobins (HbA 1): An index of glucose control in pregnancy. Biol Res Preg 3: 1980

Mansani FE, Caltabiano M, Bertoncini P (1983) HbA 1 levels in pregnancy: Evaluation of a normal range and comparative study of impaired glucose tolerance. Boil Res Preg 4: 177

Marquart FX, Poynard JP, Leutenegger M, Borel JP (1982) On the importance of a prolonged dialysis for haemoglobin A 1 c determination. Clin Chim Acta 121: 393

Marshall RN, Underwood LE, Volina SJ, Foushee DB, Van Wyk JJ (1974) Characterization of the insulin

and somatomedin-C receptors in human placental cell membranes. J Clin Endocrinol Metab 39: 283

Marshall S, Olefsky JM (1980) Effects of insulin incubation on insulin binding, glucose transport and insulin degradation by isolated rat adipocytes. J Clin Invest 66: 763

Martin AO, Simpson JL, Ober C, Freinkel N (1985) Frequency of diabetes mellitus in mothers of probands with gestational diabetes: Possible maternal influence on the predisposition to gestational diabetes. Am J Obstet Gynecol 151: 471

Matzkies F, Dauz U, Baumann M, Dorguth B (1982) Tagespläne zur diätetischen Behandlung des Diabetes mellitus. Fortschr Med 100: 266

Mayser TK, Freedman ZR (1973) Protein glycosylation in diabetes mellitus: a review of laboratory measurements and of their clinical utility. Clin Chim Acta 127: 147

Mayet N, Iialal I, Naicker RR, Moodley J, Van Middelkoop A (1983) Maternal glycosylated haemoglobin values after delivery of large infants and unexplained stillbirths. SA Mediese Tydskrif 64: 739

McDonald MJ, Shapiro R, Bleichmann M, Solway J, Bunn HF (1978) Glycosylated minor components of human adult hemoglobin. J Biol Chem 253: 2327

McFarland KF, Murtiashaw M, Bayns JW (1984) Clinical value of glycosylated serum protein and glycosylated hemoglobin levels in the diagnosis of gestational diabetes mellitus. Obstet Gynecol 64: 516

McMillan DE, Brooks SM (1982) Erythrocyte spectrin glucosylation in diabetes. Diabetes [Suppl] 3: 31, 64

Mehl TD, Wenzel SE, Russell B, et al (1983) Comparison of two indices of glycemic control in diabetic subjects: Glycosylated serum protein and hemoglobin. Diabetes Care 6: 34

Mehnert H, Standl E (1975) Ärztlicher Rat für Diabetiker. G Thieme, Stuttgart

Mehnert H (1979) Diätbehandlung des Diabetes mellitus. Monatskurse für die ärztliche Fortbildung 8: 257

Meschia G, Battaglia FC, Hay WW, Sparks JW (1980) Utilization of substrates by the ovine placenta in vivo. Fed Proc 39: 245

Mestman J (1973) Medical management of the pregnant diabetic. Contemp Obstet Gynecol 1: 61

Mestman JH (1980) Outcome of diabetes screening in pregnancy and perinatal morbidity in infants of mothers with mild impairment in glucose tolerance. Diab Care 3: 447

Mestman JH (1985) Diabetes. In: Cetrulo CL, Sbarra AJ (eds) The Problem-oriented medical record for high-risk obstetrics. Plenum, p 283

Mestman JH, Andersen GV, Burton P (1971) Carbohydrate metabolism in pregnancy. Am J Obstet Gynecol 109: 41

Mestman JH, Anderson GV, Guadalupe V (1972) Follow-up study of 360 subjects with abnormal carbohydrate metabolism during pregnancy. Obstet Gynecol 39: 421

Metzger BE, Unger RH, Freinkel N (1977) Carbohydrate metabolism in pregnancy. XIV. Relationships between circulating glucagon, insulin, glucose and amino acids in response to a "mixed meal" in late pregnancy. Metabolism 26: 151

Metzger BE, Phelps, RL, Freinkel N, Depp R, Ogata ES, Pitts T (1981) Predictive value of fasting plasma glucose in gestational diabetes (GDM). Diabetes 30 [Suppl] 1: 4A (abstract)

Metzger BE, Bybee DE, Freinkel N, et al (1985) Gestational diabetes mellitus. Correlation between the phenotypic and genotypic characteristics of the mother and abnormal glucose tolerance during the first year postpartum. Diabetes 34 [Suppl] 2: 111

Mickal A, Begneaud WP, Weese WH (1966) Glucose tolerance and excessively large infants: A twelve year follow-up study. Am J Obstet Gynecol 94: 62

Middle RFA, Bannister A, Bellingham AJ, Dean PDG (1983) Separation of glycosylated haemoglobins using immobilized phenylboronic acid. Effect of ligand concentration, column operating conditions, and comparison with ion-exchange and isoelectric focusing. Biochem J 209: 771

Miller E, Hare JW, Clogerty JP, et al (1981) Elevated maternal hemoglobin A 1 c in early pregnancy and major congenital anomalies in infants of diabetic mothers. N Engl J Med 304: 1331

Miller EC (1980) Zum Problem der Gewichtsbestimmung des Feten durch Ultraschallbiometrie. Zbl Gynäkol 102: 272

Miller EC, Steinhoff R (1982) Diabetesscreening in der Schwangerschaft. Geburtsh z Frauenheilk 42: 583

Miller JM (1983) A reappraisal of "tight control" in diabetic pregnancies. Am J Obstet Gynecol 147: 158

Mills NC, Gyves MT, Ilan J (1985) Comparisons of human placental lactogen mRNA levels from placentas of diabetics and normal term. Mol Cell Endocrinol 39: 61

Milner RDG (1971) The development of insulin secretion in man. In: Jonxis JHP, Visser HKA, Troelstra JA (eds) Metabolic processes in the fetus and newborn infant. Stenfert Kroese, Leiden, p 193

Milner RDG (1979) The role of insulin and glucagon in fetal growth and metabolism. In: Vth Nutricia symposium on nutrition and metabolism of the fetus and the infant. Martinus Nijhoff, The Hague

Milner RDG, Dinedale F, Wirdhan PK, Van Assche FA (1983) Pancreatic endocrine cell fractions in erythroblastosis fetalis. Diabetes 32: 313

Milner RDG, Hill DJ (1984) Fetal growth control: The role of insulin and related peptides. Clin Endocrinol 21: 415

Modanlou HD, Dorchester WL, Thorosian A, Freeman RK (1980) Macrosomia—maternal, fetal, and neonatal implications. Obstet Gynecol 55: 420

Modanlou HD, Komatsu G, Dorchester W, Freeman RK, Bosu SK (1982) Large-for-gestational-age neonates: Anthropometric reasons for shoulder dystocia. Obstet Gynecol 60: 417

Molnar GD, Taylor WF, Ho MM (1972) Day to day variation of continuously monitored glycemia. A further measure of diabetic instability. Diabetologia 8: 342

Molnar GD, TaylorWF, Langworthy A (1974) On measuring the adequacy of diabetes regulation: Comparison of continuously monitored blood glucose patterns with values at selected time points. Diabetologia 10: 139

Molsted-Pedersen L, Wagner L, Klebe JG, Pedersen J (1972) Aspects of carbohydrate metabolism in newborn infants of diabetic mothers. IV. Neonatal changes in plasma free fatty acid concentration. Acta Endocrinol 71: 338

Monnier VM, Cerami A (1982) Non-enzymatic glycosylation and browning of proteins in diabetes. Clin Endocrin Metab 11: 431

Montgomery D, Young M (1982) The uptake of naturally occuring amino acids by the plasma membrane of the human placenta. Placenta 3: 13

Morris MA, Grandis AS, Litton JC (1985) Glycosylated hemoglobin concentration in early gestation associated with neonatal outcome. Am J Obstet Gynecol 153: 651

Morriss FH, Makotoski EL, Meschia G, Battaglia FC (1975) The glucose/oxygen quotient of the term human fetus. Biol Neonate 25: 44

Mortensen HB, Marshall MO (1983) Effect of saline incubation on red cell content of glucosylated haemoglobins studied by iso-electric focusing. Clin Chim Acta 132: 213

Mortensen HB, Nielsen L, Soegaard, U, et al (1983) Comparison of six assays for glycosylated haemoglobin determination. Scand J Clin Lab Invest 43: 357

Muck BR, Hommel G (1977) Plasma insulin response following intravenous glucose in gestational diabetes. Arch Gynäk 223: 259

Müller-Tyl E, Salzer H (1978) Einfluß von Dexamethason und Ambroxol auf die Reifung der fetalen Lunge. Atemwegs- und Lungenkrkh [Suppl] 1: 42

Munro HN (1980) Placental protein and peptide hormone synthesis: impact of maternal nutrition. Feder Proc 39: 255

Munro HN, Pilistine SJ, Fant ME (1983) The placenta in nutrition. Ann Rev Nutr 3: 97

Murtiashaw MH, Young JE, Strickland AL, et al (1983) Measurement of nonenzymatically glucosylated serum protein by an improved thiobarbituric acid assay. Clin Chim Acta 130: 177

Mylvaganam R, Stowers JM, Steel JM, Wallace J, McHendry JC, Wright AD (1983) Insulin immunogenicity in pregnancy: Maternal and fetal studies. Diabetologia 24: 19

Naeye R (1979) Discussion on congenital syphilis. In: Hugh Philpott R (ed) Maternity services in the developing world. What the community needs. Proceedings of the seventh study group of the royal college of obstetricians and gynecologists, pp 33–34

Naeye RL, Blanc W, Paul C (1973) Effects of maternal malnutrition on the human fetus. Pediatrics 52: 494

Naeye RL, Chez RA (1981) Effects of maternal ace-tonuria and low pregnancy weight gain on children's psychomotor development. Am J Obstet Gynecol 139/2: 189

Najemnik C, Kritz H, Irsigler K (1981) Persönliche Mitteilung

Nakayama H, Manda N, Komori K, et al (1982) Studies on the determination of glucosylated albumin using affinity chromatography. J Jap Diab Soc 25: 963

Nakayama H, Manda N, Komori K, et al (1982) Measurement of glucosylated serum proteins using affinity chromatography. J Jap Diab Soc 25: 1011

Nathan DM, Avezzanoo E, Palmer JL (1982) Rapid method for eliminating labile glycosylated hemoglobin from the assay for hemoglobin A 1. Clin Chem 28: 512

Nathan DM (1983) Glycosylated hemoglobin: What it is and how to use it. Clin Diabetes 1: 2

Nathan DM, Singer, D. E. Hurxthal K, Goodson JD (1984) The clinical information value of the glycosylated hemoglobin assay. N Engl J Med 310: 341

National Center for Health Statistics (1964) Glucose tolerance tests of adults, 1960–1962. US Public Health Service Publication 1000, series II, no 2

National Diabetes Data Group (1979) Classification and diagnosis of diabetes mellitus and other categories of glucose intolerance. Diabetes 28: 1039

Nelson DM, Smith RM, Jarrett L (1978) Nonuniform distribution and grouping of insulin receptors on the surface of human placental syncytial trophoblast. Diabetes 27: 530

Nesbitt REL, Rice PA, Rourke JE (1973) In vitro perfusion studies of the human placenta. III. The relationship between glucose utilization rates and concentration. Gynec Invest 4: 243

Newman RL, Tutera G (1976) The glucose-insulin ratio in amniotic fluid. Obstet Gynecol 47: 599

Ney D, Hollingsworth DR, Cousins L (1982) Decreased insulin requirement and improved control of diabetes in pregnant women given a high-carbohydrate, high-fiber, low fat diet. Diabetes Care 5: 529

Niederau CM, Potthoff S, Gries FA, Reinauer H (1980) Zum Aussagewert von glykosilierten Hämoglobinen bei Diabetes mellitus und bei Gravidität. Lab Med 4: 9

Niederau CM, Potthoff S, Reinauer H (1981) Hyperglykämie als Risikofaktor in der Schwangerschaft. Z Geburtsh Perinat 185: 137

Niesen M (1978) Die Betreuung der Neugeborenen diabetischer Mütter. Gynäkologe 11: 92

Oats JN, Abell DA, Beischer NA, Broomhall GR (1980) Maternal glucose tolerance during pregnancy with excessive size infants. Obstet Gynecol 55: 184

Obenshain SS, Adam PAJ, King KC, Teramo K, Raivio KO, Räiha N, Schwartz R (1970) Human fetal insulin response to sustained maternal hyperglycemia. N Engl J Med 283: 566

Ogata ES, Sabbagha R, Metzger BE, et al (1980) Serial ultrasonography to assess evolving fetal macrosomia. JAMA 243: 2405

Olofsson P, Liedholm H, Sartor G, Sjoberj N, Svenningsen NW, Ursing D (1984) Diabetes and pregnancy: a 21-year Swedish material. Acta Obstet Gyn Scand [Suppl] 122

Omori Y, Minei S, Saito M, Hirata Y (1982) Insulin-receptor autoantibody detected by the human placental membrane method: Six patients with insulin-receptor autoantibody in Japan. Tohoku J Exp Med 138: 319

Oppermann W, Gugliucci C, O'Sullivan MJ, Hanover B, Camerini-Davalos RA (1975) Gestational diabetes and macrosomia. In: Camerini-Davalos RA, Cole HS (eds) Early diabetes in early life. Academic Press, New York, p 455

O'Shaugnessy R, Russ J, Zuspan FP (1979) Glycosylated hemoglobins and diabetes mellitus in pregnancy. Am J Obstet Gynecol 135: 783

Oski FA, Naiman JC (1972) Hematologic problems in the newborn, 2nd edn. WB Saunders, Philadelphia

O'Sullivan JB (1978) Gestational diabetes: Factors influencing the rates of subsequent diabetes. In: Sutherland HW, Stowers JM (eds) Carbohydrate metabolism in pregnancy and the newborn. Springer, Berlin Heidelberg New York, p 425

O'Sullivan JB (1979) Gestational diabetes: factors influencing the rates of subsequent diabetes. In: Sutherland HW, Stowers JM (eds) Carbohydrate metabolism in pregnancy and the newborn. Springer, Berlin Heidelberg New York, p 425

O'Sullivan JB (1982) Body weight and subsequent diabetes mellitus. JAMA 248: 949

O'Sullivan JB (1984) Subsequent morbidity among gestational diabetic women. In: Sutherland HW, Stowers JM (eds) Carbohydrate metabolism in pregnancy and the newborn. Churchill Livingstone, Edinburgh, p 174

O'Sullivan JB, Mahan CM (1964) Criteria for the oral glucose tolerance test in pregnancy. Diabetes 13: 278

O'Sullivan JB, Mahan CM (1964) Criteria for the oral glucose tolerance test in pregnancy. Diabetes 13: 278

O'Sullivan JB, Gellis SS, Dandrow RV, Tenney BO (1966) The potential diabetic and her treatment in pregnancy. Obstet Gynecol 27: 683

O'Sullivan JB, Charles D, Mahan CM, et al (1973) Gestational diabetes and perinatal mortality rate. Am J Obstet Gynecol 116: 901

O'Sullivan JB, Mahan CM, Charles D, Dandrow RV (1973) Screening criteria for high-risk gestational diabetic patients. Am J Obstet Gynecol 116: 895

O'Sullivan JB, Mahan CM, Charles D, Dandrow RV (1974) Medical treatment of the gestational diabetic. Obstet Gynecol 43: 817

O'Sullivan JB, Mahan CM (1980) Insulin treatment and high risk groups. Diabetes Care 3: 482

Ott WJ, Doyle S (1982) Normal ultrasonic fetal weight curve. Obstet Gynecol 59: 603

Pagano G, Cassader M, Massobrio M, Bozzo C, Trossarelli GF, Menato G, Lenti G (1980) Insulin binding to human adipocytes during late pregnancy in healthy, obese and diabetic state. Horm Metab Res 12: 177

Page MA, Williamson DH (1971) Enzymes of ketone-body utilization in human brain. Lancet ii: 66

Pajszczyk-Kieszkiewicz, T (1972) The aromatising capacity of human placenta perfunded in vitro relating to the different experimental conditions. III. The aromatising capacity of placentas from diabetic individuals. Endcrinol Pol 23: 464

Parent MA, Khadraoui, S (1979) Serum Igm of neonates in a rural population in a developing country. Child Health 25: 143

Parks DG, Ziel HK (1978) Macrosomia. A proposed indication for primary cesarian section. Obstet Gynecol 52: 407

Pastor X (1987) Doctoral thesis, unpublished data. University of Barcelona

Paterson P, Sheat J, Taft P, Wood C (1967) Maternal and foetal ketone concentrations in plasma and urine. Lancet i: 862

Paterson P, Page D, Taft P, Phillips L (1968) Study of fetal and maternal insulin levels during labour. J Obstet Gynaec Br Cwth 75: 917

Peacock J, Tattersall RB, Taylor A, Douglas CA, Reeves WG (1983) Effects of new insulins on insulin and C-peptide antibodies, insulin dose and diabetic control. Lancet i: 149

Peavy DE, Taylor JM, Jefferson LS (1978) Correlation of albumin production rates and albumin mRNA levels in livers of normal, diabetic, and insulin-treated diabetic rats. Proc Natl Acad Sci USA 75: 5879

Pedersen J (1954) Glucose content of the amniotic fluid in diabetic pregnancies. Acta Endocrinol 15: 342

Pedersen J (1977) The pregnant diabetic and her newborn. Problems and management, 2nd edn. Williams and Wilkins, Baltimore, p 211

Pedersen J, Bojen-Moller B, Poulsen H (1954) Blood sugar in newborn infants of diabetic mothers. Acta Endocrinol 15: 33

Pedersen J, Molsted-Pedersen L (1965) Prognosis of the outcome of pregnancies in diabetics. A new classification. Acta Endocrinol (Kbh) 50: 70

Pedersen JE, Molsted-Pedersen L (1981) Early fetal growth delay detected by ultrasound marks increased risks of congenital malformation in diabetic pregnancy. Br Med J 283: 269

Pedersen O (1984) Studies of insulin receptor binding and insulin action in humans. Dan Med Bull 31: 1

Pedersen O, Hjollund E, Lindskov HO, Beck-Nielsen H, Jensen J (1982) Circadian profiles of insulin receptors in insulin-dependent diabetics in usual and poor metabolic control. Am J Physiol 242: E127

Perrine SP, Greene MF, Faller DV (1985) Delay in the fetalglobin switch in infants of diabetic mothers. N Eng J Med 312: 334

Persson B, Lunell NO (1975) Metabolic control in diabetic pregnancy. Variations in plasma concentrations of glucose, free fatty acids, glycerol, ketone bodies, insulin and human chorionic somatomammotropin during the last trimester. Am J Obstet Gynecol 122: 737

Persson B, Geutz J, Lunell NO (1978) Diabetes in pregnancy. In: Scorpelli EM, Cosmi EV (eds) Reviews in perinatal medicine. Raven Press, New York, p i

Persson B, Heding LG, Lunell NO, Pschera H, Stangenberg M, Wagner J (1982) Fetal beta cell function in diabetic pregnancy. Amniotic fluid concentrations of proinsulin, insulin, and C-peptide during the last trimester of pregnancy. Am J Obstet Gynecol 144: 455

Persson B, Stangenberg M, Hansson U, Nordlander E (1985) Gestational diabetes mellitus (GDM): Comparative evaluation of two treatment regimens, diet versus insulin and diet. Diabetes 34 [Suppl] 2: 101

Persson B, Pschera H, Lunell NO, Barley J, Gumaa KA (1986) Amino acid concentrations in maternal plasma and amniotic fluid in relation to fetal insulin secretion during the last trimester of pregnancy in gestational and Type I diabetic women and women with small-for-gestational-age infants. Am J Perinatol 3: 98

Peterson CM, Jones RL (1977) Minor hemoglobins, diabetic "control" and diseases of postsynthetic protein modification. Ann Int Med 74: 489

Peterson CM, Kalan G, Jovanovic L, Jovanovic R (1979) Use of the minor hemoglobin ratio for the determination of the determination of gestational age. Am J Obstet Gynecol 135: 57

Peterson CM (eds) (1982) Proceedings of a conference on nonenzymatic glycosylation and browning reactions: their relevance to diabetes mellitus. Diabetes [Suppl] 3: 1

Peterson CM, Jovanovic L, Raskin P, Goldstein DE (1984) A comparative evaluation of glycosylated hemoglobin assays: Feasibility of references and standards. Diabetologia 26: 214

Petropoulos EA (1973) Maternal and fetal factors affecting the growth and function of the rat placenta. Acta Endocrinol 72 [Suppl] 176: 9

Pettit BR, King GS, Blau K (1977) Low glucose concentrations in amniotic fluid from anencephalic pregnancies (letter). Lancet ii: 288

Pettitt DJ, Baird HR, Aleck KA, Knowler WC (1982) Diabetes mellitus in children following maternal diabetes during gestation. Diabetes 32 [Suppl] 2: 66 A

Pettitt DJ, Baird HR, Aleck KA, Bennett PH, Knowler WC (1983) Excessive obesity in offspring of Pima Indian women with diabetes during pregnancy. N Engl J Med 308: 242

Pettitt DJ, Bennett PH, Knowler WC, et al (1985) Gestational diabetes mellitus and impaired glucose tolerance during pregnancy. Long term effects on obesity and glucose tolerance in the offspring. Diabetes 34 [Suppl] 2: 119

Phelps RL, Bergenstal R, Freinkel N, Rubenstein AH, Metzger BE, Mako M (1975) Carbohydrate metabolism in pregnancy: XIII. Relationships between plasma insulin and proinsulin during late pregnancy in normal and diabetic subjects. J Clin Endocrinol 41: 1085

Phelps RL, Honig GR, Green D, Metzger BE, Frederiksen MC, Freinkel N (1983) Biphasic changes in hemoglobion A1c concentrations during normal human pregnancy. Am J Obstet Gynecol 147: 651

Phillips AF, Dubin JW, Matty PS, Raye J (1982) Arterial hypoxemia and hyperinsulinemia in the chronically hyperglycemic fetal lamb. Pediatr Res 16: 653

Phillips AF, Rosenkrantz TS, Raye J (1985) Consequences of perturbation of fetal fuels in ovine pregnancy. Diabetes 34 [Suppl] 2: 32

Phillips L, Lumley J, Paterson P, Wood C (1968) Fetal hypoglycemia. Am J Obstet Gynecol 102: 371

Picon L (1967) Effect of insulin on growth and biochemical composition of the rat fetus. Endocrinology 81: 1419

Pictet R, Rutter W (1972) Development of the embryonic endocrine pancreas. In: Steiner DF, Freinkel N (eds) Handbook of physiology, sect 7, vol I: Endocrine pancreas. Williams and Wilkins, Baltimore, p 25

Pollak A, Widness JA, Schwartz R (1979) "Minor hemoglobins": an alternative approach for evaluating glucose control in pregnancy. Biol Neonate 36: 185

Pollak A, Brehm R (1981) Glykosyliertes Hämoglobin bei Müttern von übergewichtigen Neugeborenen. Gynäk Rdsch 21 [Suppl] 2: 164

Pollak A, Brehm R, Havelec L, Lubec G, Malamitsi-Puchner A, Simbrunner G, Widness JA (1981) Total glycosylated hemoglobin in mothers of large-for-gestational-age infants: a postpartum test for undetected maternal diabetes? Biol Neonate 40: 129

Poon P, Turner RC (1981) Glycosylated fetal hemoglobin. Br Med J 283: 469

Poon PYW, Dornan TL, Orde-Peckar C, et al (1982) Blood viscosity, glycaemic control and retinopathy in insulin-dependent diabetes. Clin Sci 63: 211

Popjack G, Beeckmans M (1950) Are phospholipids transmitted through the placenta? Biochem J 46: 99

Posner BI (1973) Insulin metabolizing enzyme activities in Human placental tissue. Diabetes 22: 552

Posner BI (1974) Insulin receptors in human and animal placental tissue. Diabetes 23: 209

Potter (1941) Zit nach Scheibenreither S, Thalhammer O (1966)

Potter JM, Reckless JPD, Cullen DR (1982) Diurnal variations in blood intermediary metabolites in mild gestational diabetic patients and the effect of a carbohydrate restricted diet. Diabetologia 19: 68

Puavilai G, Dribny EC, Domont LA, Baumann G (1982) Insulin receptors and insulin resistance in human pregnancy: Evidence for a postreceptor defect in insulin acltion. J Clin Endocrinol Metab 54: 247

Rakieten N, Rakieten M, Nedkarmi M (1963) Studies on the diabetogenic action of strepozotocin. Cancer Chemother 29: 91

Ratanasopa V, Schindler AE, Lee TY, Herrmann WL (1967) Measurement of estriol in plasma by gas-liquid chromatography. Its significance in the treatment of high-risk pregnancies. Am J Obstet Gynecol 99: 295

Rathgen GH (1980) In: Friedberg V, Rathgen GH (Hrsg) Physiologie der Schwangerschaft. G Thieme, Stuttgart New York

Reaven GM, Sageman WS, Svenson RS (1977) Development of insulin resistance in normal dogs following alloxan-induced insulin deficiency. Diabetologia 13: 459

Reiher H, Woltansky P, Hahn J (1981) Effect of glucose on human fetal pancreatic tissue in vitro. Acta Biol Med Germ 40: 61

Reiher H, Fuhrmann K, Woltanski KP, Hahn HJ (1982) Untersuchungen am humanen fetalen Pankreas — Methodenkritik zur Gewebegewinnung. Zbl Gynäkol 104: 213

Reiher H, Fuhrmann K, Noack S, Woltanski KP, Jutzi E, Hahn v. Dorsche, Hahn HJ (1983) Age-dependent insulin secretion of the endocrine pancreas in vitro from fetuses of diabetic and nondiabetic patients. Diabetes Care 6: 446

Report of the Committee on Statistics of the American Diabetes Association (1969) Standardization of the oral glucose tolerance test. Diabetes 8: 299

Reusens-Billen B, Remacle C, Daniline, J, Hoet JJ (1984) Cell proliferation in pancreatic islets of rat fetuses and neonates from normal and diabetic mothers. An in vitro and in vivo study. Horm Metab Res 16: 565

Rinderknecht E, Humbel RE (1976) Polypeptides with nonsuppressible insulin-like and cell-growth promoting activities in human serum: Isolation, chemical characterization, and some biological properties of forms I and II. Proc Natl Acad Sci USA 73: 2365

Rizvi J, Gillmer MDG, Oakley NW, Beard RW (1980) Evaluation of plasma glucose control in pregnancy complicated by chemical diabetes. Br J Obstet Gynaecol 87: 383

Robb SA, Hytten FE (1976) Placental glycogen. Br J Obstet Gynecol 83: 43

Robbers H, Traumann KJ (1980) Diätbuch für Zuckerkranke. G Thieme, Stuttgart New York

Roberts AB, Baker JR, Court DJ, et al (1983) Fructosamine in diabetic pregnancy. Lancet ii: 998

Robertson AF, Sprecher H (1968) A review of human placental lipid metabolism and transport. Acta Paediatr Scand [Suppl] 183: 1

Rollmann J (1987) Glykohämoglobinbestimmungen bei makrosomen Neugeborenen und deren Müttern postpartum. Inaugural-Dissertation, Marburg/L

Rosanelli K, Lichtenegger W, Weiss PAM (1982) Über die Wirkung der Tokolyse durch Beta-Mimetika auf den Fet und das Neugeborene. Z Geburtsh Perinat 186: 93

Rosenkranz A, Sedlak W, Ogris E (1979) Kohlenhydratstoffwechseluntersuchungen bei Kindern diabetischer Mütter. Wien Klin Wochenschr 91: 563

Rosso P (1980) Placental growth, development, and function in relation to maternal nutrition. Fed Proc 39: 250

Rost J, Freude G, Fuchs G, Bali Ch, Kriz H, Irsigler K, Leodolter S (1982) Frühgeburtlichkeit bei schwangeren Diabetikerinnen. Gynäk Rdsch 22 [Suppl] 1: 180

Roux JF, Green R (1968) Lipid metabolism by the human placenta. Obstet Gynecol 29: 446

Roversi GD, Gargiulo M, Nicolini U, Pedretti E, Marini A, Barbarani V, Peneff P (1979) A new approach to the treatment of diabetic pregnant women. Report of 479 cases seen from 1963 to 1975. Am J Obstet Gynecol 135: 567

Roversi GD, Gargiulo M, Nicolini U, Ferazzi E, Pedretti E, Gruft L, Tronconi G (1980) Maximal tolerated insulin therapy in gestational diabetes. Diabetes Care 3: 489

Rubinstein P, Walker M, Krassner J, et al (1981) HLA antigens and islet cell antibodies in gestational diabetes. Human Immunol 3: 271

Rudolf MCJ, Sherwin RS (1983) Maternal ketosis and its effects on the fetus. Clin Endocrin Metabol 12: 413

Ruge S, Andersen T (1985) Obstetric risks in obesity. An analysis of the literature. Obstet Gynecol Survey 40: 57

Rule AH, Sliogeris V, Farber M, Britton G, Vandervoorde J (1981) Relation of glucose to alphafetoprotein in amniotic fluid. Obstet Gynecol 57: 310

Ruzycki SM, Kelley LK, Smith CH (1977) Placental amino acid uptake. IV. Transport by microvillous membrane vesicles. Am J Physiol 234: C 27

Sabata V, Wolf H, Lausmann S (1968) The role of free fatty acids, glycerol, ketone-bodies and glucose in the energy metabolism of the mother and fetus during delivery. Biol Neonate 13: 7

Sachs L (1973) Angewandte Statistik, 4. Aufl. Springer, Berlin Heidelberg New York

Sadik W, Kovacs L, Pretnar-Darovec A, de Acosta M, Toddywalla VS, Ng CSA, Holck S, Belsey E, Pinol A, Hall P (1985) A randomized double-blind study of the effects of two low-dose combined oral contraceptives on biochemical aspects. Contraception 32: 223

Sadovski E, Jaffe H, Polishuk WZ (1974) Fetal movement monitoring in normal and pathological pregnancies. Int J Gynecol Obstet 12: 75

Saintonge J, Cotge R (1983) Intrauterine growth retardation and diabetic pregnancy: Two types of fetal malnutrition. Am J Obstet Gynecol 146: 194

Sakurai T, Takagi H, Hosoya N (1969) Metabolic pathways of glucose in human placenta. Changes with

gestation and with added 17-Beta-estradiol. Am J Obstet Gynecol 105: 1044

Salem HT, Seppala M, Ranta T, Bohn H, Chard T (1981) The effects of protamine on serum levels of placental protein S (PPS) in normal and abnormal pregnancy: A possible relation to coagulation abnormalities. Br J Obstet Gynaecol 88: 367

Salzberger M, Liban E (1975) Diabetes and antenatal fetal death. Isr J Med Sci 11: 623

Salzberger M, Sharon A, Liban ED (1975a) Significance of the oral glucose tolerance test performed on the third day after delivery for the diagnosis of diabetes in pregnancy. Isr J Med Sci 11: 629

Samaan NA, Vassilopoulou-Sellin R, Schultz PN, Rivera ME, Held B (1985) Nonsuppressible insulin-like activity and somatomedin C levels in normal pregnant women, in pregnant women with gestational diabetes, and in umbilical cord blood of mature and premature infants. Am J Obstet Gynecol 153: 457

Samaja M, Melotti D, Carenini A, Pozza G (1982) Glycosylated haemoglobins and the oxygen affinity of whole blood. Diabetologia 23: 399

Sarles MD, Adamsons K (1978) Diabetes: New management concepts. Perinatal Care 2: 13

Schatz H (1977) Biosynthese von Insulin. Dtsch Med Wochenschr 102: 734

Scheibenreiter S, Thalhammer O (1966) Kindlicher Diabetes mellitus infolge fetaler Inselschädigung. Dtsch Med Wochenschr 91: 216

Schernthaner G (1980) Neue Aspekte in der Pathogenese und im Krankheitsverlauf des Typ-I-Diabetes mellitus. Wien Klin Wochenschr [Suppl] 114: 1

Schernthaner G, Müller MM, Prager R, Mühlhauser I (1980) Die klinische Bedeutung des Glykohämoglobins (HBA 1). Wien Klin Wochenschr 92 [Suppl] 115: 1

Schindler AE, Meyfort J (1979) Übergewichtigkeit von Mutter und Kind und Glucosetoleranztest im Wochenbett. Geburtsh Frauenheilk 39: 593

Schlichtkrull J, Munk O, Jersild M (1965) The M-value, an index of blood-sugar control in diabetics. Acta Med Scand 177: 95

Schlichtkrull J (1977) The absorption of insulin. Acta Paediatr Scand [Suppl] 270: 97

Schmid AI (1983) Morphometrische Untersuchungen an Placentazotten bei mütterlichem Diabetes mellitus und Rhesus-Inkompatibilität. Thesis, Ulm

Schmid R, Pollak A, Vycdilik, W, et al (1984) Determination of glucitollysine for the quantitation of non-enzymatic glucosylation by ion exchange chromatography and reverse phase liquid chromatography. Clin Chim Acta 139: 119

Schneider H, Dancis J (1974) Amino acid transport in placenta slices. Am J Obstet Gynecol 120: 1092

Schneider H, Challier JC, Dancis J (1981) Transfer and metabolism of glucose and lactate in the human placenta studied by a perfusion system in vitro. Placenta [Suppl] 2: 129

Schönhardt RE (1981) Perinatale Morbidität und Mortalität — eine EDV-Analyse der Geburten der Jahre 1973—1977 an der UFK Bonn. Thesis, Medical Faculty, University of Bonn

Schuhmann R (1986) Morphologie und Pathomorphologie der Placenta. In: Wulf KH, Schmidt-Mathiessen H (Hrsg) Klinik der Frauenheilkunde und Geburtshilfe. Urban & Schwarzenberg, München

Schwartz HC, Widness J, Thompson D, Tsuboi KK, Oh W, Schwartz R (1980) Glycosylation and acetylation of hemoglobin in infants of normal and diabetic mothers. Biol Neonate 38: 71

Second International Workshop—Conference on Gestational Diabetes Mellitus (1985) Summary and Recommendations. Diabetes 34 [Suppl] 2: 123

Seeds AE, Leung LS, Talbor MW, Russell PT (1979) Changes in amniotic fluid glucose, beta-hydroxybutyrate, glycerol, and lactate concentration in diabetic pregnancy. Am J Obstet Gynecol 135: 887

Senft HH, Foedisch HJ, Bellmann O (1986) Placentafunktionsstörungen bei Gestationsdiabetes — eine morphometrische Analyse. Ber Gynäkol Geburtsh 122: 797

Sepe SJ, Connell FA, Geiss LS, Teutsch SM (1985) Gestational diabetes. Incidence, maternal characteristics, and perinatal outcome. Diabetes 34 [Suppl] 2: 13

Service FJ, Molnar GD, Rosevear JW, Ackerman E, Gatewood LC, Taylor WF (1970) Mean amplitude of glycemic excursions, a measure of diabetic instability. Diabetes 19: 644

Service FJ, Nelson RL (1980) Characteristics of glycemic stability. Diabetes Care 3: 58

Shambaugh GE, Mrozak SC, Freinkel N (1977) Fetal fuels. I. Utilization of ketones by isolated tissues at various stages of maturation and maternal nutrition during late gestation. Metabolism 26/6: 623

Shambaugh GE, Korehler RA, Freinkel N (1977) Fetal fuels II. Contributions of selected carbon fuels to oxidative metabolism in rat conceptus. Am J Physiol 233/6: E 457

Shannon K, Davis JC, Kitzmiller JL, Fulcher SA, Koenig HM (1986) Erythropoesis in infants of diabetic mothers. Pediatr Res 20: 161

Shelly J, Basset JM, Milner RDG (1975) Control of carbohydrate metabolism in the fetus and newborn. Br Med Bull 31: 37

Shen SJ, Wang CY, Nelson KK, Jansen M, Ilan J (1986) Expression of insulin-like growth factor II in human placentas from normal and diabetic pregnancies. Froc Natl Acad Sci USA 83: 9179

Shenouda FS, Cockram CS, Baron MD (1982) Importance of short-term changes in glycosylated haemoglobin. Br Med J 284: 1084

Shepard MJ, Richards VA, Berkowitz RL, Warsof SL, Hobbins JC (1982) An evaluation of two equations for predicting fetal weight by ultrasound. Am J Obstet Gynecol 142: 47

Sherwood LM, Burstein Y, Schechter I (1979) Primary structue of the NHZ-terminal extra piece of the precursor to human placental lactogen. Proc Natl Acad Sci USA 76: 3819

Sicree RA, Hoet JJ, Zimmet P, King HOM, Coventry JS (1986) The association of non insulin dependent diabetes with parity and stillbirth occurrence amongst five pacific populations. Diab Res Clin Practice 2: 113

Silverman W, Anderson DH (1956) A controlled chemical trial of effects of water mist on obstructive respiratory signs, death rate and necropsy findings among premature infants. Pediatrics 17: 1

Simpson HCR, Lonsley S, Geekie M, Simpson RW, Carter RD, Hickaday TDR, Mann JJ (1981) A high carbohydrate leguminous fibre diet improves all aspects of diabetic control. Lancet i: 1

Sing MM, Anthony F (1979) The relation between pregnancy specific fetal glycoprotein (SPI) levels and abnormal glucose tolerance in pregnancy. Br J Obstet Gynecol 86: 458

Singer DB (1984) The placenta in pregnancies complicated by diabetes mellitus. Perspect Pediat Pathol 8: 199

Singer K, Chernoff AI, Singel L (1965) Studies on abnormal hemoglobins: their demonstration in sickle cell anemia and other hematologic disorders by means of alkali denaturation. Blood 6: 413

Skouby SO (1982) Low dosage oral contraception in women with previous gestational diabetes. Obstet Gynecol 59: 325

Skouby SO (1986) Update on metabolic effects of oral contraceptives. J Obstet Gynecol 6: 104

Skouby SO, Kühl C, Mosted-Pedersen L, Petersen K, Christensen MS (1985) Triphasic oral contraceptives: Metabolic effects in normal women and those with previous gestational diabetes. Am J Obstet Gynecol 153: 495

Skouby SO, Kühl C, Andersen O (1986) Oral contraceptives and insulin receptor binding in normal women and those with previous gestational diabetes. Am J Obstet Gynecol 155: 802

Skyler JS, O'Sullivan MJ, Robertson EG (1980) Blood glucose control during pregnancy. Diabetes Care 3: 69

Smith CH, Adcock EW, Teasdale F, Mesehina G, Battaglia FC (1973) Placental amino acid uptake. Tissue preparation, kinetics and the preincubation effect. Am J Physiol 224: 558

Smith CJ, Wejksnora PJ, Warnher JR, Rubin CS, Rosen OM (1979) Insulin-stimulated protein phosphorylation in 3 T 3-Li preadipocytes. Proc Natl Acad Sci USA 76: 2725

Smith NC, Brush MG (1978) Preparation and characterization of human syncytiotrophoblast plasma membrane. Med Biol 56: 272

Smythe AR, Sakakini J (1980) Maternal metabolic alterations secundary to terbutaline therapy for premature labor. Obstet Gynecol 57: 566

Snoeck A, Remacle C, Reusens B, Hoet JJ (1987) Effect of protein deprivation during pregnancy on the fetal endocrine pancreas in the rat. Acta Clin Belg (in press)

Sodoyez-Goffaux F, Sodoyez JCl, De Vos CJ (1982) Maturation of liver handling of insulin in rat fetus. Diabetes 31: 60

Sosenko IR, Kitzmiller JL, Loo SW, Blix P, Rubenstein AH, Gabbay KH (1979) The infant of the diabetic mother. Correlation of increased cord C-peptide levels with macrosomia and hypoglycemia. N Engl J Med 301: 859

Sosenko JM, Kitzmiller JL, Fluckinger R, Loo SWH, Younger DM, Gabbay KH (1982) Umbilical cord glycosylated hemoglobin in infants of diabetic mothers: relationships to neonatal hypoglycemia, macrosomia, and cord serum C-peptide. Diabetes Care 5: 566

Spellacy WN (1976) Carbohydrate metabolism in male infertility and female fertility-control patients. Fertil Steril 27: 1132

Spellacy WN (1982) Carbohydrate metabolism during treatment with estrogen, progestogen, and low-dose oral contraceptives. Am J Obstet Gynecol 142: 732

Spellacy WN, Buhi WC, Birk SA (1972) The effect of estrogens on carbohydrate metabolism: Glucose, insulin and growth hormone studies on one hundred and seventy one women ingesting premarin, mestranol and ethinyl estradiol for six months. Am J Obstet Gynecol 114: 378

Spellacy WN, Buhi WC, Bradley B, Holsinger KK (1973) Maternal, fetal and amniotic fluid levels of glucose, insulin and growth hormone. Obstet Gynecol 41: 323

Spellacy WN, Buhi WC, Birk SA (1976) Carbohydrate and lipid metabolic studies before and after one year of treatment with ethynodiol diacetate in "normal" women. Fertil Steril 27: 900

Spellacy WN, Buhi WC, Birk SA (1977) Vitamin B 6 treatment of gestational diabetes mellitus. Am J Obstet Gynecol 127: 599

Spellacy WN, Buhi WC, Birk SA (1979) Carbohydrate metabolism prospectively studied in women using a low-estrogen oral contraceptive for six months. Contraception 20: 137

Spellacy WN, Buhi WC, Birk SA (1980) Carbohydrate metabolism in women with a twin pregnancy. Obstet Gynecol 55: 688

Spencer GSG, Hill DJ, Garsson GJ, Mac Donald AA, Colenbizander B (1983) Somatomedin activity and growth hormone levels in body fluids of the fetal pig: effect of chronic hyperinsulinemia. J Endocrinol 28: 1058

Sperling MA (1982) Integration of fuel homeostasis by insulin and glucagon in the newborn. Monogr Pediatr 16: 39

SPSS User's guide (1966) 2nd edn. McGraw-Hill, New York

Srikanta S, Rabizadeh MAK, Eisenbarth GS (1985) Assay for islet cell antibodies. Diabetes 34: 300

Steel JM, Thomson P, Johnstone F, Smith AF (1981) Glycosylated haemoglobin concentrations in mothers of large babies. Br Med J 282: 1357

Steel RB, Mosley JD, Smith CH (1979) Insulin and

placenta: Degradation and stabilization, binding to microvillous membrane receptors, and amino acid uptake. Am J Obstet Gynecol 135: 522

Stehbens JA, Baker GL, Kitchel M (1977) Outcome at ages 1, 3 and 5 years of children born to diabetic women. Am J Obstet Gynecol 127: 408

Stein Z, Susser M, Saenger G (1975) Famine and human development. The dutch hunger winter of 1944–1945. Oxford University Press

Steldinger R, Weber B (1981) Dauerketose und Kindesentwicklung. Kasuistik einer Gestationsdiabetikerin mit Atkins-Diät während der Schwangerschaft. Gynäk Rdsch 21 [Suppl] 2: 165

Steldinger R, Weber B, Kneer J, Hättig H (1986) Kindesentwicklung nach ketogener Reduktionsdiät in der Schwangerschaft. In: Weiss PAM (Hrsg) 1. Internationales Grazer Symposium: Kohlenhydratstoffwechselstörungen und Schwangerschaft. Maudrich, Wien München Bern (Probl perinat med 15)

Stonestreet BS, Piasecki GJ, Oh W, Jackson BT (1984) Cardiovascular responses to insulin infusions in the ovine fetus. Pediatr Res 18: 130 A

Stowers JM, Sutherland HW, Kerridge DF (1985) Long-range implications for the mother. The Aberdeen experience. Diabetes 34 [Suppl] 2: 106

Straus DS (1984) Growth-stimulatory actions of insulin in vitro and in vivo. Endocrine Rev 5: 356

Summary and recommendations of the second international workshop-conference on gestational diabetes mellitus (1985). Diabetes 34 [Suppl] 2: 123

Susa JB, Naeve C, Sehgal P, Singer DB, Zeller WP, Schwartz R (1984) Chronic hyperinsulinemia in the fetal rhesus monkey. Effects of physiologic hyperinsulinemia on fetal growth and composition. Diabetes 33: 565

Susa JB, Schwartz R (1985) Effects of hyperinsulinemia in the primate fetus. Diabetes 34 [Suppl] 2: 36

Sybulski S (1969) In vitro estrogen biosynthesis from testosterone by homogenates of placentas from normal pregnancies and pregnancies complicated by intrauterine fetal malnutrition. Am J Obstet Gynecol 105: 1055

Szabo AJ, Cole HS, Grimaldi RD (1970) Glucose tolerance in gestational diabetic women during and after treatment with a combination-type oral contraceptive. N Engl J Med 282: 646

Szabo AJ, Grimaldi RD (1970) The effect of insulin on glucose metabolism of the incubated human placenta. Am J Obstet Gynecol 106: 75

Szabo AJ, De Lellis R, Grimaldi RD (1973) Triglyceride synthesis by the human placenta. I. Incorporation of labeled palmitate into placental triglycerides. Am J Obstet Gynecol 115: 257

Takano K, Hall K, Fryklund L, Holmgren A, Sieevertsson H, Uthne K (1975) The binding of insulin and somatomedin A to human placental membranes. Acta Endocrinol 80: 14

Tallarigo L, Giampietro O, Penno G, Miccoli R, Gregori G, Navalesi R (1986) Relation of glucose tolerance to complications of pregnancy in nondiabetic women. N Engl J Med 315: 989

Tamas Gy Jr, Baranyi E Gy, Petranyi Jr, Dimeny Z, Banyai Z, Kerenyi Z (1981) Improvement of metabolic control: criteria for normoglycaemia in diabetic pregnancy. In: Irsigler K, et al (eds) New approaches to insulin therapy. MTP Press, Falcon House, Lancester, England

Tanner JM (1970) Standards for birth weight or intrauterine growth. Pediatrics 46: 1

Tchobroutsky G, Heard I, Tchobroutsky C, Eschwege E (1980) Amniotic fluid C-peptide in normal and insulin-dependent diabetic pregnancies. Diabetologia 18: 289

Teasdale F (1981) Histomorphometry of the placenta of the diabetic women: class A diabetes mellitus. Placenta 2: 241

Teramo KA, Widness JA, Clemons GK, Voutilainen P, McKinlay S, Schwartz R (1986) Relationship of fetal erythropoietin to amniotic fluid C-peptide levels in diabetic pregnancy. Proceedings of the society for gynecologic investigation. Toronto, March 1986

Teramo KA, Widness JA, Clemons GK, Voutilainen P, McKinlay S, Schwartz R (1987) Amniotic fluid erythropoietin correlates with umbilical plasma erythropoietin in normal and abnormal pregnancy. Obstet Gynecol (in press)

Thalme B, Edström K (1974) Intravenous glucose tolerance test and its relation to a scoring system for the degree of diabetic fetopathy in newborn infants. J Perinat Med 2: 233

Theile U, Rückel E, Hust U (1985) Kinder diabetischer Eltern — Zur Frage von Entwicklungsstörungen während der Schwangerschaft mit besonderer Berücksichtigung des sogenannten Riesenkindes. Z Geburtsh Perinat 189: 79

Thomas G, Martin-Perez J, Siegmann M, Otto AM (1982) The effect of serum, EGF, PGF 2 alpha and insulin on 56 phosphorylation and the initiation of protein and DNA synthesis. Cell 30: 235

Thomsen K, Lieschke G (1958) Untersuchungen zur Placentamorphologie bei Diabetes mellitus. Acta Endocrin 29: 602

Thomson AM, Billewicz WZ, Hytten FE (1968) The assessment of fetal growth. J Obstet Gynaec Br Cwlth 77: 903

Thorsson AV, Hintz RL (1977) Insulin receptors in the newborn. Increase in receptor affinity and number. N Engl J Med 297: 908

Tibi L, Young RJ, Smith AF (1982) Clinical implications of labile HbA 1 as assayed by the electrophoretic method. Clin Chim Acta 126: 257

Tiitinen A, Ylinen K (1986) Circulating levels of placental protein 10 (PP 10) in diabetic pregnancy. Acta Obstet Gynecol Scand 55: 709

Toyoda N (1982) Insulin receptors on erythrocytes in normal and obese pregnant women during the follicular and luteal phases. Am J Obstet Gynecol 144: 679

Travis LB, Hürter P (1980) Einführungskurs für Kinder und Jugendliche mit Diabetes mellitus. Bund diabetischer Kinder (Gerhards u Co OHG, Frankfurt a M, Hrsg)

Trivelli LA, Ranney HM, Lai HT (1971) Hemoglobin components in patients with diabetes mellitus. N Engl J Med 284: 353

Truman P, Ford HL (1984) The brush border membrane of the human term placenta. Biochim Biophys Acta 779: 139

Tsibris JCM, Raynor LO, Buhi WC, Buggie J, Spellacy WN (1980) Insulin receptors on circulating erythrocytes and monocytes from women on oral contraceptives or pregnant women near term. J Clin Endocrinol Metab 51: 711

Turner RC, Harris E, Bloom SR, Uren C (1977) Relation of fasting plasma glucose concentration to plasma insulin and glucagon concentrations. Studies in latent diabetics and women who have produced large-for-dates babies. Diabetes 26: 166

Turner RC, Harris E, Ounstead M, Ponsford C (1979) Two abnormalities of glucose-induced insulin secretion: dose-response characteristics and insulin sensitivity. Acta Endocrin 92: 148

Turtle JR, Kipnis DM (1967) The lipolytic action of human placental lactogen on isolated fat cells. Biochim Acta 144: 583

Tyson JE (1971) Obstetric management of the pregnant diabetic. Med Clin North Am 55: 961

Tyson JE, Hock RH (1976) Gestational and pregestational diabetes: Approach to therapy. Am J Obstet Gynecol 125: 1009

Uher J (1969) Metabolism of the trophoblast in tissue culture under normal and modified conditions. In: Diczfalusy E (ed) The feto-placental unit. WB Saunders, Philadelphia

Ukena T, Merrill E, Morgan C (1982) An analysis of the importance of the "labile" fraction of glycosylated hemoglobin as determined by a minicolumn method. Am J Clin Path 78: 724

Underwood LE, D'Ercole AJ (1984) Insulin and insulin-like growth factors/somatomedins in fetal and neonatal development. Clin Endocrinol Metab 13: 69

Upton GV (1983) The phasic approach to oral contraception: The triphasic concept and its clinical application. Int J Fertil 28: 121

Urban G, Baumgarten K, Baumung H, Beck A, Fröhlich H, Gruber W, Seidl A (1972) Die Beeinflussung des Glukosetoleranztestes bei normalen Schwangerschaften mit Ritotrine und Th 1165a. In: Proceedings of the international symposium on the treatment of foetal risks, Baden, Austria

Van Assche FA (1970) The foetal endocrine pancreas. A quantitative morphological approach. Thesis, University of Leuven

Van Assche FA, De Prins F, Aerts L, Verjans L (1977) The endocrine pancreas in small-for-dates infants. Br J Obstet Gynecol 84: 751

Van Assche FA, Aerts L, Gepts W (1982) The different

celltypes in the endocrine pancreas (human). Diabetologia 16: 151

Van Assche FA, Aerts L, De Prins F (1983) Degranulation of the insulin-producing beta-cells in an infant of a diabetic mother. Case report. Br J Obst Gynaec 90: 182

Van Assche FA, Aerts L (1985) Long-term effect of diabetes and pregnancy in the rat. Diabetes 34 [Suppl] 2: 116

Van Beek (1939) Zit. nach Scheibenreither S, Thalhammer O (1966)

Van Kreel BK, Van Dijk JP, Pijnenburg AMCM (1982) Placental transfer and metabolism of purines and nucleosides in the pregnant guinea pig. Placenta 3: 127

Van Leusden H, Villes CA (1965) The de novo synthesis of sterols and steroids from acetate by a preparation of human term placentas. Steroids 6: 31

Van Lierde M, Buysschaert M, De Hertogh R (1982) Administration intraveineuse de ritodrine chez le diabetique insulino dependante enceinte repercussions metaboliques. J Gynecol Obstet Biol Reprod 11: 869

Verhaeghe J, Bouillon R, Nyomba BL, Lissens W, Van Assche FA (1987) Vitamin D and bone mineral homeostasis during pregnancy in the diabetic BB rat. Endocrinology 118 (in press)

Villar-Palasi C, Larner J (1961) Insulin treatment and increased UDPG-glycogen transglucosylase activity in muscle. Arch Biochem Biophys 94: 436

Villee CA (1953) The metabolism of human placenta in vitro. J Biol Chem 205: 113

Villee CA (1962) Metabolism of the Placenta. Am J Obstet Gynecol 84: 1684

Vlassara H, Brownlee M, Cerami A (1982) Assessment of diabetic control by measurement of urinary glycopeptides. Diabetologia 23: 252

Vohr BR, Lipsitt LP, Oh W (1980) Somatic growth of children of diabetic mothers with reference to birth size. J Pediatr 97: 196

Vorherr H (1982) Factors influencing fetal growth. Am J Obstet Gynecol 142: 577

Waine H, Frieden EH, Caplan HI, Colt T (1963) Metabolic effects of Enovid in rheumatoid patients. Arthritis Rheum 6: 796

Wakhloo AK, Beyer J, Dietrich C, Schulz G (1984) Einfluß von Nahrungsfett auf Blutzuckerspiegel und Insulinverbrauch nach Einnahme verschiedener Kohlenhydratträger bei Typ I Diabetikern am künstlichen Pankreas. Dtsch Med Wochenschr 42: 1589

Walinder O, Ronquist G, Fager PJ (1982) New spectrophotometric method for the determination of hemoglobin A1 compared with a microcolumn technique. Clin Chem 28: 96

Walker DG, Lea MA, Rossiter G, Addison MEB (1954) Glucose metabolism in the placenta. Arch Biochem Biophys 120: 646

Wang T (1980) Amniotic epithelium in diabetes mellitus. Light and electron microscopic examination. Virch Arch A 387: 185

Ward WK, Johnston CLW, Beard JC, Benedetti TJ, Halter JB, Porte Jr D (1985) Insulin resistance and impaired insulin secretion in subjects with histories of gestational diabetes mellitus. Diabetes 34: 861

Ward WK, Johnston CLW, Beard JC, et al (1985) Abnormalities of islet B-cell function, insulin action, and fat distribution in women with histories of gestational diabetes: relationship to obesity. J Clin Endocr Metab 61: 1039

Warsof SL, Gohari P, Berkowitz RL, Hobbins JC (1977) The estimation of fetal weight by computer-assisted analysis. Am J Obstet Gynecol 128: 881

Warth MR, Knopp RH (1977) Lipid metabolism in pregnancy. V. Interactions of diabetes, body weight, age, and high carbohydrate diet. Diabetes 26: 1056

Wassermann L, Shiesinger H, Abramovici A, Goldman JA, Allalouf D (1980) Glycoseaminoglycan patterns in diabetic and toxemic term placentas. Am J Obstet Gynecol 138: 769

Weare JA, Brinkman C, Nix N, Roosevelt TS, Seegan GW (1984) Personal communication

Weber B, Tech J, Schmidt H, Oberdisse U (1978) Antibody formation and insulin requirements in diabetic children during treatment with purified commercial pork insulins. Europ J Ped 128: 89

Weidinger H, Mohr D, Haller K, Hiltmann WD, Vogel M (1976) Zeitlicher Verlauf der Blutglucose des immunoreaktiven Insulins und der Kaliumionen bei Neugeborenen nach langzeitiger und akuter Gabe von Partusisten mit und ohne Isoptin. Z Geburtsh Perinat 180: 258

Weiss PAM (1979) Die Überwachung des Ungeborenen bei Diabetes mellitus an Hand von Fruchtwasserinsulinwerten. Wien Klin Wochenschr 91: 293

Weiss PAM (1986) Diskussionsbeitrag. 1. Internationales Grazer Symposium Kohlenhydratstoffwechselstörungen und Schwangerschaft

Weiss PAM, Lichtenegger W, Pürstner P (1975) Insulin im Fruchtwasser bei gesunden und diabetischen Schwangeren. VII. Akademische Tagung deutschsprechender Hochschullehrer in der Gynäkologie und Geburtshilfe. Kongreßband

Weiss PAM, Winter R, Pürstner P (1978) Insulin levels in amniotic fluid. Management of pregnancy in diabetes. Obstet Gynecol 51: 292

Weiss PAM (1979) Die Überwachung des Ungeborenen bei Diabetes mellitus an Hand von Fruchtwasserinsulinwerten. Wien Klin Wochenschr 91: 293

Weiss PAM, Hofmann H, Winter R, Pürstner P, Lichtenegger W (1984) Gestational diabetes and screening during pregnancy. Obstet Gynecol 63: 776

Weiss PAM, Hofmann HMH, Pürstner P, Winter R, Lichtenegger W (1984a) The fetal insulin balance: Gestational diabetes and postpartal screening. Obstet Gynecol 64: 65

Weiss PAM, Hofmann HMH (1984b) Intensified conventional insulin therapy for the pregnant diabetic patient. Obstet Gynecol 64: 629

Weiss PAM, Pürstner P, Winter R, Lichtenegger W (1984c) Insulin levels in amniotic fluid of normal and abnormal pregnancies. Obstet Gynecol 63: 371

Weiss PAM, Hofmann H (1985) Diabetes mellitus und Schwangerschaft. In: Burghardt E (Hrsg) Spezielle Gynäkologie und Geburtshilfe. Springer, Wien New York, S 337

Weiss PAM, Winter R, Pürstner P, Lichtenegger W (1985) Amniotic fluid glucose values in normal and abnormal pregnancies. Obstet Gynecol 65: 333

Weiss PAM (1986) Diskussionsbeitrag. 1. Internationales Grazer Symposium Kohlenhydratstoffwechselstörungen und Schwangerschaft

Weiss PAM, Hofmann HMH, Winter R, Lichtenegger W, Pürstner P, Haas J (1986) Diagnosis and treatment of gestational diabetes according to amniotic fluid insulin levels. Arch Gynecol 239: 81

Weiss PAM, Hofmann HMH (1986) Indikation zur Insulintherapie bei Gestationsdiabetes. In: Weiss PAM (ed) Probl perinat med 15. Maudrich, Wien München Bern, p 95

Weiss PAM, Hofmann HMH, Kainer F, Haas JG (1987) Fetal outcome in gestational diabetes with elevated amniotic fluid insulin levels. Dietary versus insulin treatment. Diab Res Clin Pract (in press)

Weith HL, Wiebers JL, Gilham PT (1980) Synthesis of cellulose derivatives containing the dihydroxyboryl group and a study of their capacity to form specific complexes with sugars and nucleic acid components. Biochem 9: 4396

White P (1949) Pregnancy complicating diabetes. Am J Med 7: 609

White P (1959) Pregnancy complicating diabetes. In: Joslin EP, et al (eds) The Treatment of diabetes mellitus. Lea and Febinger, Philadelphia, p 690

White P (1974) Diabetes mellitus in pregnancy. Clin Perinat 1: 331

White P (1978) Classification of obstetric diabetes. Am J Obstet Gynecol 130: 228

White NH, Skor DA, Cryer PhE, Levandosky LA, Bier DM, Santiago JV (1983) Identification of type I diabetic patients at increased risk for hypoglycemia during intensive therapy. N Engl J Med 308: 485

Whitelaw A (1977) Subcutaneous fat in newborn infants of diabetic mothers: An indication of quality of diabetic control. Lancet i: 15

Whitsett JA (1980) Specializations in plasma membranes of the human placenta. J Pediatrics 96: 600

Whitesett JH, Lessard JL (1978) Characteristics of the microvillous brush border of human placenta: Insulin receptor localization in brush border membranes. Endocrinol 103: 1458

WHO expert committee on diabetes mellitus (1980) Second report. Techn Report Series 646: 7

Widness JA, Schwartz HC, Thompson D, et al (1978) Hemoglobin A1c (glycohemoglobin) in diabetic pregnancy: an indicator of glucose control and fetal size. Br J Obstet Gynecol 57: 652

Widness JA, Schwartz HC, Thompson D, Kahn CB, Oh W, Schwartz R (1978) Hemoglobin A1c (Gly-cohaemoglobin) in diabetic pregnancy: an indicator of glucose control and fetal size. Br J Obstet Gynecol 85: 812

Widness JA, Schwartz HC, Kahn C, Oh W, Schwartz R (1980) Glycohemoglobin in diabetic pregnancy: A sequential study. Am J Obstet Gynecol 136: 1024

Widness JA, et al (1981) Increased erythropoiesis and elevated erythropoietin in infants born to diabetic mothers and in hyperinsulinemic Rhesus monkeys. J Clin Invest 67/3: 637

Widness JA, Schwartz HC, Zeller WP, Oh W, Schwartz R (1981) Glycohemoglobin in postpartum women. Obstet Gynecol 57: 414

Widness JA, Clemons GK, Garcia JF, et al (1984) Increased immunoreactive erythropoietin in cord serum after labor. Am J Obstet Gynecol 148: 194

Widness JA, Teramo KA, Clemons GK, Coustan DR, Cavalieri RL, Oh W, Welch GP, Schwartz R (1985) Correlation of the interpretation of fetal heart rate records with cord plasma erythropoietin levels. Br J Obstet Gynecol 92: 326

Widness JA, Cowett RM, Coustan DR, Carpenter MW, Oh W (1985) Neonatal morbidities in infants of mothers with glucose intolerance in pregnancy. Diabetes 34 [Suppl] 2: 61

Widness JA, Piasecki GJ, Clemons GK, Oh W, Jackson BT (1986) Graded hyperinsulinemia in fetal sheep: effects on oxygenation and plasma erytropoietin. Proceedings of the society for gynecologic investigation. Toronto, March 1986, p 186

Widness JA, Teramo KA, Clemons GK, et al (1986) Temporal response of immunoreactive erythropoietin to acute hypoxemia in fetal sheep. Pediatr Res 20: 15

Wieland OH (1983) Late diabetic damage and nonenzymatic glucosylation of proteins. Med Klin 78: 107

Wilcox AJ (1981) Birth weight, gestation, and the fetal growth curve. Am J Obstet Gynecol 139: 863

Williams PF, Turtle JR (1979) Purification of the insulin receptor from human placental membranes. Biochem Biophys Acta 579: 367

Williams PR, Sperling MA, Racasa Z (1979) Blunting of spontaneous and alanine-stimulated glucagon secretion in newborn infants of diabetic mothers. Am J Obstet Gynecol 133: 51

Williams RL, Creasy RK, Cunnhingham GC, Hawes WE, Norris FD, Tashiro M (1982) Fetal growlth and perinatal viability in California. Obstet Gynecol 59: 624

Wingrave SJ, Kay CR, Vessy MP (1979) Oral contraceptives and diabetes mellitus. Br Med J 1: 23

Winick M, Noble A (1967) Cellular growth in human placenta. II. Diabetes mellitus. J Pediatr 71: 216

Winter R, Hofmann H (1985) Pränatale Diagnose genetischer Defekte. In: Burghardt E (Hrsg) Spezielle Gynäkologie und Geburtshilfe. Springer, Wien New York, S 296

Winter WE, MacLaren NK, Riley WJ, Clarke DW, Kappy MS, SpillarRP (1987) Maturity onset diabetes in black americans. N Engl J Med 316: 285

Wissenschaftliche Tabellen Geigy (1982) Teilband Somatometrie und Biochemie, 8. Aufl. Ciba-Geigy, Basel

Wladimiroff JW, Bloemsma CA, Wallenburg HCS (1978) Ultrasonic diagnosis of the large-for-dates infant. Obstet Gynecol 52: 285

Wolff F, Jung C, Bolte A (1981) Ist der postpartale Glucosetoleranztest bei Müttern makrosomer Neugeborener sinnvoll? Gynäk Rdsch 21 [Suppl] 2: 160

Wood, GP, Sherline DM (1975) Amniotic fluid glucose: A maternal, fetal, and neonatal correlation. Am J Obstet Gynecol 122: 151

Worth R, Ashworth L, Home PD, Gerrard J, Lind T, Anderson J, Alberti GKMM (1983) Glycosylated haemoglobin in cord blood following normal and diabetic pregnancies. Diabetologia 25: 482

Worth R, Potter JM, Drury J, Fraser RB, Cullen DR (1985) Glycosylated haemoglobin in normal pregnancy: a longitudinal study with two independent methods. Diabetologia 28: 76

Wright AD, Nicholson HD, Pollack A, et al (1983) Spontaneous abortion and diabetes mellitus. Postgraduate Med J 59: 295

Wulf KH (1981) Das Placenta-Insuffizienzsyndrom. Z Geburtsh Perinat 185: 2

Wynn W, Doar JWH (1966) Some effects of oral contraceptives on carbohydrate metabolism. Lancet ii: 715

Wynn V, Adams PW, Godsland I, Niththyananthnan R, Adams PW, Melrose J, Oakley NW, Seed M (1979) Comparison of different combined oral contraceptive formulations on carbohydrate and lipid metabolism. Lancet i: 1045

Yatscoff RW, Braidwood JL (1982) Comparison of column chromatographic, colorimetric and electrophoretic methods for determination of glycosylated hemoglobin (HbA). Clin Biochem 15: 302

Yatscoff RW, Mehta A, Dean H (1985) Cord blood glycosylated (glycated) hemoglobin: correlation with maternal glycosylated (glycated) hemoglobin and birth weight. Am J Obstet Gynecol 152: 861

Yen SSC (1964) Abnormal carbohydrate metabolism and pregnancy. Am J Obst Gynecol 40: 468

Yen SSC (1978) Metabolic homeostasis during pregnancy. In: Yen SSC, Jaffe RB (eds) Reproductive endocrinology. WB Saunders, Philadelphia, p 537

Ylinen K, Raivia K, Teramo K (1981) Haemoglobin A l c predicts the perinatal outcome in insulin dependent diabetic pregnancies. Br J Obstet Gynaecol 88: 961

Yoshicke T, Roux JF (1972) In vitro metabolism of palmitic acid in human fetal tissue. Pediatr Res 6: 675

Younoszai MK, Haworth JC (1969) Chemical composition of the placenta in normal preterm, term, and intrauterine growth-retarded infants. Am J Obstet Gynecol 103: 262

Yudilevich DL, Sweiry JH (1985) Transport of amino acids in the placenta. Biochim Biophys Acta 822: 169

Yue DK, McLennan S, Church DB, Turtle JR (1982) The measurement of glycosylated hemoglobin in man and animals by aminophenylboronic acid affinity chromatography. Diabetes 31: 701

Zahn V, Zach HP, Sigmund R (1978) Über die Möglichkeit der pränatalen Behandlung des Atemnotsyndroms bei Frühgeburten mit Ambroxol. Atemwegs- u Lungenkrkh 4 [Suppl] 1: 35

Zanjani ED, Gordon AS (1971) Erythropoietin production and utilization in fetal goats and sheep. Isr J Med Sci 7: 850

Zeller WP, Susa JB, Widness JA, et al (1983) Glycosylation of hemoglobin in normal and diabetic mothers and their fetuses. Pediat Res 17: 200

Zilker Th, Paterek K, Ermler R, Bottermann P (1977) Untersuchungen zur Frage einer Autoregulation der Insulinsekretion. Klin Wochenschr 55: 475

Zuppinger K, Wiesmann U, Siegrist HP, Schäfer T, Sandru L, Schwartz HP, Herschkowitz N (1981) Effect of glucose deprivation on sulfatide synthesis and oligodendrocytes in cultured brain cells of newborn mice. Pediatr Res 15: 319

Subject Index

Acknowledgements and Grants

Bellmann Otto et al.:
The authors are grateful to Mrs. I. Maßen and Mrs. H. Pilgermann for excellent technical assistance.

Desoye Gernot:
The author is greatly indebted to Drs. Weiss P. A. M. and Motter W. for careful reading of the manuscript and valuable discussion. The secretarial assistance of A. Klimpfinger is gratefully acknowledged.

Domenech Pedro et al.:
This work has been supported in part by a R 5/85 grant from Hospital Clinic and an AIUB research grant for Barcelona University.

Kühl Claus and **Andersen Ole:**
The expert technical assistance of Marie Louise Borgen, Connie Kühl, and Lene Poulsen is gratefully acknowledged. The studies have been supported by grants from the Danish Medical Research Council, the Danish Hospital Foundation for Medical Research, Region of Copenhagen, the Faroe Islands and Greenland, the Danish Diabetes Foundation, and the Novo Foundation.

Mestman Jorge H.:
The author is grateful to Robert Nakamura PhD., for statistical analysis.

Oats Jeremy et al.:
This study was supported by the Mercy Maternity Hospital Research Foundation. We wish to thank our consultant colleagues for their cooperation in reviewing their patients, and Mrs. J. Walstab for expert statistical advice.

Pastor Xavier et al.:
This work has been supported in part by a R 5/2/85 grant from Hospital Clinic and an AUIB Research grant from the University of Barcelona.

Weiss Peter A. M.:
Professor Donald R. Coustan's immense efforts and collaboration in the editorial work of this book are herewith gratefully acknowledged and also his invaluable advice and suggestions are highly appreciated. The editor is also much indebted to H. Rosegger, M.D. for the coverdrawing and to Mrs. Hiptmayer O. for her extensive work and secretarial assistance as well as to K. Tamussino, M.D., for linguistic support.